3rd International Conference on Energy and Environmental Science (ICEES2019)

IOP Conference Series: Earth and Environmental Science
Volume 291

Seoul, South Korea
26 – 29 January 2019

ISBN: 978-1-5108-9078-7
ISSN: 1755-1307

Printed from e-media with permission by:

Curran Associates, Inc.
57 Morehouse Lane
Red Hook, NY 12571

CURRAN ASSOCIATES INC. proceedings.com

Some format issues inherent in the e-media version may also appear in this print version.

Institute of Physics
Dirac House, Temple Back
Bristol BS1 6BE UK

Phone: 44 1 17 929 7481
Fax: 44 1 17 920 0979

techtracking@iop.org

Additional copies of this publication are available from:

Curran Associates, Inc.
57 Morehouse Lane
Red Hook, NY 12571 USA
Phone: 845-758-0400
Fax: 845-758-2633
Email: curran@proceedings.com
Web: www.proceedings.com

TABLE OF CONTENTS

ENVIRONMENTAL QUALITY MONITORING AND MANAGEMENT

ESTIMATED EFFECTS OF AIR QUALITY CONTROL MEASURES ON MORTALITY REDUCTION AND ECONOMIC BENEFITS DURING THE 2014 NANJING YOUTH OLYMPIC GAMES 1
S Y Yu, Y Qian

CROSS-CITY PM2.5 PREDICTIONS WITH RECURRENT NEURAL NETWORK 8
Rh Zong, Ty Zhang, Z Chen, Y Zhu

TACKLING ENVIRONMENTAL POLLUTION THROUGH PARTICIPATORY APPROACH IN NAMBO, SOUTH TANGERANG, INDONESIA 14
Oc Dewi, Tu Atmoko, Ha Alrasyid, D Priyandana, Mbv Dewi, D Siska

CONSTRUCTING A FRAMEWORK FOR ECONOMIC VALUE OF URBAN RENEWAL IN TAIWAN 20
Kuo-Wei Hsu, Jen-Chih Chao

THE DISCUSSION ON THE RELATIONSHIP AMONG INDOOR AIR QUALITY, ENVIRONMENTAL COGNITION AND USER'S BEHAVIOUR 29
Xue-Min Xu, Sheng-Jung Ou

ACCESS THE CHANGE OF WATER QUALITY IN THE SPONGE CITY OF LCTIP 35
N Li, P F Du, X C Huang, Z P Shu

PARTICIPATION PATROLLING EFFORTS BY LOCAL PEOPLE: CASE OF NAM TIEN FOREST STATION IN PU HU NATURE RESERVE, VIETNAM 41
L. K. Dong, S. Sinutok, P. Manop, K. Techato

ESTIMATING THE IMPACT OF THE MRT ELEVATED RAILWAY SYSTEM ON CARBON REDUCTION IN TAICHUNG, TAIWAN 48
Kuo-Wei Hsu, Jen-Chih Chao, Hsiao-Sung Chan, Pei-Chen Wu

SEWAGE TREATMENT AND WASTE MANAGEMENT

THE ROAD AHEAD FOR MUNICIPAL SOLID WASTE MANAGEMENT IN THE 21ST CENTURY: A NOVEL-STANDARDIZED SIMULATED PARADIGM 55
M S S Danish, H Zaheb, N R Sabory, H Karimy, A B Faiq, H Fedayi, T Senjyu

FEASIBILITY STUDY OF MUNICIPAL SOLID WASTE FOR PRODUCING REFUSE-DERIVED FUEL (RDF): A CASE STUDY OF BANGKOK, THAILAND 60
N Eaktasang, Y Suma, S Kingkeaw

IN SITU AEROBIC BIODEGRADATION IN SOIL CONTACT USING THE STANDARD TEST METHOD OF ASTM D 5988-12 65
S. M. Al-Salem, A. Al-Nasser, H. Sultan, H. Karam

THE APPLICATION OF ORGANIC SOLVENTS AND THERMAL PROCESS FOR ELIMINATING EVA RESIN LAYER FROM WASTE PHOTOVOLTAIC MODULES 70
Wei Sheng Chen, Yen Jung Chen, Yu An Chen

PRELIMINARY STUDY OF DYE REMOVAL FROM AQUEOUS SOLUTION USING ELEPHANT DUNG ACTIVATED CARBON 76
Y Suma, N Pasukphun, N Eaktasang, P Laor

EFFECT OF SALINITY, SINGLE AND BINARY IONIC COMPOUNDS' LOW SALINITY WATER ON WETTABILITY ALTERATION IN CARBONATE ROCKS 82
C Ho, M A Ayoub

EVALUATION OF ANTIFOULING COATING PERFORMANCE WITH COMMERCIAL AND LEMON NATURAL BIOCIDE ON COMPOSITE FIBERGLASS SURFACE 89
Asmalina Mohamed Saat, Muhammad Irfan Sabri, Hanisah Johor, Fatin Zawani Zainal Azaim, Mohd Iqbal Lokman

ENVIRONMENTAL ECOLOGICAL CONSTRUCTION AND SUSTAINABLE DEVELOPMENT

A NEW PREDICTIVE MODEL FOR PLANTS PHOTOSYNTHESIS INFLUENCED BY MAJOR CLIMATIC CONDITIONS ... 94
Zs Liu, Wy Yang, Xl Yu

TAIWAN BROWNFIELD REDEVELOPMENT & ECOLOGICAL RESTORATION INDICATOR ANALYSIS ... 101
S J Feng, N C Chao, Y J Chen

KEY STRATEGIES OF SUSTAINABLE AND ENERGY-SAVING DEVELOPMENT FOR GREEN UNIVERSITIES ... 108
Yi-Xin Cai, Sheng-Jung Ou, Hsuan-Hung Chen, Chien-Chen Chiu

AN ENVIRONMENTAL IMPROVEMENT STRATEGY FOR AN URBAN GREEN FARM GARDEN: A CASE STUDY OF DONGGUANG GREEN GARDEN ROAD, TAICHUNG CITY ... 114
Meng-Li Lin, Rui-Xing Yang

RENEWABLE ENERGY

APPLICATION OF ECOVILLAGE FOR THE REDEVELOPMENT STRATEGY OF RURAL COMMUNITIES IN TAIWAN. A CASE STUDY OF JIANGUO MILITARY DEPENDENTS' VILLAGES ... 119
Jhong-Ping Xie, Kuo-Wei Hsu

A STUDY ON APPLICATION OF ECOLOGICAL ENGINEERING METHODS TO A RIVER POLLUTION REMEDIATION – CASE STUDY OF LIUCHUAN RIVER ... 124
Che-Yu Hsu, Sheng-Jung Ou

HYBRID-APPROACH FOR SINKHOLE OCCURRENCE RISK MITIGATION IN URBAN AREAS ... 129
A A Malinowska, A Guzy, R Hejmanowski, P. Ulmaniec

LAND USE EVOLUTION AND SIMULATION PREDICTION OF PYONGYANG MUNICIPALITY IN NORTH KOREA FROM 2000 TO 2020 ... 140
Qinchen Zhang, Pingyu Zhang, Dongyan Wang

A CUSTOMIZATION-ORIENTED CARBON FOOTPRINT SERVICE FOR MECHANICAL PRODUCTS ... 148
Yajun Zhang, Jianjun Yi, Zhuoran Wang, Liang He

APPLICATION AND THERMAL ECONOMIC ANALYSIS OF PLATE LOW PRESSURE ECONOMIZER IN FLUE GAS WASTE HEAT RECOVERY ... 154
X Xu, G D Ju

APPLICATION OF BUILDING INTEGRATED PHOTOVOLTAIC IN HOT HUMID CLIMATE. CASE STUDY: OFFICE BUILDING IN INDONESIA ... 161
N A Ardiani, Suhendri, M D Koerniawan, D Y N Naimah

MULTI-CRITERIA ANALYSIS FOR RENEWABLE ENERGY GENERATION: A CASE OF TOKYO ... 166
Dinna Fitriyana, Mutia Prabawati, Frank Tanuwijaya, Dintani Y N Naimah

HOUSE-SCALE RENEWABLE ENERGY SYSTEM SIZING IN LIVERPOOL ... 172
Suhendri, Tri Pratiwi Handayani

HYDROGEN PRODUCTION POSSIBILITY USING MONGOLIAN RENEWABLE ENERGY ... 178
A Dorjgotov, B Ulziidelger, J Yunseong, K Oh Chan, J Ok Sung, S Yong-Gun

ENERGY ENGINEERING

A SIMULATION METHODOLOGY ON GRID DISPLACED PHOTOVOLTAIC (PV) PROSPECTIVE FOR EXISTING TERRACE HOUSE SUPPLY IN MALAYSIAN PENINSULAR ... 184
Ahn Abdul-Razak, N. A Ahmad, S. S. Ahmad, S. N Haron, A. K Ermeey

ENERGY USE PREDICTION WITH INFORMATION THEORY AND MACHINE LEARNING TECHNIQUE ... 189
Yw Tong, Wy Yang, Dl Zhan

THE IMPACT OF ECONOMIC GROWTH, ENERGY CONSUMPTION AND DEMOGRAPHIC DYNAMICS ON CARBON EMISSIONS: A CROSS COUNTRY ANALYSIS IN 133 NATIONS ... 195
P T T Trang, N T X Hanh, B P P Thanh

A REVIEW ON ENERGY PERFORMANCE IN MALAYSIAN UNIVERSITIES THROUGH BUILDING INFORMATION MODELLING (BIM) ADAPTATION 201
A F Latif, N A Ahmad, M R Abdullah, A Ismail, A A A Ghani

BIOMASS ENERGY FROM AGRICULTURE IN CHINA: POTENTIAL AND EVOLUTIONARY TREND 207
Liangzhen Zang, Caihong Zhang

CONNECTING SOLAR GENERATED ELECTRICITY DURING FLOOD DISASTER IN THE FELDA COMMUNITIES OF MALAYSIA: THE PUBLIC PERCEPTION 212
S J T Jamil, N A Ahmad, J J Jamian

PALM EMPTY FRUIT BUNCH GASIFIER CHARACTERIZATION USING EULERIAN-LAGRANGIAN CFD MODELLING 217
Priyambodo Nur Ardi Nugroho, Shuichi Torii

THERMODYNAMIC PERFORMANCE COMPARISON BETWEEN SINGLEPRESSURE AND DUAL-PRESSURE EVAPORATION ORGANIC RANKINE CYCLES FOR HEAT SOURCES WITH OUTLET TEMPERATURE LIMIT 223
Jian Li, Jingze Yang, Zhong Ge, Yuanyuan Duan, Zhen Yang

CONVERSION OF CARBON DIOXIDE INTO STORABLE SOLAR FUELS USING SOLAR ENERGY 230
Houda Ennaceri, Bernd Abel

ENERGY AND ELECTRICAL ENGINEERING

PERFORMANCE ANALYSIS OF A TWO-STAGE GRAVITATIONAL WATER VORTEX TURBINE 235
Taqi Ahmad Cheema, Rizwan Ullah, Abdul Samad Saleem

DEVICE SELECTION OF THE POTENTIAL WAVE ENERGY SITE IN INDONESIAN SEAS 240
B Triasdian, Y S Indartono, N S Ningsih, D Novitasari

DESIGN AND DEVELOPMENT OF INTEGRATED SAVONIUS AND DARRIEUS SMALL SCALE VERTICAL AXIS WIND TURBINE FOR POWER GENERATION 247
Edison E Mojica, Cian Marvince Fabay, Felix Kehinde, John Lambert Tenorio

EVALUATING THE SUSTAINABILITY IMPACT OF GREENING, WATER MANAGEMENT, VENTILATION AND INTERIOR LIGHTING ON CAR PARKING IN TAIWAN 254
Shu-Hui Hsieh, Chien Yu-Chen, Lee Yi-Chien

IMPLEMENTATION OF A UAV-SENSORY-SYSTEM-BASED HAZARD SOURCE ESTIMATION IN A CHEMICAL PLANT CLUSTER 259
Dan Li, Feiran Chen, Yiping Wang, Xiaodong Wang

TAICHUNG CITY'S MANAGEMENT STRATEGY FOR LEISURE CYCLEWAY MAINTENANCE: A CASE STUDY OF GAOMEI CYCLEWAY 265
Yu-Chen Chien, Sue Yee Yew

A DESIGN OPTIMIZATION INDEX FOR TWO TYPES OF CYCLEWAYS 270
Yu-Chen Chien, Cheng-Hung Wu

NEW ENERGY MATERIALS AND EQUIPMENT

SYNTHESIS AND PHYSICOCHEMICAL PROPERTIES OF EPOXIDIZED OLEIC ACID-BASED PALM OIL 275
Mohd Jumain Jalil, Mohd Saufi Md Zaini, Aliff Farhan Mohd Yamin, Intan Suhada Azmi, Siu Hua Chang, Norhashimah Morad, Abdul Hadi

START-UP AND DYNAMIC PROCESSES SIMULATION OF SOFC-MGT HYBRID POWER SYSTEM 281
Kangjie Deng, Chen Yang

PERFORMANCE PREDICTION OF ZERO HEAD TURBINE AT DIFFERENT WATER LEVELS 289
I. A. Masud, S. Yusuke, Y. Suwa

THE USE OF FABA-BEAN CROPPING AS A SUSTAINABLE AND ENERGY SAVING TECHNOLOGY – A NEW PROTEIN SELF-SUFFICIENCY OPPORTUNITY FOR EUROPEAN AGRICULTURE? 294
R Lybæk, H Hauggaard-Nielsen

Author Index

IOP Conf. Series: Earth and Environmental Science **291** (2019) 011001 doi:10.1088/1755-1315/291/1/011001

Preface

This volume of the proceedings collects the papers accepted for publication and presentation on 2019 3rd International Conference on Energy and Environmental Science held in Seoul, South Korea on January 26-29, 2019. With the successful experience of the last two years in Thammasat University, Thailand and University of Malaya, Malaysia, ICEES 2019 achieved a success and provide a better platform for all the participants.

The goal of ICEES is to discuss the latest research and results of scientists and engineers from academic side and industrial side related to energy and environmental science topics. ICEES2019 had submissions from 19 countries and regions including Kuwait, Thailand, UK, Saudi, Arabia, China, Malaysia, Viet Nam, Taiwan, Indonesia, Poland, India, Pakistan, South Korea, Mongolia, Japan, Philippines, Germany, and Denmark, and accepted 51 high quality papers related to energy engineering, environmental quality monitoring and protection, power system and power generation technology, environmental and ecological engineering, environment and urban construction, clean energy and renewable energy and so on. All the papers were subjected to peer-review by conference committee members and international reviewers.

The conference is highlighted with the three invited speakers who gave an excellent speech: Prof. Hee-Je Kim from Pusan National University, Korea delivered a speech entitled "Various Solar+ESS (Energy Storage System) Applications Controlled by Smart-phone"; Prof. Saad Mekhilef (IET Fellow) from University of Malaya, Malaysia focuses on "Transformer-less Online Uninterruptible Power Supply (UPS) System" as well as Prof. Hyungkee Yoon from Korean Maritime University, South Korea gave a talk on "Energy Transition toward Green and Sustainable KOREA".

The proceedings give an exciting and wide-ranging discussion of the topics presented at ICEES2019. The peer-reviewed and accepted papers have been edited as conference proceedings to be published with IOP Conference Series: Earth and Environmental Science (EES).

IOP Conf. Series: Earth and Environmental Science **291** (2019) 011001 doi:10.1088/1755-1315/291/1/011001

In the end, these proceedings could not have been published without the dedication and hard work of the ICEES committee. We appreciate all the authors who have contributed to this conference and also to the reviewers, speakers, chair persons and all the conference participants for their selfless time and efforts to ICEES 2019. Hope to see you in ICEES2020 again!

Best Regards.

Prof. Dr. Saad Mekhilef,
Conference Chair of ICEES2019
University of Malaya, Malaysia

Statement of Peer Review

All papers published in this volume of IOP Conference Series: Earth and Environmental Science (EES) has been peer reviewed through processes administered by the proceedings Editors. Reviews were conducted by expert referees to the professional and scientific standards expected of a proceedings journal published by IOP Publishing.

IOP Conf. Series: Earth and Environmental Science **291** (2019) 011002 doi:10.1088/1755-1315/291/1/011002

COMMITTEES

Conference Chairs

Prof. Chul H. Jo, Inha University,Incheon, South Korea
Prof. Dr. Saad Mekhilef, University of Malaya, Malaysia

Program Chairs

Prof. Yuanyuan Duan,Tsinghua University, China
Prof. Hazlie Mokhlis, University of Malaya, Malaysia
Prof. Hyungkee Yoon, Korean Maritime University, South Korea

Technical Committee

Asst. Prof. Jennifer L. Santos, Centro Escolar University, Philippines
Assoc.Prof. Sandro Nizetic, FESB Faculty, University of Split, Croatia
Assoc. Prof. Thanakom Soontornchainacksaeng, King Mongkut's University of Technology North Bangkok. Thailand
Prof. Yee-Chang Feng, National Cheng Kung University, Taiwan
Prof. Yogesh Pahariya, Lakshmi Narain College of Technology, India
Assit. Prof. Varun Goel, National Institute of Technology, India
Assit. Prof. Kumar K, SV College of Engineering, India
Dr.Tsung-Mou Huang,Taiwan power company,Taiwan
Dr.Jiahao Liu, University of Science and Technology of China,China
Dr.Lau Woei Jye, Universiti Teknologi Malaysia, Malaysia
Lecturer Brabha Hari Nagaratnam, Universiti Utara Malaysia, Malaysia
Dr.Umarin Sangpanich, Kasetsart University, Thailand
Lecturer Unggul Wibawa, Universitas Brawijaya, Indonesia
Prof. Ali CHEKNANE, Amar Telidji University of Laghouat, Algeria
Dr. Alireza Ghasempour, University of Applied Science and Technology, Iran
Dr. Song He, Xiamen University, China
Assoc.Prof. R. K. Saket, Indian Institute of Technology (Banaras Hindu University), India
Dr. G. Kumaresan, Anna University, India
Dr. Sandip Singh Bhatti, Guru Nanak Dev University, India
Dr. Lau Woei Jye, Universiti Teknologi Malaysia, Malaysia
Dr.Umarin Sangpanich, Kasetsart University,Thailand
Prof. Manish Bhardwaj, SRM University, India
Asst. Prof.Sohail Ahmed, University of Haripur, Pakistan
Prof. Abdelbaki Benamor, Qatar University, Qatar
Yang (Leo) DU, Illinois State Geological Survey, USA
Dr. Di Song, Praxair Inc.,USA
Dr.Qing Zhu, US Lawrence Berkeley National Laboratory, USA
Dr. SITI AKHTAR BINTI MAHAYUDDIN, UiTM Cawangan Perak, Seri Iskandar Campus, Malaysia
Senior Lecturer Nuruol Syuhadaa Mohd, University of Malaya, Malaysia
Assoc.Prof. Dr. K. J. Sreekanth, Kuwait Institute for scientific research (KISR), Kuwait

IOP Conf. Series: Earth and Environmental Science **291** (2019) 011002 doi:10.1088/1755-1315/291/1/011002

Peer review statement

All papers published in this volume of *IOP Conference Series: Earth and Environmental Science* have been peer reviewed through processes administered by the proceedings Editors. Reviews were conducted by expert referees to the professional and scientific standards expected of a proceedings journal published by IOP Publishing.

IOP Conf. Series: Earth and Environmental Science **291** (2019) 012001 doi:10.1088/1755-1315/291/1/012001

Estimated effects of air quality control measures on mortality reduction and economic benefits during the 2014 Nanjing Youth Olympic Games

S Y Yu[1], Y Qian[1,2]

[1] State Key Laboratory of Pollution Control and Resource Reuse, School of the Environment, Nanjing University, Nanjing 210093, China

[2] Corresponding author: yqian@nju.edu.cn

Abstract. During the 2014 Youth Olympic Games in Nanjing, the local government implemented a series of emission control measures. Fixed-effect model and generalized additive model were conducted to evaluate the association between control measures and health benefits. Daily non-accidental mortality decreased from 71 deaths per day before taking control measures to 61 during the under-control period. Interim measures reduced SO2, NO2, PM2.5, PM10 by 6.63%, 12.96%, 8.72% and 10.30%, while comprehensive measures further reduced pollutants by 12.19%, 8.89%, 25.05% and 30.86%. A 10μg/m3 decrease in SO2, NO2, PM2.5, PM10 was associated with decreases in all-cause mortality of 2.91%(95%CI:1.57%,4.24%), 1.36%(95%CI:0.62%,2.11%), 0.55%(95%CI:0.23%,0.87%) and 0.39%(95%CI: 0.15%,0.62%). According to the calculation, the game specifically reduced about 1000 acute deaths from all cause and produced about 500 million USD of benefits, demonstrating a statistically significant association between air quality controls and health benefits. The result also shows that relevant controls to deal with NO2 and PM should be taken more stringently in heavily polluted cities.

1. Introduction

The 2nd Summer Youth Olympic Games(YOG) was held from 17 to 28 August 2014 in Nanjing, where suffers from severe air pollution. In order to realize the promise of "Green YOG", the local government took a series of measures to reduce emissions of air pollutants. A large number of studies have shown that air pollution can cause significant human health risks [1]. Most researches on air control measures for major events focus on the evaluation of air quality. Huang et al.[2] found that the concentration of NO_2 columns in Beijing and surrounding areas decreased significantly during the 2014 Beijing APEC compared with the previous three years. Integrated emission reduction measures reduced 66.8%, 51.3%, 21.5% and 17.1% of the SO_2, NO_2, PM_{10} and $PM_{2.5}$ concentration during the Guangzhou Asian Games[3]. The three pollution control measures implemented at the 2015 2nd World Internet Conference (WIC) in Jiaxing cut $PM_{2.5}$, inhalable particles, NO_2 concentration by 62.1%, 47.1%, 31.2% respectively[4].

In this study, an integrated air quality and health benefit model was applied to evaluate the health-based benefits for YOG. The first step was to investigate the specific effects of control measures on the concentration changes of air pollutants. The second epidemiological question was to estimate the short-term health effects of changes in pollutants caused by control measures. Finally, we measured air pollution-associated health outcome in terms of mortality and monetization, which is used to provide a basis for the formulation of control plans of environmental protection administrative departments.

2. Material and methods

2.1 Integrated air quality and health benefit framework

The study integrates three models, including Fixed Effect model to evaluate air quality, GAM[5] model to evaluate health impact and VOSL(Value of Statistic Life) model[6] of health economic cost of air pollution, to identify the health and economic benefits of air pollution due to emission control measures during YOG. The air quality model outputs the quantitative influence of control measures on the current period of pollutants based on the grid meteorological data and air pollutant data. The health effects of air quality changes including the mortality of all-cause, cardiovascular, respiratory, stroke, ischemic heart disease(IHD), chronic obstructive pulmonary disease(COPD) are calculated in the health impact model using the exposure-response functions which is drawn from an annual time-series analysis. Finally, Value Of Statistics Life(VOSL) Approach is used to monetize health benefits.

2.2 data collection and scenarios setting

The data consist of ground observations of air pollutants, meteorological parameters and basic population data from 1 January 2014 to 31 December 2014. Air quality data including $PM_{2.5}$, PM_{10}, NO_2 and SO_2 was collected from Qingyue Open Environmental Data Center (https://data.epmap.org), which has been providing air quality information of state controlling air monitoring sites in Nanjing. Weather data was drawn from the European Centre For Medium Range Weather Forecasts(ECMWF) Project[7] and China ground international exchange station(V3.0) (http://data.cma.cn/data/cdcdetail/dataCode/SURF_CLI_CHN_MUL_ DAY_CES_V3.0.html).The population data was selected as the exposed population, according to the Nanjing statistics yearbook (2014) (Statistics Bureau of Nanjing City, 2011). Daily mortality data was obtained from Jiangsu Provincial Center for Disease Control and Prevention, providing total and specific-cause mortality. Causes of deaths were coded by the Tenth Revision of the International Classification of Diseases[8][9], which included total natural causes (A00 – R99), cardiovascular diseases (I00 – I99), respiratory diseases (J00 – J98), stroke (I60 – I69), ischemic heart disease(IHD:I20-I25) and chronic obstructive pulmonary disease (COPD: J41 – J44).

According to “the 2014 Nanjing Youth Olympic Games environmental quality interim control plan”, we divide the implementation stage of control measures into four periods. The air quality control measures in different situations are shown in table 1.

Table 1. Air quality control measures in different stages of YOG.

Period	Stage of implementation	Specific control measures
January 1st to June 30th	Without control	
July 1st to July 31st	Interim control	Coal-fired power plants control: Coal enterprises used high-quality low-sulfur coal with sulphur content less than 0.7 percent.
August 1st to August 31st	Comprehensive implementation	Industrial point sources control: A total of 298 enterprises in Nanjing were suspended production and restricted production. The construction site control: 2,952 construction sites in Nanjing were closed. Road moving sources control: 30% of business vehicles were suspended; yellow label cars were banned all day; highly polluting vehicles were strictly restricted to enter into the main city; A total of 6277 heavy or medium vehicle transport vehicles were suspended. Regional joint prevention and control: Temporary controls on coal burning and other industries were imposed in 23 key cities around Nanjing.
September 1st to December 31st	Without control	

2.3 Statistical analysis

We estimated the effect of air quality control measures of YOG on air pollutant concentrations using a fixed-effect model with instrumental variables. We introduced the lag period of 2 days to pollutants for regression by student's t test and R square.

$$Q_t = \alpha_0 + \alpha_1 * Q_{t-1} + \alpha_2 * Q_{t-2} + \alpha_3 * pressure_t + \alpha_4 * temperature_t + \alpha_5 * DOW + \alpha_6 * CON1 + \alpha_7 * CON2 + \varepsilon_t \quad (1)$$

where Q represents the concentration of pollutant, pressure, temperature and DOW represent the surface pressure, 24-hour daily average temperature and indicator for day of week (regarding weekends as 1). CON is an indicator of control variable, as the enforce in August, we set CON1 as 1 in July and August, CON2 as 1 in August to differentiate the impact. t indicates the time, α is the coefficient of the model. ε_t *is* unobservable disturbances.

Poisson additive models with over-dispersion were used to estimate the associations between mortality and air pollutants.

$$\log[E(Y_t)] = \alpha_0 + s(temperature, k = 6) + s(humidity, k = 5) + s(pressure, k = 4) + lag(rain, 1) + \alpha_1 * Q_t + \alpha_2 * DOW \quad (2)$$

Where $E(Y_t)$ is the expected mortality count, humidity is relative humidity, rain is daily precipitation and we lag the period for 1 day. Following the previous study, we determine the degrees of freedom k by AIC and Anova criterion, this resulted 6df for temperature, 5df for humidity and 4df for pressure. Also, we estimate the moving average for previous 3 day's concentration of the air pollutant.

The mortality reduction can be estimated using the following formula (3). Our economic benefit assessment was based on the VOSL method (5) [10][11].

$$N = -\text{population} * (baseline\ incidence * [\exp(-\alpha_1 * \Delta C) - 1]) \quad (3)$$

$$\Delta C = Q_t * CON4 + Q_t * CON5 \quad (4)$$

$$VOSL = VOSL_{CQ} \times (I_{NJ}/I_{CQ})^e \quad (5)$$

$$EC_{\mathrm{al}} = N \times VOSL \quad (6)$$

Where N is the cases of health effects, α_1 is the coefficient, ΔC is the change in air pollutant concentration. Where VOSL and $VOSL_{CQ}$ were the values of a statistical life of Nanjing residents and Chongqing residents in 2014, respectively. I_{NJ} and I_{CQ} were the incomes of Nanjing and Chongqing in 2014, respectively. VOSL represents an individual's willingness to pay(WTP) for a marginal reduction in risk of death. e is the elastic coefficient of WTP and is assumed to be 1.0. EC_{al} is the economic benefit of human impacts linked to air pollution during YOG in Nanjing.

3. Results

Descriptive statistics of the data used in the statistical analysis are presented in Table 2. A total of 24728 total deaths were recorded. Daily non-accidental mortality decreased from 71 deaths per day before taking control measures to 61 during the under-control period, a 13.27% overall decrease. Daily total death toll rose to 66 when control measures were lifted after the game. Daily cardiovascular, respiratory, stroke, IHD and COPD disease mortality decreased by 18.32%, 23.99%, 17.96%, 11.36% and 31.81% and rose by 5(18.57%), 0.31(1.64%). 2.13(24.26%), 0.46(11.32%) and 0.38(14.66%). Compared to the WHO air quality guidelines of 25 μg/m^3, 50 μg/m^3, 25 μg/m^3 for $PM_{2.5}$, PM_{10}, SO_2[12], the corresponding average concentration during under-control period were still not all up to standard, especially PM_{10} and $PM_{2.5}$ were 1.5 to 2 times higher than the WHO guideline. However, the recovery of fine particle concentration after the cessation of the measures was weaker than that of other pollutants.

Table 2. Overall summary statistics for daily air pollution, mortality and weather variables in Nanjing, 2014.

Variables	Mean(SD)			% change	
	Without-control period[a]	Under-control period [b]	Without-control period[c]	Under-control period**	Without-control period***

Moratality					
Total	70.90(13.24)	61.48(7.77)	66.26(12.06)	-13.27%	7.77%
Cardiovascular	30.59(8.40)	24.98(5.14)	29.62(6.90)	-18.32%	18.57%
Respiratory	8.40(3.85)	6.39(2.57)	6.49(2.88)	-23.99%	1.64%
Stroke	10.73(3.76)	8.81(2.97)	10.94(3.89)	-17.96%	24.26%
IHD	4.59(2.49)	4.06(1.98)	4.52(2.23)	-11.36%	11.32%
COPD	3.78(2.23)	2.58(1.50)	2.96(1.78)	-31.81%	14.66%
Weather					
Temperature (℃)	14.36(8.33)	26.22(2.89)	14.77(7.30)	82.63%	-43.68%
Humidity (%)	70.64(14.78)	85.39(6.94)	73.08(14.04)	20.88%	-14.41%
Pressure (Pa)	10131.87(87.03)	10026.19(38.91)	10169.07(68.35)	-1.04%	1.43%
Rain (mm)	24.64(65.15)	68.11(142.17)	18.27(57.54)	176.48%	-73.18%
Air pollutants($\mu g/m^3$)					
SO_2_24h	28.51(14.21)	12.63(5.38)	20.51(9.97)	-55.68%	62.31%
NO_2_24h	55.85(19.54)	33.84(10.00)	56.29(21.16)	-39.42%	66.36%
$PM_{2.5}$_24h	84.93(46.19)	54.93(28.19)	67.41(32.80)	-35.32%	22.72%
PM_{10}_24h	138.92(68.24)	78.49(37.10)	120.97(49.15)	-43.50%	54.13%

[a] The period from January 1st to June 30th
[b] The period from July 1st to August 31st
[c]The period from September 1st to December 31st

Table 3. The associations between control measures and air pollutants.

Pollutants	Lag 1	Lag 2	CON1(SD)	95%CI	CON2(SD)	95%CI	R^2
log_SO_2	0.5537***	0.0134	-0.0663**(0.0187)	(-0.1029,-0.0297)	-0.1219***(0.0226)	(-0.1663,-0.0776)	0.5027
log_NO_2	0.4468***	0.0575***	-0.1296***(0.0137)	(-0.1564,-0.1029)	-0.0889***(0.0165)	(-0.1213,-0.0565)	0.4468
log_$PM_{2.5}$	0.6468***	-0.1736***	-0.0872***(0.0181)	(-0.1226,-0.0517)	-0.2505***(0.0226)	(-0.2947,-0.2063)	0.4005
log_PM_{10}	0.6234***	-0.1419***	-0.1030***(0.0169)	(-0.1361,-0.0698)	-0.3086***(0.0212)	(-0.3502,-0.2670)	0.4573

*p<0.10, **p<0.05, ***p<0.01

Table 3 shows Interim control measures reduced concentrations of SO_2, NO_2, $PM_{2.5}$, PM_{10} by 6.63%, 12.96%, 8.72% and 10.30%, respectively. In August, further control measures reduced air pollutant concentrations by 12.19%, 8.89%, 25.05% and 30.86%. It can be seen that the greater the intensity of air quality control measures, the more air pollutant concentrations will be cut. At the same time, panel data provided sufficient data to determine the difference in air pollutant concentration caused by the policy, which supported us to accurately estimate the environmental benefits of regulatory measures.

We observed decrements of 2.91%(95% CI: 1.57%,4.24%), 1.36%(95% CI: 0.62%,2.11%), 0.55%(95% CI: 0.23%,0.87%) and 0.39%(95% CI: 0.15%,0.62%) in all-cause mortality for a 10 $\mu g/m^3$ changes in SO_2, NO_2, $PM_{2.5}$, PM_{10} concentrations(**Table 4**). The results indicated that the influence of pollutants on mortality is delayed. $PM_{2.5}$ showed significant influence on mortality and especially useful in Stroke, SO_2 was also a good indicator of air condition. In multi-pollutant models(**Table 5**) the relationship was no longer statistically significant(**Table 6**), so we used the single-pollutant model to estimate health and economic effects.

Table 4. Estimated percent decrease in mortality for a 10 $\mu g/m^3$ change of pollutants in single-pollutant model.

Pollu tant	La g	All cause	Cardiovascular	RPT	Stroke	IHD	COPD
		Increase(95% CI)	Increase(95% CI)	Increase(95% CI)	Increase(95% CI)	Increase(95% CI)	Increase(95% CI)

SO_2	0	2.91***(1.57,4.24)	3.86***(1.89,5.84)	7.31***(3.52,11.09)	3.15(-0.24,6.54)	5.49**(0.56,10.43)	8.01***(2.48,13.55)
	0-2	4.70***(2.97,6.44)	5.29***(2.77,7.81)	12.32***(7.58,17.07)	2.21(-2.16,6.59)	7.98**(1.70,14.25)	11.91***(4.59,19.23)
NO_2	0	1.36***(0.62,2.11)	2.09***(0.97,3.20)	1.99(-0.19,4.17)	1.00(-0.90,2.90)	3.76***(1.04,6.49)	1.44(-1.67,4.54)
	0-2	1.58***(0.54,2.62)	2.79***(1.23,4.34)	3.53**(0.48,6.57)	1.27(-1.34,3.89)	3.72(-0.05,7.49)	2.83(-1.52,7.19)
$PM_{2.5}$	0	0.55***(0.23,0.87)	0.69***(0.21,1.16)	1.59***(0.69,2.50)	0.33***(-0.49,1.16)	1.65***(0.50,2.80)	2.06***(0.76,3.37)
	0-2	0.77***(0.36,1.18)	0.90***(0.29,1.50)	2.02***(0.91,3.12)	0.38***(-0.66,1.42)	2.07***(0.63,3.51)	2.49***(0.88,4.10)
PM_{10}	0	0.39***(0.15,0.62)	0.59***(0.24,0.94)	1.05***(0.38,1.72)	0.27(-0.33,0.88)	1.31***(0.46,2.15)	1.01**(0.03,1.99)
	0-2	0.62***(0.31,0.93)	0.85***(0.40,1.31)	1.29***(0.44,2.14)	0.41(-0.38,1.19)	1.66***(0.57,2.74)	1.24(-0.01,2.50)

Table 5. Estimated percent decrease in mortality for a 10 $\mu g/m^3$ change of pollutants in multi-pollutant model.

Mortality	SO_2 Increase(95% CI)	NO_2 Increase(95% CI)	$PM_{2.5}$ Increase(95% CI)	PM_{10} Increase(95% CI)	R^2
All-cause	2.04** (0.124,3.96)	0.64(-0.31,1.58)	0.12(-0.71,0.94)	-0.03(-0.63,0.58)	0.331
Cardiovascular	2.61*(-0.27,5.48)	1.07(-0.34,2.49)	-0.65(-1.88,0.58)	0.55(-0.35,1.44)	0.348
RPT	6.04**(0.38,11.71)	-1.16(-3.96,1.64)	1.20(-1.21,3.61)	0.28(-2.04,1.49)	0.199
Stroke	3.88(-0.99,8.76)	0.16(-2.25,2.57)	-0.53(-2.65,0.16)	0.16(-1.37,1.68)	0.169
IHD	0.62(-7.84,6.60)	2.10(-1.38,5.58)	0.06(-2.98,3.09)	0.98(-1.23,3.20)	0.066
COPD	5.67(-2.77,14.11)	-1.47(-5.55,0.03)	4.63**(0.92,8.33)	-2.57*(-5.36,0.22)	0.198

Table 6. Spearman correlations between air pollutants.

	SO_2	NO_2	$PM_{2.5}$	PM_{10}
SO_2	-	-	-	-
NO_2	0.666***	-	-	-
$PM_{2.5}$	0.611***	0.519***	-	-
PM_{10}	0.713***	0.642***	0.885***	-

We estimated the case of health effects and present the significant result in Table 7.The reduction of air pollutants concentration will decrease 977-1697 deaths. We obtained the per capita disposable income of Nanjing in 2014 as 6069 USD. According to the VOSL of Chongqing as 34458 USD[13], we calculated the VOSL of Nanjing as 426787 USD in 2014. Table 7 represents the economic Benefit of control measure as 553.12 million USD on average of all air pollutant, 452.77 million USD on average of SO_2 and $PM_{2.5}$. Although NO_2 was cut down less than other pollutants, it was actually going to have a larger population health effect. From the point of view of maximizing health benefits, the control measures should concentrate on reducing SO_2, but when taking air quality in Nanjing, reducing NO_2, $PM_{2.5}$ and PM_{10} is more emergency.

Table 7. The cases and economic benefits of health effects between air pollutants and mortalities.

Air Pollutant	the Reduction of Death in Statistic life						Economic Benefit(USD, Million)
	All cause	Cardiovascular	RPT	Stroke	IHD	COPD	
SO_2	977.3	528.2	255.3	-	122.1	113.1	417.11
NO_2	1697.0	1055.7	-	-	309.6	-	724.27
$PM_{2.5}$	1144.4	581.7	206.0	99.5	227.9	107.7	488.43
PM10	1365.2	839.4	227.0	-	302.9	88.6	582.68

4. Discussion

The study has presented a new comprehensive evaluation method of the economic assessment of the health effects related to the air quality control measures during YOG. The key aims of this study were to (1) estimate quantitative impact of control measures on air quality at different stages. (2) use the exposure-response relationship to calculate and monetize the health effects of control measures by affecting air quality.

Meanwhile, Limited by the length of the data, we failed to consider the seasonal trend of pollutant concentration in the model, which may lead to errors in the estimation results due to changes in climatic conditions. We should supplement data for more years especially exposed population demographic data in further research. In addition, our hypothesis about the relationship between the concentration of air pollutants and mortality(Eq.(1)) was validated by the model only from a statistical point of view, and the pathological correlation was not further explored. Evidence about the relative toxicity of different particles is ambiguous, and the synergistic effects of different particles do exist from a pathological point of view. Similar to most papers, we can only abandon this relationship and infer from the results of the single pollutant model because the results of this study cannot show significant differences.

5. Conclusion

An integrated air quality, health and economic model was established to evaluate the mortality reduction from non-accidental causes and from cardiovascular, respiratory, IHD, COPD and stroke diseases. In summary this event specifically reduced about 1000 acute deaths from all cause and produced about 500 million USD for any of the four pollutants. According to the model, we found that air quality control measures during YOG made a great contribution to population health effects and economic benefits. The results are of great reference value to similar studies in developing countries with rapid economic development, and the model can be applied to other major international events where air control measures are implemented.

6. References

[1] Mannucci P M, Harari S, Martinelli I, et al. Effects on health of air pollution: a narrative review[J]. Internal & Emergency Medicine, 2015, 10(6):657-662.

[2] Huang K, Zhang X, Lin Y. The "APEC Blue" phenomenon: Regional emission control effects observed from space[J]. Atmospheric Research, 2015, 164-165:65-75.

[3] Liu H, Wang X, Zhang J, et al. Emission controls and changes in air quality in Guangzhou during the Asian Games[J]. Atmospheric Environment, 2013, 76(5):81-93.

[4] Shen L, Wang H, Sheng L, et al. Influence of pollution control on air pollutants and the mixing state of aerosol particles during the 2nd World Internet Conference in Jiaxing, China[J]. Journal of Cleaner Production, 2017.

[5] Dominici F, Mcdermott A, Zeger S L, et al. On the use of generalized additive models in time-series studies of air pollution and health.[J]. American Journal of Epidemiology, 2002, 156(3):193-203.

[6] Zhang M, Song Y, Cai X, et al. Economic assessment of the health effects related to particulate matter pollution in 111 Chinese cities by using economic burden of disease analysis.[J]. Journal of Environmental Management, 2008, 88(4):947-954.

[7] Palmer T N, Barkmeijer J, Buizza R, et al. The ECMWF Ensemble Prediction System[J]. Meteorological Applications, 2010, 4(4):301-304.

[8] Shaughnessy W J, Venigalla M M, Trump D. Health effects of ambient levels of respirable particulate matter (PM) on healthy, young-adult population[J]. Atmospheric Environment, 2015, 123:102-111.

[9] Organization W H. Tenth revision of the International Classification of Diseases chapter V (F: mental, behavioural and developmental disorders, clinical descriptions and diagnostic guidelines[J]. World Health Organization, 1988.

[10] Du Y, Li T. Assessment of health-based economic costs linked to fine particulate (PM 2.5) pollution: a case study of haze during January 2013 in Beijing, China[J]. Air Quality Atmosphere & Health, 2016, 9(4):439-445.
[11] Zhang M, Song Y, Cai X, et al. Economic assessment of the health effects related to particulate matter pollution in 111 Chinese cities by using economic burden of disease analysis.[J]. Journal of Environmental Management, 2008, 88(4):947-954.
[12] Organization W H. Air quality guidelines: Global update 2005. Particulate matter, ozone, nitrogen dioxide and sulfur dioxide.[J]. Indian Journal of Medical Research, 2007, 4(4):492-493.
[13] Wang H, Mullahy J. Willingness to pay for reducing fatal risk by improving air quality: A contingent valuation study in Chongqing, China[J]. Science of the Total Environment, 2006, 367(1):50-57.

Acknowledgements
This research was financially supported by the National Key R&D Plan (2016YFC0207603) and National Natural Science Fundation (71573116) of China. We would like to thank Qingyue Open Environmental Data Center, Jiangsu Provincial Center for Disease Control and Prevention and China ground international exchange station for data support.

IOP Conf. Series: Earth and Environmental Science **291** (2019) 012002 doi:10.1088/1755-1315/291/1/012002

Cross-city PM2.5 predictions with recurrent neural network

RH Zong[1], TY Zhang[2], Z Chen[3], Y Zhu[4]

[1] School of Computer Science, Sichuan University, Chengdu, Sichuan, China
[2] School of Statistics, Jiangxi University of Finance and Economics, Qingshanhu, Nanchang, China
[3] School of Life Science and Technology, University of Electronic Science and Technology of China, Chengdu, Sichuan, China
[4] School of Electronic Engineering, Beijing University of Posts and Telecommunications, Beijing, China

Email: 2016141093007@stu.scu.edu.cn

Abstract. PM2.5 is inhalable particulate with a diameter less than 2.5 μM that easily enters the lungs and causes diseases and non-accidental death. The generation and dissipation of PM2.5 are strongly affected by a variety of environmental factors, thus the concentration of PM2.5 is presumably predictable with the observations of environmental conditions. This paper used multi-year meteorological and PM2.5 concentration data across multiple megacities in China (Beijing, Chengdu, and Shenyang) and sought for a universal predictive model. Our results showed that data-driven machine-learning model was able to not only capture PM2.5 dynamics at the city where the model was trained but also could be generalized to predict PM2.5 concentrations over other cities. Therefore, the modeling results indicated a universally existing predictive relationship between PM2.5 source-sink dynamics and the environmental drivers.

1. Introduction

PM2.5 are particulate matters having a diameter of fewer than 2.5 micrometers. These fine particulate matters are the critical pollutant, which could cause adverse impacts on the human health system. The size feature of PM2.5 makes it difficult to be protected from. They are small in diameter and the large surface areas that are capable of suspending in the atmosphere for a long time and carrying a large number of toxic substances. In addition, the small size enables itself to pass throw the nose hair and impair the function of lung and other parts of the body. A growing number of researches revealed that particulate matter enhanced the relevant diseases and mortality rate. For example, "Harvard Six Cities Study" demonstrates that there is a linear dependence between PM2.5 and non-accidental mortality [1].

Observational evidence came from Beijing Normal University site, from 2001 to 2006, revealed that atmospheric PM2.5 were black carbon crustal elements, nitrates, ammonium salts, and sulfates [2], which indicated that the PM2.5 concentrations were source emission controlled mainly from industrial emission, soil dust, coal combustion, secondary aerosol, and biomass burning [2]. Besides, the evolution of atmospheric PM2.5 concentration is also dependent on the variability of boundary layer depth, wind speed, air humidity, and many other environmental drivers that control the sink strength of PM2.5.

Predictive modeling of PM2.5 concentrations across multiple cities is challenging. Traditional process-based modeling framework of PM2.5 source-sink dynamics is comprised of atmospheric transport and chemical models, such as WRM-Chem and CMAQ [3]. These models can be used to predict multiple atmospheric pollutants over the whole region, but their accuracy is highly dependent on the prescribed source emissions that are very difficult to generate [3]. Data-driven times series based data mining models are the alternative numerical framework to predict atmospheric PM2.5 concentrations, including e.g., Auto Regression Moving Average, Stochastic Volatility, Stock-Watson Model, Support Vector Machine, and artificial Neural Network [4]. These frequently used machine-learning methods vary significantly by its complexity and prior assumption about the modeled system. However, none considers the influence of the system memory (previous moment) on the future dynamics. Furthermore, each model is subject to the specific condition in the city where the model is trained. Thus, it is hard to be generalized to other cities. In this study, we employ Recurrent Neural Network, to tackle the system memory of PM2.5 dynamics and aim to generate a universal machine learning models suitable to predict PM2.5 concentrations over multiple megacities in China.

2. Methodology

2.1 Data

We used the meteorology data and PM2.5 concentration data recorded in Beijing, Chengdu, and Shenyang in China from 2010 to 2015 with the hourly resolution [5]. The density distribution of the data is shown in Figure 1. The red, green and blue curves correspond to Beijing, Chengdu, and Shenyang, respectively.

The mean values of PM2.5 concentration in Beijing, Chengdu, and Shenyang are 95.83, 84.66, and 75.08, and the variances are 91.80, 57.96, and 65.96, respectively. Among them, Beijing has the highest average value, but its variance is also the highest, indicating that it is likely to be more serious when it is polluted. The mean value of Chengdu is medium, but the lowest variance indicates that its fluctuation is the smallest, revealing that PM2.5 pollution event could last longer (sink slowly) than the other two cities.

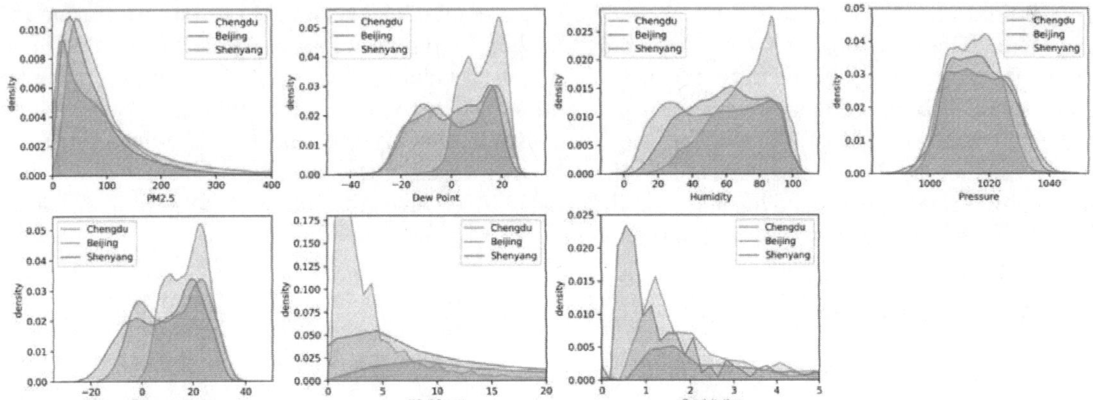

Figure 1. Density distribution of various different drivers and PM2.5 concentrations at Beijing, Chengdu, and Shenyang.

2.2 LSTM Model

We used memory-enabled Recurrent Neural Network (RNN) to predict PM2.5 concentrations across Beijing, Chengdu, and Shenyang at the same time. RNN maintains and updates the model hidden state h_t by a nonlinear activation of a linearly combining of input matrix X, previous hidden state h_{t-1} with weight parameters *W* and error *b*. Then, the output prediction matrix y is computed by multiplying the value of the hidden layer by its corresponding weighted matrix (Eq. 1, Eq. 2).

$$h_t = \sigma(W_{ih}x_t + W_{hh}h_{t-1} + b_h) \quad (1)$$

$$y_t = W_{ho}h_t + b_o \quad (2)$$

Specifically, we used Long Short-Term Memory Networks (LSTM) [6] implementation of RNN with memory cells that can add the cell value of the last time to the new model with a certain weight. LSTM can process data that is highly dependent on historical state or currently input by optimizing the value of the input gate and the forget gate.

$$i_t = \sigma(W_{xi}x_t + W_{hi}h_{t-1} + W_{ci}c_{t-1} + b_i) \quad (3)$$

$$f_t = \sigma(W_{xf}x_t + W_{hf}h_{t-1} + W_{cf}c_{t-1} + b_f) \quad (4)$$

$$c_t = f_t c_{t-1} + i_t \tanh(W_{xc}x_t + W_{hc}h_{t-1} + b_c) \quad (5)$$

$$o_t = \sigma(W_{xo}x_t + W_{ho}h_{t-1} + W_{co}c_t + b_o) \quad (6)$$

$$h_t = o_t \tanh(c_t) \quad (7)$$

W in the equations is the corresponding weight matrix, and σ is activation function. In the following equations, i, f, and o are the input gate, the forget gate, and the output gate, all of which are scaled value between 0 and 1 that are typically activated by the sigmoid function. New input matrix and hidden layer matrix of the last time are activated by tanh function and multiply by the input gate i (Eq. 3, Eq. 5). New cell activation c (Eq. 5) is computed by linearly combining the product of the last c and the forget gate f (Eq. 4), and the activated input value multiplied by the input gate. The output gate o (Eq. 6) is computed by linearly combining the input matrix X, the hidden layer matrix h of the last time, and the cell activation value. Then the memory cell value activated by tanh function is multiplied by the output gate to calculate the new hidden layer matrix h (Eq. 7).

We aim to use the data from three cities (Beijing, Chengdu, and Shenyang) to build and train the LSTM model and to predict the respective PM2.5 concentration across all three cities. The input matrix X is meteorological features, and specifically includes dew point, humidity, pressure, temperature, wind speed, and precipitation. The output matrix y is the predicted PM2.5 concentration at the city where LSTM is trained. After the respective predictions, the models of the three cities are used to predict the PM2.5 concentration of the other two cities to test whether the model is scalable. We find the most representative of the models that best predict other cities. It may be possible to extend it to a wider range of forecasts in the future.

3. Results and Discussion

3.1 Optimize model performance at a single city

Since the various important hyper-parameters have significant impacts on the overall model performance, we test and compare the effects of those parameters on the model predictability. The batch size was tested with six values: 100, 200, 1000, 2000, 5000, 10000. We found that the value for the best test results was 100, while 10000 was significantly inferior to other results. As the value increases, the Pearson coefficients become smaller and smaller, which is consistent with the empirical knowledge that training fewer data at a time can lead to more accurate results.

We tested six epoch values: 10, 20, 40, 60, 80, 100, and the results are shown in the figure below (Figure 2) as an example of the experiments in different parameters. The results corresponding to 10 and 20 were significantly worse than the other results, while the results of the other values of epoch increased slightly as the value increases, which was consistent with training more times can get better results. Nevertheless, the result corresponding to 100 in the model trained in Chengdu had a large error, which may be due to over-fitting caused by excessive training. Therefore, epoch size 80 was considered the optimal parameter for selection in this experiment.

We also tested the number of nodes per layer: 10, 20, 30, 40, 50. The difference in prediction results is less obvious than the first two parameters, and there are single values do not match the overall trend. However, in general, the prediction result is more accurate as the number of nodes increases.

We also conducted leave-one-out experiment to generate insights on which input feature was more important in modeling PM2.5 dynamics. The correlation coefficients of the prediction results were not much different, but the mean square error was the smallest when the feature 'pressure' was removed, which confirmed the potential importance of pressure fields in controlling the local atmospheric transport of pollutants.

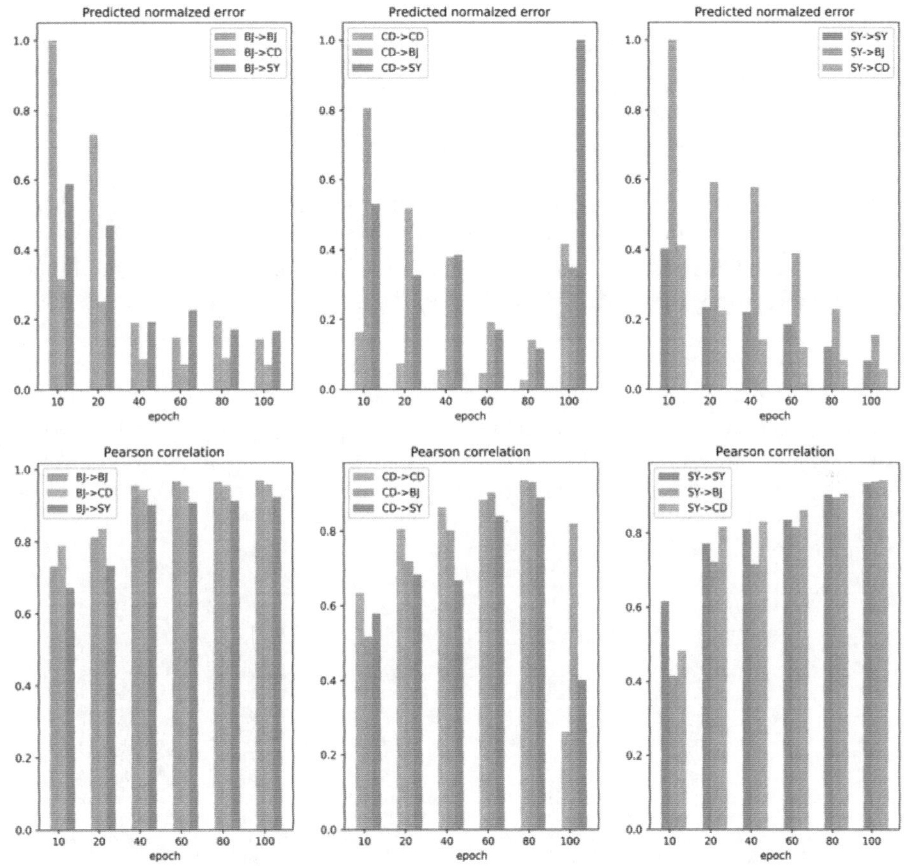

Figure 2. Example of hyper-parameter tuning of RNN PM2.5 modeling in three cities.

In Beijing, Chengdu, and Shenyang, the Pearson correlation coefficients of the predicted and tested values of the PM2.5 predictions using the best parameters were 0.975, 0.962, and 0.959, respectively. Among them, Beijing has the best results, thus, we showed the distribution of the observed versus modeled PM2.5 concentrations in Beijing as an example (Figure 3).

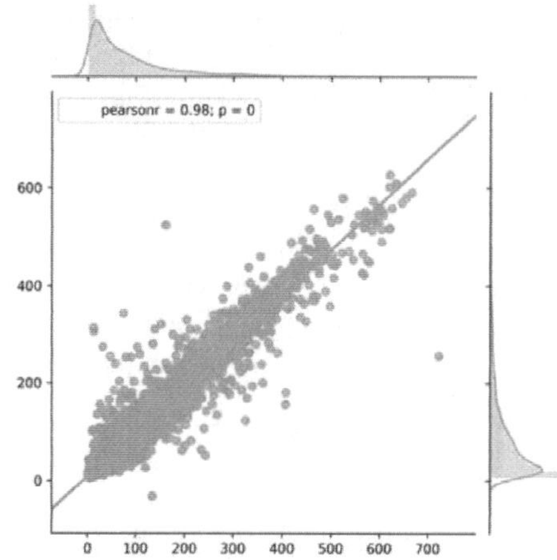

Figure 3. Best model performance in Beijing.

3.2 PM2.5 predictions across multiple cities

The second goal of this study is to evaluate the generality of RNN models from one city to another. The results of the cross-city PM2.5 predictions with those best hyper-parameters (section 3.1) across cities indicated that individual RNN trained in one city could be generalized to the other two cities (Figure 4). Specifically, using the model trained by Beijing's data, the Pearson correlation coefficients of the predicted and tested values of Beijing, Chengdu, and Shenyang were 0.975, 0.962, 0.936, respectively. With the model trained by Chengdu's data, the Pearson coefficients of the predicted and tested values of Chengdu, Beijing, and Shenyang were 0.962, 0.963, 0.940, respectively. With the model trained by Shenyang's data, the Pearson coefficients of the predicted and tested values of Shenyang, Beijing, and Chengdu were 0.959, 0.965, 0.964, respectively. Among them, RNN trained in Shenyang was the most representative model because it had better results both in Beijing and Chengdu.

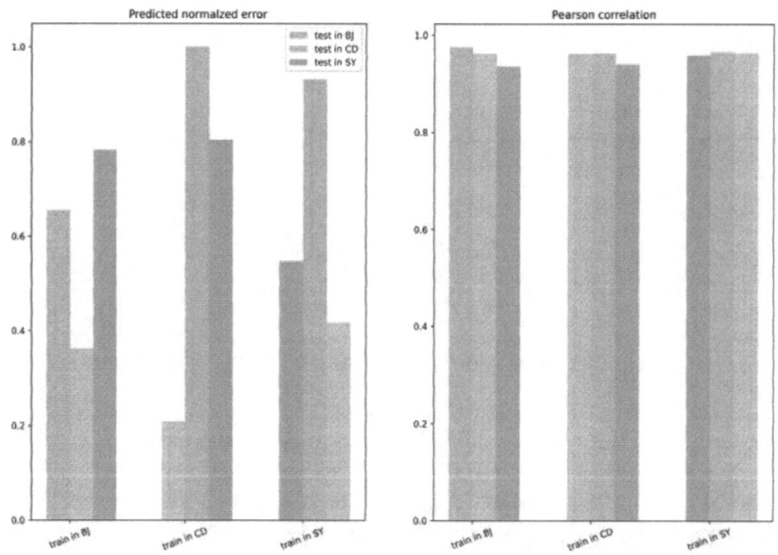

Figure 4. Across-city model performances shown by normalized error and Pearson correlation.

4. Conclusion

Particular matter pollution over megacities is increasing severely due to the rapid industrial and economic development. This paper aimed to use the advanced memory-enabled neural network to model PM2.5 dynamics across Beijing, Chengdu, and Shenyang. Our results demonstrated that multiple hyper-parameters significantly contributed to the overall performance of the model at each individual city. More importantly, the machine-learning model trained in one city could be effectively generalized to other cities, without significantly losing model performance. It implied that the internal relationship between PM2.5 source-sink dynamics and the environmental drivers were closely linked and the linkage might be universal across cities.

5. References

[1] Dockery D, Pope C, Xu X, Spengler J and Ware J 1993 An association between air pollution and mortality in six U.S. cities *NEJM*, 329(24): 1753

[2] Wang H, Zhuang Y, Wang Y, et al 2008 Long-term monitoring and source apportionment of PM2.5/PM10 in Beijing, China *Journal of Environmental Sciences*, 20 (11) :1323-1327

[3] Xia X, Zhao W, Rui X, et al 2016 A comprehensive evaluation of air pollution prediction improvement by a machine learning method *IEEE International Conference on Service Operations and Logistics*, 176-181

[4] Shen J 2012 PM2.5 concentration prediction using times series based data mining *City*, 2013, 2014

[5] Liang X, Li S, Zhang S, Huang H, Chen SX, et al 2016 PM2.5 data reliability, consistency, and air quality assessment in five Chinese cities[J] *Journal of Geophysical Research Atmospheres*, 2016, 121

[6] Gers FA, Schraudolph NN, et al 2003 Learning precise timing with LSTM recurrent networks *JMLR*, 3 (1) :115-143

IOP Conf. Series: Earth and Environmental Science **291** (2019) 012003 doi:10.1088/1755-1315/291/1/012003

Tackling Environmental Pollution through Participatory Approach in Nambo, South Tangerang, Indonesia

OC Dewi[1], TU Atmoko[1], HA Alrasyid[2], D Priyandana[3], MBV Dewi[4] and D Siska[4]

[1] Lecturer, Universitas Indonesia, Depok, Indonesia
[2] Director, Bina Ekonomi Sosial Terpadu NGO, South Tangerang, Indonesia
[3] Head, Building and Spatial Agency, South Tangerang Indonesia
[4] Postgraduate Student, Universitas Indonesia

E-mail: ova.candra@gmail.com

Abstract. A participatory approach is well-known for its time-consuming and no-instant approach. Tackling environmental pollution directly from its source in an urban area by participatory approach is very critical. Universitas Indonesia's Community Engagement Program together with the City Without Slum or Kotaku Program in South Tangerang City, Banten Province, Indonesia, conducted in Kampung Nambo. This joint program carried out in two lines through the local government and the community by a bottom-up approach. The inhabitants of the neighbourhood involved through a *charrette* process. The outcome of the *charrette* is the guideline of the new master plan development of Kampung Nambo that responding to tackling the pollution created in the kampung. This master plan will use as a parameter for the Kampung Nambo development program. A team of experts that consists of the local government, the university scholars, the NGO, and the inhabitants joined together for the implementation plan under the intensive and an interactive designated workshop. It concluded that this approach is valuable to integrate the unplanned urban areas development, which happened quite vast in some Indonesian cities and towns. Thus, it helps to reduce the urgency of environmental degradation in the kampung, through awareness and educational campaign. This approach is challenging yet it is a long-term investment.

1. Introduction

South Tangerang city is one of the rapid growing cities in Indonesia. The city established in 2008 but its density is almost as dense as Jakarta, the capital city. There are 3 giant developers invested in the city and thus start-up very rapid urban development in the city. However, there are many slum areas that still exist in the city. Those areas have insufficient access to many services such as clean water, sanitation, and electricity [1].

In the case of towns and cities of Indonesia, slums are created from an unplanned or spontaneous settlement, where majority of urban poor live. To prevent the formation of slum within the kampung areas, Universitas of Indonesia together with BEST (*Bina Ekonomi Sosial Terpadu* or Integrated Social and Economic Development Institution) NGO, and Local Government of South Tangerang City, initiated the program called "Livable City". This program aims to prevent the foundation of slum within the unplanned areas of the city and to integrate the kampung with the city development. Particularly on Kampung Nambo, 03/04 neighborhood. Kampung Nambo is adjacent to Cipeucang landfill and has some threats to become an example of an unplanned area of the city due to its proximity to the other urban infrastructures yet in an affordable land price. Living near to the landfills area is chosen for the implementation of this new model approach. Nonetheless, Kampung Nambo has various potentials, with the natural environment and unique contoured land.

2. Materials and Methods

A broad array of public policy and anthropological research has demonstrated how the slum environment physically and symbolically omits its residents from society. Women in particular use the slum environment as a tool for personal and social redefinition. Women commented on how their techniques to see value in poor environmental conditions, could result in physical changes to environment and body. It reflects their spiritual capacities as "know-ers" of the environmental field, able to identify potential in an inadequate environment [2].

2.1 Charette Method

This study engaged the residents through a charette method, known as an intensive and multi-disciplinary workshop design [3]. This method is commonly used to facilitate discussions between stakeholders on a project or a program. A charrette, a group of experts come together with the community groups and related stakeholders. The program introduces and educates the community in Kampung Nambo about the concept and the goal of Livable City, as well as examples of well-organized villages/ kampung and unwell-organized villages/ kampung in Indonesia. Then, the joined team explore the potentials as well as the problems occurring in the Kampung, and planning the area to mitigate the problems, including mitigating the pollutions caused by the households.

2.2 Stakeholder Formation and The Roles Identification

The program is a joined program from several designated program. Those are Community Engagement Program Universitas Indonesia, City Without Slum or "Kotaku" South Tangerang City, The Polycentric Approach on Urban Waters from Bremen Overseas and Development Association (BORDA) and BEST NGO Group, and also Community-based Sanitation Program from Islamic Development Bank. Stakeholders have their own role and contribution to the program (Table 1).

Table 1. Stakeholder Contribution

No.	Stakeholder	Designated Program	Main Roles	Contribution
1	Universitas Indonesia Team	Community Engagement	Experts for Planning	Master Plan
2	South Tangerang Local Government Team	City without Slum or *"Kotaku"*	Experts for Implementation and Decision Makers	Funding and External Cooperation and Program Integration
3	BEST-BORDA NGO Team	Polycentric Approach on Urban Waters	Experts for Community Development and Sanitation	Household Sanitary Landfill (HSL)
4	Islamic Development Bank Team	Community Based Sanitation Program	Expert of Sanitation	Wastewater Treatment Plant (WWTP)
5	Kampung Nambo Community	Neighbourhood Association 03/04	Expert of the Actual Condition	Reporter and Users

3. Finding and Discussion

3.1 Problems and Opportunities

Obviously, the major problems which are occurring in Kampung Nambo are the odors from Cipeucang landfills which located near the community housing. However, 9 out of 120 households in the area are piping their wastewater directly to the river and polluted clean water source due to overcrowded and the un-proper septic tank constructions. More details from the problems are as follows:

3.1.1 Insufficient public facilities

There is an insufficient public facility in Kampung Nambo. There is only one mosque or Musholla within the kampung. It is the only public facility in the kampung which is used not only for praying but also meetings and discussions. The community does not have a place to do social activities, such as joint workshop and woman empowerment program. Whilst the community has many great potentials to enhance their abilities and to develop their village.

3.1.2 All day with odors

Odors from Cipeucang landfill has become part of the life of the Nambo community. The TPA is in the southwest side of the kampung Nambo, it causes the smell effortlessly carried by the wind. During the rainy season, it gets worst. Plants along the border of TPA and kampung Nambo do not act as buffer zones properly. Even though the community used to it, it is less comfortable for other visitors.

3.1.3 Un-proper sanitation system

Among all the environmental degradation happens in the kampung, un proper sanitation system found significantly in some houses. There are 9 houses out of 120 don't have the septic tank. These houses throw their black water into the river and water bodies. This certainly has a negative impact on the environment because it pollutes river water and reduces groundwater quality.

3.1.4 Unmanaged drainage

In some points in the Kampung, many unmanaged drainages can be found. They are sometimes due to the un-fitted size thus gets the wastewater overloaded or rather there is no proper drainage. Even in some locations, there is no drainage. So that household wastewater flows into the road and causes unpleasant odors. In some locations the road conditions damaged due to the water runoff. This is certainly also due to the limited land and the narrow roads.

3.1.5 Lack of lighting intensity during night time

During night time, many spots are very dark. These spots found almost in alleys in the lower part of the kampung. So, there is not much activity going on in the area at night. In addition, although the safety matter in the village of Nambo sufficiently maintained, residents still feel worried about the dark road conditions.

Aside from the problems, there are also some potentials, such as the abundances of greeneries and water bodies which become the habitat of birds and fish, and the prosperous social life which shown by the proximity and the friendliness among the inhabitants. They know each other houses locations as well as the types and the location of each clean water sources and septic tanks within the settlement. They are also knowledgable about the terrain of their Kampong, the location of the greeneries and water bodies, the width of the streets, as well as the lands' owner, whether they are belong to private or local government. During the intervention, they are also acknowledged by the need of open spaces for socializing and recreation and ways to fulfil those needs.

3.2 Design Interventions

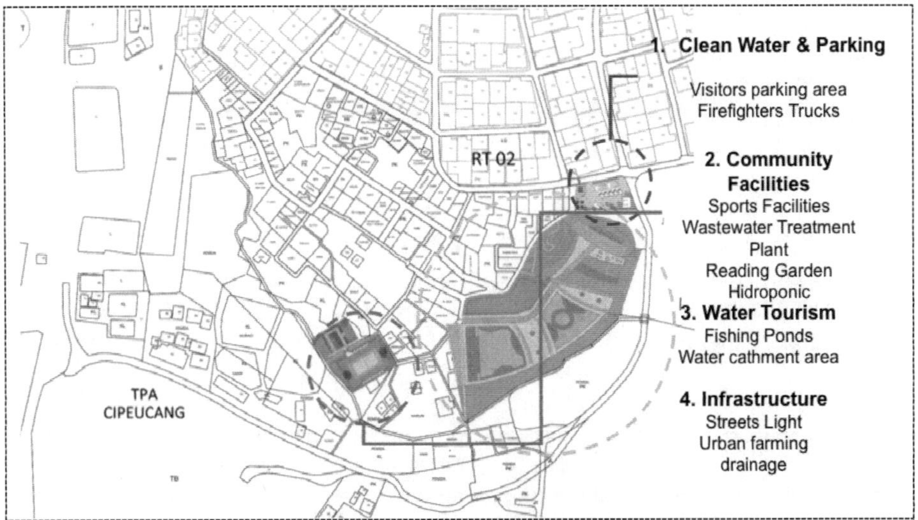

Figure 1. Design Interventions

Based on the problems and potentials, it is clear that they need interventions, mainly on those 5. Through charrette and discussions that have taken place several times, there are various lists of community needs that necessarily applied in the village of Nambo. Some of them are the buffer zone implementation to minimize the bad smell coming from the landfill, a well-managed wastewater treatment plant and clean drinking water sources. For safety reason, the residents also state that they are in need of parking spaces for their cars and fire trucks in case of emergency. Also, it is necessary to fulfill the needs of the children, by suggesting a reading garden to improve the knowledge of the children, and soccer field where the children can play futsal and volleyball during their free time so that they can keep their health as well (Figure 1). The list of needs is then reconciled and developed together with all relevant parties, to produce several criteria regarding design interventions. Such as the placement, dimension, amount, ownership, development plan, construction plan, and maintenance.

3.2.1 Community Facilities and Wastewater Treatment Plant

The Millennium Development Goals (MDG) calls for a 50 percent reduction in the proportion of people without sustainable access to safe drinking water and sanitation. Provision of these services by city authorities is often absent in slum settlements in densely populated countries, nevertheless in Indonesia [4]. That is why the joined program aims to provide communal wastewater plant to make sure that the wastewater is clean enough to released to the river or reused for watering plants, and to keep the source of clean water unpolluted by wastewater. Moreover, a community deep well and pipe water is developed to keep the clean water free from any pollution and manageable.

Because the land price is very high, the communal wastewater plant incorporated within the community park. On top of the communal wastewater plant, there is a playground area for children which also adjacent to a reading spot. WWTP placed in the lowest plane of the village to facilitate wastewater distribution. The reading space can also be used as a workshop for the women to improve their abilities. Later the products of the workshop and the people's handicrafts sold to visitors who come to the village. Behind the reading spot, there is a hydroponic garden where the residents can plant a variety of herbs and vegetables. Adjacent to the park, there is a soccer field where the residents can play futsal or volleyball during their free time to keep them healthy.

3.2.2 Water Recreation Park

Urban kampung tourism has evolved from being practiced at only a limited number of places into a truly global phenomenon, which now performed on five continents [5]. Kampung Nambo blessed with a natural spring, and so that we provide a water recreation park as a source of tourism for Kampung Nambo. The water recreation park consists of clean water reservoir in which the clean water is going to be channeled to the drinking water well above, a skywalk bridge to connect the parking spaces and the water recreation park, a children playground, an artificial river with stepping-stones for children to play and get across, a sitting area for the parents waiting on their children, a dancing fountain, a hamster ball where the children can get inside and walk on the water, a fishing pond for the fish that the residents cultivate, and a buffer zone which consists of greenery to minimize the smell from Cipeucang landfill. In the buffer zone area which is quite extensive, the scheme to reduce odor is by planting trees that can reduce air pollution, such as tanjung trees, then trees that are able to filter groundwater so that it is not polluted by water from landfill, such as bamboo trees and flower trees to provide a fragrance. In addition, the planting of broadleaf trees is also needed to direct the wind naturally.

3.2.3 Clean Water Source, Visitors and Firefighters Trucks Parking Area

The drinking water well located just above the clean water reservoir. The parking space provides the firefighters' trucks, tourists' cars, and motorcycles. There is also a gateway to welcome the tourists, with some stalls nearby which functioned as a place to sell the souvenirs. The residents process the gift such as foods made from fishes, hydroponic vegetables and compost products. Alongside the cliff, there are gazebos where visitors can sit and eat the local cuisine while enjoying the view to the water recreation park below, as well as services area for public toilets and a security post. The master plan implementation, at least for some of the items executed this year by the municipality, including the communal waste treatment plant construction and the deep well construction.

3.2.4 Urban Infrastructure Improvement

The kampung master plan also includes the existing roads and lanes improvement within the community. It provides the guidance for new land/property subdivision to create a proper road and lanes to allow the light and fresh air penetration to the buildings. Moreover, the mediation on organic solid waste directed by community- and home-based composting through Household Sanitary Landfill (HSL). It is a simple method of composting at a source, supervised by BEST NGO. It guarantees the waste at its best quality. Further, the compost will used by the urban farming in the area. The interventions are intended to make the settlement pattern more organized. Thus, the kampung environment becomes more healthy, comfortable, and livable.

4. Conclusion

Tackling the environmental pollution can be done by many approaches; by involving the community is the most effective way of doing it. The participatory approach in Nambo has pulled in many stakeholders to contribute to the livable city program. Improving the impromptu areas and preventing it from slum transformation is critical for the city and especially for the people that possess in the area. The process of the kampung improvement as a pilot project that involves the community, the experts including the NGO, and the municipal agencies whose response to urban infrastructures, proved very fruitful and colourful. The procedure of intervention took some time and the outcome came gradually but evolving. Community approach in Kampung Nambo is very significant to build the sense of ownership. Thus, it empowers the community and keeps the local wisdom alive. Finally, kampung moves toward a livable kampung and becomes a part of a livable city, live, life and prospers.

5. References:

[1] Galiani, S., Undurraga, R., & Gertler, PJ. (2018). The Half-line of Happiness Hedocni Adaptation in the Sujective Well-being of Poort Slum Dwellers to Satisfaction of Basic Housing Needs *Journal of the European Economic Association EEA 16* (4), 1189-1233.

[2] Adubofour, K., Obiri-Danso, K., & Quansah, C. (2012). Sanitation Survey of Two Urban Slum Muslim Communities in the Kumasi Metropolis, *Ghana International Institute for Environment and Development IIED 25* (1), 189-207.
[3] Town paper. (2017). What is Charrette. Availabe from http://www.tndtownpaper.com/what_is_charrette.htm
[4] Smell-Roof, C. (2013). To Know the Field: *Shaping the Slum Environment and Cultivation and Self. Journal of the Society for Psychological Anthropology ETHOS 2013 41* (3), 271-291.
[5] Frenzel, & Koens, K. (2012). Slum Tourism: *Developments in a Young Field of Interdiciplinary Toursim Research, Tourism Geographies TG 2012 14* (2), 195-212.

Acknowledgments
Authors would like to express gratitude to The City Major of South Tangerang, The Research and Community Engagement Directorate Universitas Indonesia (DRPM UI), the community of Kampung Nambo RT 03/RW 04, Agency of Housing, Settlement, and Land South Tangerang, Agency of Building and Spatial Planning South Tangerang, City without Slum or Kotaku Program, and Environmental Agency South Tangerang, Bina Ekonomi Sosial Terpadu NGO. Our sincere gratitude goes to Architecture Department Faculty of Engineering Colleagues and Students for extensive support: Evawani Ellisa, PhD., Intan Chairunisa, Aisyah, Cut Sannas, Erina, Rizka Felly, Intan Findanavy for their worthy contribution and support during the program.

IOP Conf. Series: Earth and Environmental Science **291** (2019) 012004 doi:10.1088/1755-1315/291/1/012004

Constructing a Framework for Economic Value of Urban Renewal in Taiwan

Kuo-Wei Hsu[1] and Jen-Chih Chao[2]

[1]Department of Landscape and Urban Design, Chaoyang University of Technology, Taiwan
[2]PhD program in Architecture and Urban Design, Department of Architecture, Chaoyang University of Technology, Taiwan

E-mail: kwhsu@cyut.edu.tw[1]; jenchihchao@gmail.com[2]

Abstract. Following a phase of early prosperity, Taiwan's urban centers have slowly declined, resulting in the hollowing-out of business districts and declining populations, and prompting widespread interest in urban redevelopment. To improve the quality of urban environments, not least in terms of disaster prevention, the central government's urban-renewal policy is aimed at re-inspiring local development via the introduction of funds and dynamism from the private sector. However, progress has been slow because potential private-sector partners tend to measure projects' merit and performance chiefly in financial terms, and the economic value of urban renewal remains notoriously difficult to quantify. Accordingly, this study proposes a complete economic-value framework for urban renewal in Taiwan through secondary-data methods. Specifically, this framework consists of three facets and six factors, of which the land-use value facet (comprising the land and real-estate value factors) relates to the strength of the local economy; the energy-conservation value facet's two factors are energy-use and carbon-emissions reductions; and the disaster-prevention value facet consists of reducing the likelihood of disasters, and reducing their impact.

1. Introduction

The highly developed cities of early Taiwan experienced slow decline beginning in the 1970's, with urban business districts shrinking and populations migrating to the suburbs. Beginning in the late 1990s, urban-renewal policies for older urban areas have aimed to check or reverse these processes, and have been widely discussed by scholars, policymakers and the general public. Urban renewal is an inherently complex and difficult process involving a wide range goals, including the improvement of living quality, the promotion of sustainable development, and preparation for and prevention of disasters. Through the Urban Renewal Policy Act, passed in 1998, Taiwan's government has sought to accelerate the promotion and implementation of urban renewal by harnessing the power and funds of the private sector: specifically, via measures such as volume awards and tax concessions, and incentivization of green architecture and the provision of open space.

However, private-sector participants and potential participants in urban-renewal projects often use decision-making models that measure the merit and success of such projects simply in terms of increases to real-estate prices. However, the economic value created by urban renewal is both broader and more complex than such models would suggest: including factors such as quality of life, aesthetics, disaster resilience, and what might be called the morale of local people and businesses, all of which have a direct and/or indirect impact on the economic value of the city as a whole. Therefore, this study proposes a comprehensive framework for measuring the effectiveness of urban renewal in terms of its wider economic value.

As Chou has shown, the most controversial aspect of the economic assessment of urban renewal is the measurement of external costs and benefits: for example, the intangible value of historic districts and emotions in community networks, post-renewal increases in local commercial activities, or environmental pollution caused by increased numbers of vehicles. Some of these ripple effects of urban renewal are simply unquantifiable in economic terms, while others are quantifiable but diffused across a complex array of public and private stakeholders [1]. Based on findings that urban renewal positively affected the environmental quality of the surrounding area, Wu proposed that such quantification adopt all or more of the following three orientations: 1) health, e.g., ventilation, sunshine and air quality; 2) safety, e.g., traffic accidents and disaster preparedness; and 3) convenience, e.g., the presence and quality of public facilities, open space, and parking, as well as streetscape aesthetics and crowding [2]. Li noted that the price of housing will be affected by its age, layout and transfer area, as well as by its distance from transportation facilities and green space [3]; and Chen indicated that economic value was accurate as an evaluation index for historical re-use investment [4]. Accordingly, it is suggested that no matter what mode of development is pursued, post-restoration development of historically significant cultural assets should take account of the economic value of their reuse. As such, research on urban renewal should explore its financial impacts and risks, its effects on rights, the role of public/private partnerships, implementation methods, floor-area incentives, capacity transfers, options, and trust funds. Anyone who implements an urban-renewal business plan must confront multi-faceted social, cultural, economic, and ecological problems. Private-sector decision-making about whether to invest in such plans is closely tied to floor-area incentives and capacity transfers, as well as financial costs/benefit analyses and execution-time projections. However, in the view of the Taiwanese government and general public, the promotion of urban renewal is not merely about the generation of economic value, but also about redevelopment in a broader sense, including reduction of the impact of natural disasters on cities in the face of extreme climate events.

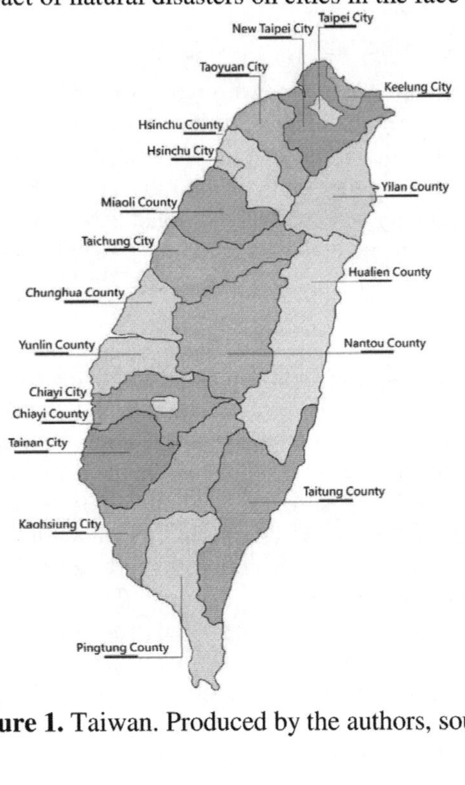

Figure 1. Taiwan. Produced by the authors, sources: [5]

Nevertheless, rates of return on investment continue to dominate private-sector decision-making about urban-renewal investment, which ignore the wider interrelationships between such investment and its less tangible consequences, good and bad alike. The same myopia affects academic research on urban renewal, which likewise tends to prioritize economic value as its chief conceptual orientation. Therefore, the key aim of the current study is to establish a completely novel framework for evaluating urban renewal projects' using the secondary-data method. The geographic scope of this research is depicted in Figure 1.

2. Literature Review

2.1 Definitions of Urban Renewal

Article Three of the Urban Renewal Law of Taiwan defines urban renewal as falling within the scope of urban planning, and as consisting of the implementation of reconstruction, rehabilitation or maintenance measures. "Old" urban renewal, according to Article Seven of the same law, is defined as taking place in areas where old buildings are dense, dilapidated, unsightly, and affect public safety, and therefore must be demolished, rebuilt, or specially maintained.

2.2 Definitions of Economic Value

In economic-value theory, "total economic value" includes "use value" and "non-use value" (Turner), with use value relating to the direct or indirect use of a certain thing, the benefits of which can be divided into current (direct), current (indirect), and future use value. Non-use value, meanwhile, is inherent in the existence of the thing – i.e., is neither called into existence nor changed by any other thing(s) – and can be divided into existential value and bequest value. The former means that even if people have little or no chance of using the resource, they are willing to pay a certain price to maintain its existence. The bequest value, on the other hand, refers to the total value people are willing to pay to protect the resource for the use of future generations, regardless of whether they are also using it themselves (Turner) [6].

3. Research Methods

Discussion of the economic value of urban renewal involves a wide range of orientations. The current study proposes an urban-renewal economic-value framework and factors based on an exploration of urban-renewal law and the related literature that adopts an economic-value perspective. Li [7], Feng [8], and Yeh [9]have all used the secondary-data method for indicator construction and argumentation; and this study also adopts it to collect, summarize and organize the essence of previous studies, before proceeding to a complete discussion and elaboration of the framework and factors mentioned above.

4. Results and Discussion

Urban-renewal implementers must cooperate with Taiwan's government and promote urban-renewal policy in accordance with the relevant laws and regulations if they are to win volume awards. The areas targeted for urban renewal are often old neighborhoods, developed in their wider urban settings' early history, and representing the essence of such settings. Therefore, land-use efficiency is improved after the land is re-planned to enhance the land use value via the urban-renewal process.

From prior discussions of the real essence economic value of urban renewal, it is clear that implementers also conform to urban-renewal law during the planning process, as evidenced by their plans' inclusion of public facilities and green buildings, which not only serve their agenda of gaining volume awards, but also enhance the value of real estate.

In addition to improving the quality of the urban environment and its functionality and safety, prevention of disasters is an important goal for urban renewal. In line with the principles of eco-city development and eco-city planning, planners plan various systems for urban public facilities, open space, mass transit, human-oriented transportation environments, green transportation and urban water resources, as well as formulating overall development strategies. Urban renewal's non-essence economic value includes the use of new technologies and building technologies in housing, transportation and environmental planning, which can reduce energy use and carbon emissions. On the other hand, the addition of public facilities, open space and certain other types of change can lead to setbacks in public safety, and enhance the potential impacts of disasters.

The Urban Renewal Act is aimed in part at enhancing urban disaster-prevention and energy-conservation measures. Thus, planning and design processes conducted under the urban-renewal umbrella conform to the green building code and are more likely to secure volume awards. They also reduce energy consumption by buildings; legally reserve and demarcate open space and space for public facilities; enhance the environment for non-motorized transportation, e.g., walking and cycling; reduce carbon emissions; and improve urban evacuation and disaster-prevention functions. In the demarcation phase of urban-renewal areas, it is generally held to be possible to comprehensively assess the disaster-prevention systems of buildings and the overall environment in such areas. When updates are subsequently made, moreover, regional-level planning of disaster prevention systems can be improved.

This study's framework for the economic value of urban renewal, as governed by the Urban Renewal Act, therefore incorporates the real essence and non-essence economic values of urban renewal, re- integration of land, and land-use planning; the gaps between urban areas, fire-protection structures or adjacent buildings, and the seismic strength of buildings; and the potential benefits of urban renewal to local development. The economic value of urban renewal is reviewed pursuant to the Urban Renewal Act, and in addition to major reconstruction takes account of improvements to the quality of the urban environment and urban disaster resilience; adherence to eco-city principles; and energy conservation. Urban-disaster prevention and climate-change adaptability concerns also mandate land-use reviews as part of urban-renewal projects. Increases in public space allow for the installation of permeable pavement layers that contribute to reducing rainwater runoff and storm impacts, while at the same time enhancing the amount of shelter space that is available in the event of earthquakes and fires. Lastly, indicators related to green buildings will be included in the framework, as they have been shown to reduce energy use and enhance resource reuse, thereby reducing urban carbon emissions and improving environmental quality. To sum up, then, this study explores urban renewal's economic- value structure in terms of three main aspects: land-use value, energy-conservation value, and urban disaster mitigation value. The specifics of its economic-value framework are shown in Figure 2 and Table 1.

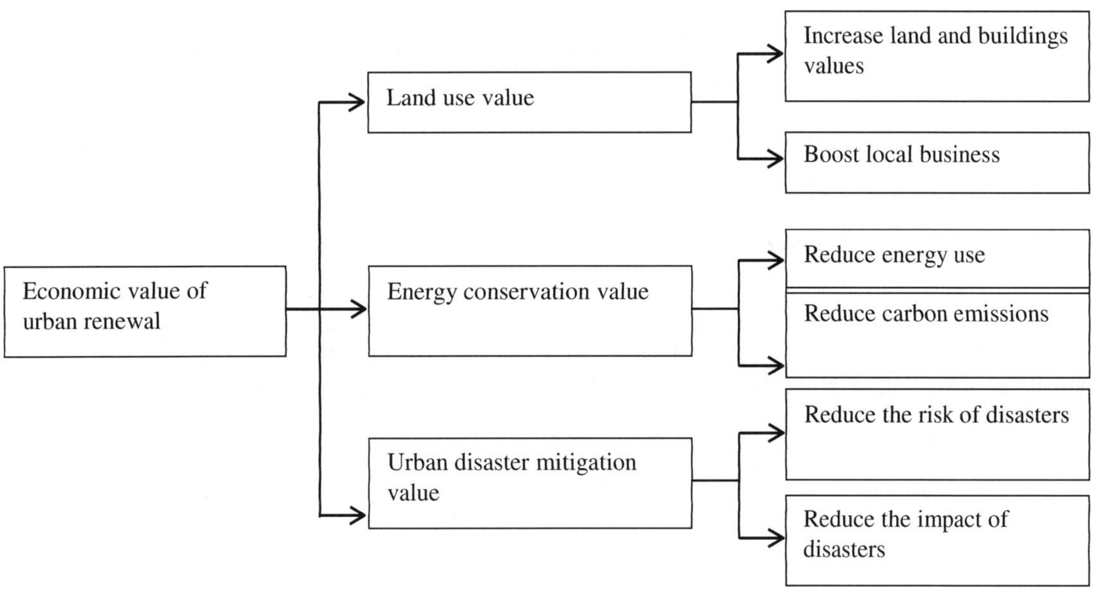

Figure 2. Economic-value framework for urban renewal

Table 1. Impact factors, economic value of urban renewal

Facet	Impact factors	Narrative explanation
Land use value	Increase land and buildings values	Green space improves the urban environment, protects environmental quality, establishes sound public facilities and open-space systems, mitigates the impact of disasters, boosts the markets in land and residential buildings, and increases commercial retail rents and occupancy rates.
	Boost local business	Comprehensive regional planning and good-quality residential can attract new residents and inward investment, improve local environmental quality, and increase the attractiveness of retail centers and tourist destinations.
Energy conservation value	Reduce energy use	Urban renewal can introduce green building design; reduce energy use and increase resource reuse, to obtain volume awards.
	Reduce carbon emissions	Construction for urban renewal uses new technologies, materials and commodities with low carbon footprints to reduce carbon emissions, improve urban air quality, and enhance urban carbon neutrality.
Urban disaster mitigation value	Reduce the risk of disasters	During the re-planning and consolidation of an area, fireproofing of its buildings and the adjacent buildings are enhanced, and the overall amount of open space is increased, thus reducing surface runoff, thus reducing both the incidence and the impact of fires and flooding.
	Reduce the impact of disasters	During the reconstruction of buildings, their seismic strength is increased, and open spaces and public facilities are planned with disaster-related evacuation, relocation and shelter in mind, thus enhancing the city's disaster resilience.

4.1. Land Use Value

Urban development tends to spread outward slowly from a city's core. With the passage of time and changes to the environment, the building density, infrastructure, public facilities and open space of early urban planning have proved unable to cope with residents' needs. Urban environments are more at risk than ever before from extreme climate-related natural disasters, and the intensity their preparations for such events is generally insufficient, making urban-renewal projects potentially critical to residents' safety. On the other hand, re-planning of urban land use can also enhance the cityscape and residents' quality of life, not least by prioritizing open space. As such, urban renewal can indirectly promote the reinvigoration of local economic development.

4.1.1 Increasing land and buildings values

As noted in the previous section, the process of urban renewal tends improve the urban environment's aesthetics, environmental quality, and land-use efficiency. Harvey showed that in real estate-led reconstruction, the timing of a given project depends on three basic issues: the value of the available land resources as currently used; their value if re-assigned to the best possible alternative uses; and the cost of such re-assignment [10]. Effective use of land and buildings, and thus urban-development efforts more generally, have been found especially effective if the methods adopted include restoring and upgrading existing structures and infrastructures (Malta Planning Authority) [11]. In addition, Jang observed that the greater positive social and economic impacts of urban renewal include changes in business practices, social values, architectural aesthetics, the presence of and demand for recreational facilities, shopping convenience, land-use patterns, land prices, and rents. Urban renewal provided a comfortable environment and higher security buildings, which promoted the value of land and property and improved land use efficiency in the land relocation and planning, as well as the re-construction of buildings [12].

4.1.2 Boosting local business

As any city develops, changes in business practices and consumer tastes, and the passage of time itself, inevitably result in decay; and urban renewal is rightly seen as a means of reversing this process by re-inspiring local development. Andersen's research indicated that urban renewal's transformation of sites can serve as a catalyst for further investment and create a range of benefits for local communities[13]. The retail industry in particular provides numerous opportunities and hires from a wide array of socioeconomic groups, leading some to argue that retail-led regeneration provides significant economic opportunities through direct employment and additional investment while at the same time helping to rebuild strained emotional connections between districts and their inhabitants (Calxton and Siora) [14]. However, urban-renewal initiatives should also incorporate sufficient public resources to address the issue of urban stagnation head-on, while working to increase social and economic benefits and promote entrepreneurial activities (Calxton and Siora; Hamnet; Andersen) [13-15]. Nor should it be forgotten that the process of urban renewal, if successful, will tend to promote not only local employment and other local economic activity, but also the government's tax take.

4.2. Energy Conservation Value

Older urban-renewal projects often failed to create connections between well-established walking environments and mass transit systems. Increasingly, however, efficient and interconnected mass transit, green transportation, and pedestrian zones are important indicators of urban environments' resident-friendliness. Urban-renewal schemes must of course be compliant with regulations and congruent with other government policies, and this is perhaps especially prominent in the case residential and transportation-based carbon emissions. Newly built buildings use energy-efficient equipment to reduce their energy-consumption, and urban-renewal, planning and design decisions – notably, regarding the creation of external space – can also help reduce energy use. The Sustainable Neighborhood Retrofit Action Plan observed that gas and electricity consumption where be reduced through residents' and businesses' participation in old-community and urban-renewal environmental action activities [16].

4.2.1 Reducing energy use

As previously noted, urban-renewal schemes involving the erection of new buildings are required to comply with the Green Building Code, which emphasizes the use of energy-efficient materials and design features that facilitate energy efficiency, energy-use reduction, and water-saving. Specifically, this can involve solar-energy systems, lighting technology, ventilation control, rainwater harvesting and roof greening, among other elements. Such approaches to building represent the most cost-effective approach to CO2 reduction; and urban renewal also provides a useful opportunity to switch to sustainable fuel sources, as changes in energy infrastructure can be linked across multiple buildings, thus reducing costs (Sunikka) [17]. Moreover, urban-renewal efforts generally tend to increase the amount of pedestrian-friendly space and reduce the use of personal motor vehicles, thus lowering carbon emissions and having a broad positive impact on the welfare and health of the inhabitants.

4.2.2 Reducing carbon emissions

As noted above, residential buildings and transportation are the two main sources of urban greenhouse-gas emissions, which cause urban temperatures to rise, often to uncomfortable or even dangerous levels. Ideally, therefore, new buildings constructed as part of urban-renewal schemes should comply with green building codes, improve pedestrian-friendliness and the proportion of green transportation, and use clean and renewable energy to reduce urban carbon emissions and create a low-carbon urban environment. As Mehdipanah et al. has noted, most urban-renewal projects result in improved walking capacity, construction of new public spaces, and increases in the number of community projects that have important positive impacts on residents' overall well-being [18]. Urban renewal's role in CO2 reduction, in particular, is also important in terms of policy outcomes and regulatory compliance (Sunikka) [17].

4.3. Urban Disaster Mitigation Value

4.3.1 Reducing the risk of disasters

Thomas suggested that being prepared for reconstruction represented an opportunity to replace any substandard buildings, resolve disaster-relevant land-use conflicts, and remove environmental nuisances [19]. Reconstruction can also help make the most of the land's potential, and help communities solve various problems (Sigworth and Wilkinson) [20] such as residential basement flooding (SNAP) [16]. Ideally, other infrastructure, and the environment should all be improved by urban-renewal initiatives, in a manner supported by all social partners (Adair et al.) [21]. The selection and prioritization of urban-renewal areas should be based on comprehensive consideration (i.e., at the regional as well as local scales) of how to reduce the incidence of preventable disasters, in what Shi has referred to as active disaster management [22].

4.3.2 Reducing the impact of disasters

During the re-planning and consolidation of an area, urban-renewal projects can upgrade its buildings' fire safety (and that of adjacent buildings in other districts), and add open space that reduces surface runoff and improves the drainage system, thus reducing the impact of non-preventable disasters involving fires, storm surges, and flooding caused by heavy rain. As well as reducing their immediate negative consequences, such adaptations to natural disasters reduce post-disaster recovery times and costs. Moreover, many old buildings in Taiwan remain in use beyond what might be called the end of their natural lifespan: a problem that can best be dealt with via government regulation aimed at ensuring that more buildings meet certain safety standards. The disaster-prevention and mitigation strategies used in the field of urban disaster management, combined with urban renewal's reconstruction methods, could usefully develop assessment indicators for the disaster-risk factors of buildings and their surrounding environments in the contexts of earthquakes, fires, and floods, singly or in combination. Indeed, active promotion of urban-renewal measures has become the main strategy for improving Taiwan's urban disaster Mitigation functions (Deng, 2011) [23]. As we have seen, urban renewal promotes compliance with regulations and governmental policies, including those aimed specifically at smoothing post-disaster recovery and reconstruction processes at both the micro (building) and macro (regional) scales.

5. Conclusion

Urban-renewal areas in Taiwan are considered priority planning areas, and are selected in part according to various disaster-preparedness factors. According to the principles of eco-city planning and urban-design guidelines, a key development criterion for urban renewal is whether it will benefit sustainable urban development, slow the heat-island effect, and enhance overall environmental quality via the provision or more open space, green space, and water features. Urban renewal is expected to achieve both disaster reduction (in the case of man-made disasters such as fires) and disaster mitigation (in the case of both man-made and natural disasters), regionally as well as locally. The present study has explored the economic value of urban renewal in Taiwan through an analysis of the

IOP Conf. Series: Earth and Environmental Science **291** (2019) 012004 doi:10.1088/1755-1315/291/1/012004

relevant laws and scholarly literature, and established a useful framework for the economic valuation of urban-renewal projects that goes beyond the usual, simplistic method of simply measuring post-project increases in property values. Rather, this new framework – in addition to property values – takes account of projects' tangible and intangible impacts on local business activity; reductions in energy use and carbon emissions; and disaster prevention and mitigation.

6. References:

[1] Chou, Chao-Pei (2010) Model Analysis on Cost-Benefit and Economic Evaluation of Urban Renewal. Master's Thesis, Department of Accounting, Soochow University, Taipei, Taiwan.
[2] Wu, I-An (2014) A study on the impacts of living quality of urban renewal projects on the surrounding area from local residents' perspective. Master's Thesis, Department of Urban Planning, National Cheng Kung University, Tainan, Taiwan.
[3] Li, Wen-Chun (2015) Estimating the House Economic Values of Cultural Heritage Attributes in Tainan City. Master's Thesis, Department of Sociology, National Donghua University, Hualien, Taiwan.
[4] Chen, Ya-Wen (2013) Research on the re-use of historic site economic value: A Case Study of the Ciaotou Sugar Refinery, Master's Thesis, Department of Creative Design and Architecture, National University of Kaohsiung, Kaohsiung, Taiwan.
[5] Taiwan Map ONLINE. Available online: http://www.safetourbus.com (accessed on 1 August 2018).
[6] Turner, R.K., & Pearce, D., & Bateman, I.(1994) Environmental economics: An elementary introduction. London: Harvester Wheatsheaf.
[7] Li, Po-Tsang (2015) Assessment Index Construction of Brownfield Redevelopment in Taiwan-A Case Study of Formosa Chemicals & Fibre Corporation in the Changhua City Expanded Planning Area. Master's Thesis, Graduate institute of Architecture and Urban Design Chaoyang University of Technology, Taichung, Taiwan.
[8] Feng, Shih-Jen (2015) Building Social Network of Creative Clusters - A Case Study of Urban Regeneration Station in Taipei City. Ph.D. Thesis, PhD Program in Architecture and Urban Design, Department of Architecture, Chaoyang University of Technology, Taichung, Taiwan.
[9] Yeh, Yi-Cheng (2013) Cultural and Creative Industry and Place identity in Kinmen. Master's Thesis, Department of Tourism, Ming Chuan University, Taipei, Taiwan.
[10] Harvey, J. (1992) Urban Land Economics. Macmillan, Basingstoke.
[11] Malta Planning Authority (1999) The Malta Council for Science and Technology. State of the Environment Report for Malta 1998. Environment Protection Department.
[12] Jang, Yuan-Ru (1994) A Study on the Social-Economic Impact Assessment on Land Development--The First square Urban Renewal Project, Master's Thesis, Department of Land Management, Feng Chia University, Taichung, Taiwan.
[13] Skifter Andersen, H. Urban Sores (2003) On the interaction between segregation, urban decay and deprived neighbourhoods. Aldershot: Ashgate.
[14] Ruth Claxton and Georgia Siora (2008) Retail-led regeneration: why it matters to our communities. DTZ Consulting & Research, London: DTZ.
[15] Hamnet, C. (2000) Gentrification postindustrialism, and industrial and occupational restructuring in global cities', in G. Bridge and S. Watson (eds) A Companion to the city, Oxford: Blackwell: pp. 331-341.
[16] Sustainable neighborhood Retrofit Action Plan (SNAP) Projects (2013) Toronto and Region Conservation Authority.
[17] Minna Sunikka (2006) Energy efficiency and low-carbon technologies in urban renewal, Journal Building Research & Information. 34(6): 521-533.
[18] Mehdipanah, R, Malmusi, D, Muntaner, C, Borrell, C. (2013) An evaluation of an urban renewal program and its effects on neighborhood resident's overall wellbeing using concept mapping. Health Place. 23: 9-17.

[19] Thomas, K. (1977) The impact of renewal on small firms. The Planner.69(3): 48-9.
[20] Sigsworth, E.M. and Wilkinson, R.K. (1967) Rebuilding or renovation. Urban Studies. 4(2): 109-21.
[21] Adair A., McGreal S., Smyth A., Cooper J. and Ryley T. (2000) House prices and accessibility: The testing of relationships within the Belfast Urban Area. Housing Studies. 15: 699-716.
[22] Shih, Yi-Cheng (2008) The Establishment of Urban Redevelopment Assessment Index with Mitigation Concept: The Case of Wanhua District in Taipei City. Master's Thesis, Department of Urban Planning and Disaster Management. Ming Chuan University, Taipei, Taiwan.
[23] Teng, Kai-wen (2011) The Study on Disaster Prevention Evaluation Indicators Created on Delimit of Urban Renewal-The Case Study of Urban Renewal Based in Datong District, Taipei City. Master's Thesis, Departments of Urban Planning and Development, Chinese Culture University, Taipei, Taiwan.

The Discussion on the relationship among indoor air quality, environmental cognition and user's behaviour

Xue-Min Xu[1,2,a], **Sheng-Jung Ou**[3,b]

[1]PhD program in Architecture and Urban design, Department of Architecture, Chaoyang University of Technology, Taiwan
[2]Department of Environmental Art and Design, Shanghai Art and Design Academy, China
[3]Department of Landscape and Urban Design, Chaoyang University of Technology, Taiwan

E-mail: [a] 452780491@qq.com; [b] sjou@cyut.edu.tw

Abstract. Indoor Air quality inspection is an important environmental assessment work before the newly decorated space is open to use. Air quality is a dynamic factor affecting environmental health. It mainly evaluates the concentration of harmful volatile gases such as formaldehyde (HCHO), volatile organic compounds (VOCs) in the air, and the concentration of dust and other indicators. The release of these harmful substances changes with the physical factors of the environment that many studies focus on and explore the complex correlations among them. But as a user in the environment, does the behaviour also affect the air quality? In view of this, the study takes the Master's Research Office of Chaoyang University of Technology in Taichung as an example and tries to explore if the release of formaldehyde (HCHO) and volatile organic compounds (VOCs) has any correlation with the cognitive and behavioural activities of people in their environment. In addition to the measurement of the concentration of formaldehyde (HCHO) and volatile organic compounds (VOCs) under the established conditions in the process of the study, the questionnaire survey was used to know the subject's cognition of the environment, such as the freshness of the air, the ventilation of the space, etc. Through statistical analyses, However, the results showed that there were no significant differences in environmental cognition between the before and after measurements. The results also revealed that there were no change of having bad psychological feeling and only a slight change of having bad physiological feeling between the before and after measurements. The study concludes that the assessment of indoor air quality needs to be based on consideration of people's perceptions of the environment and behavioural activities, not just the level of harmful substances in the air. Through the cognition of the environment and the effective intervention in the behaviour process, the harm caused by the air quality problem to the health of the human being can be alleviated to some extent.

1. Introduction

1.1. Background

The inspection of indoor air quality is an important step to evaluate the health of indoor air before a newly decorated space is open to use. According to the air quality standards set by the local environmental department, they are usually being used as the basis for judging whether the space can be opened for use. It has been proved that volatile organic compounds (VOCs), formaldehyde (HCHO) and suspended dust are harmful to human health in the air of newly decorated space [1]. So, in the air quality measurement, the values of formaldehyde (HCHO), volatile organic compounds (VOCs) and PM (particulate matter) 2.5 are the main targets. Studies have confirmed that air pollutants, formaldehyde (HCHO) and volatile organic compounds (VOCs) are mainly derived from the release of composite panels used in space [2]. According to the annual climate of Taichung City, the average temperature range from January to December is 18 °C to 30 °C [3], which is a favorable condition for promoting the continuous volatilization of formaldehyde (HCHO) and volatile organic compounds (VOCs) [4]. After the newly decorated space has been passed the test of air quality standards and put into use, it is not uncommon for users to become physically ill due to environmental changes. As air quality is one of the main factors affecting environmental health, related department needs to consider the rationality of its assessment methods and various affecting factors.

1.2. Motivation

WELL Building Standard™（WELL）[5] was launched by The International WELL Building Institute (IWBI) in 2014, compared with the green building energy conservation and environmental protection standards established by countries in the past to save the earth's resources, many indicators about the people-oriented health have been added in WELL. WELL's seven concepts, including air quality, not only comprehensively address the energy-efficient design and operation of buildings, but also promote and influence human behaviour associated with health and well-being. 90% of human life is spent in indoor space, which has a good air quality environment and keep the indoor space healthy through the user's maintenance and management is the way of sustainable coexistence between human and environment.

In the process of air quality monitoring in space, dynamic data is a comprehensive reflection of the influence of disturbance factors such as environmental space size, congestion level, ventilation, material safety, temperature and humidity on air quality. In addition to the above influencing factors, previous studies may ignore the existence of human being as the leading role of behaviour in space, and the service object of air quality is the user in the environment. Therefore, this study attempts to investigate the correlation between the air quality indicators in space and human environmental cognition or behaviour by taking the space of the master's research office of Chaoyang University of Technology, which was newly completed in September 2018.

1.3. The image of the study

The master's research office of Chaoyang University of Technology completed the interior decoration on September 16, 2018. In the entire space, the main surrounding wall material is painted. The floor and ceiling materials are composite panels. The interior and exterior materials of the desks and cabinets are also composite panels. In addition to the above work area, the space also includes two conference areas and two leisure areas. The actual scene of the office is shown in Figure 1.

Figure 1. Scene of the master's research office of Chaoyang University

2. Research methods

2.1. Measurement experiment
This study uses the instruments to measure the indoor air quality of the object space. The research office which completed decoration and cleaning work was sealed for more than 1 hour before tests. Under the established temperature conditions, the concentration of formaldehyde (HCHO) and volatile organic compounds (VOCs) in the space height of 80cm to 150cm was measured twice before and after a week using sensor-type air monitors. At the same time, the formaldehyde (HCHO) and volatile organic compounds (VOCs) concentrations were measured in the internal space of the door cabinet. For the same desk surface, the formaldehyde (HCHO) and volatile organic compounds (VOCs) concentration values with direct sunlight one hour or not were also measured. The measurements have been implemented twice in terms of before and after. The final value is obtained by averaging one hour of data acquisition for each measurement.

2.2. Questionnaire survey
A questionnaire survey was used to understand the daily behaviours such as learning, rest and communication in the experimental space of the users (N=11, male: 5, female: 6, average age: 22), as well as the cognition of the indoor environment such as air freshness and spatial ventilation, etc.

3. Results and discussion

3.1. Air quality and analysis
Through the measurement by several groups of formaldehyde (HCHO) and total volatile organic compounds (TVOC) concentrations under specific conditions in the experimental environment, the corresponding values were obtained and the comparative analysis was attempted.

Table 1. Measurements of HCHO and TVOC of indoor air.

Date	Temperature（℃）	HCHO（ppm)	TVOC（ppm)
18-Sep-18	28	0.03	0.26
25-Sep-18	28	0.04	0.38

Note: Air quality standards in Taiwan shows: HCHO⩽0.08（ppm), TVOC⩽0.56（ppm).

The above data shows that the HCHO and TVOC values measured in two experiments have met the air quality standards in Taiwan [6]. The value differences are quite small between the two tests. This may be explained by large space, stable ambient temperature, and insufficient measurement time interval.

Table 2. Measurements of HCHO and TVOC of air in cabinet with door closed.

Date	Temperature（℃）	HCHO（ppm)	TVOC（ppm)
18-Sep-18	28	0.09	7.76
25-Sep-18	28	0.17	9.98

Note: Air quality standards in Taiwan shows: HCHO⩽0.08（ppm), TVOC⩽0.56（ppm).

The second set of data shows that the air in the cabinet is seriously exceeded the standard of air quality in Taiwan. It indicates that the harmful gases are easy to exhibit the cumulative effect when the space is small and closed. At the same time, the HCHO and TVOC values in the cabinet (Table 1.) are much higher than the values of space air (Table 2.), because of the composite panels material for the cabinets which is the main release source of the formaldehyde (HCHO) and volatile organic compounds (VOCs).

Table 3. Measurements of HCHO and TVOC of the desk surface with sunlight or not.

Date	Sunlight（h）	HCHO（ppm)	TVOC（ppm)
25-Sep-18	0	0.07	0.51
25-Sep-18	1	0.12	0.67

The last set of data (Table 3.) shows that direct sunlight causes the desktop temperature to rise and promotes the accelerated release of the formaldehyde (HCHO) and volatile organic compounds (VOCs). The desks are layout around the surrounding walls in the room, 60% of the desks without sunlight all day and 40% of the desks with 2-3 hours of direct sunlight every day.

3.2. User's behaviour

A questionnaire survey on the behaviour of the research office was conducted for the users (N=11) of the experimental space. The results and analysis are as follows:

3.2.1. The using behaviour of cabinets. According to the behavioural survey, cabinets are primarily used to store documents, personal items, and learning tools, each of which has the potential to be opened or repeatedly opened every day. At the same time of opening, high concentrations of formaldehyde (HCHO) and volatile organic compounds (VOCs) which are trapped in the cabinets for a certain period of time are released into the space.

3.2.2. The sitting behaviour. According to the behavioural survey, the user is in a sitting position by the desk or in the sofa for about 93% of the time in the research office, and the height of the breathing apparatus in the adult sitting position by desk or in sofa ranges from 90 cm to 140 cm. Studies have confirmed that the height range of 80 cm to 150 cm in space is a dense suspension area of formaldehyde (HCHO) and volatile organic compounds (VOCs) in the air. Therefore, users in the environment spend most of their time breathing the most polluted air.

3.2.3. The breathing behaviour. According to the behavioural survey, the height of the desktop which is also the major source of formaldehyde (HCHO) and volatile organic compounds (VOCs) is 75 cm. Combined with its release to the final suspension area, the height is 80 cm to 150 cm. In the closed space without obvious influence of ventilation on air flow, the users sit in front of the desk and breathe the volatile formaldehyde (HCHO) and volatile organic compounds (VOCs) from the desk nearly.

3.2.4. The using behaviour on the desks. The desktop temperature is positively correlated with the release rate of formaldehyde (HCHO) and volatile organic compounds (VOCs). Besides direct sunlight, according to the behavioural survey, users usually sit in front of their desks for 78% of the time, and their body contact with the desk for quite long causes heat transfer from body to the desk. Users use laptops for learning and entertainment on their desks for 71% of the time, and laptops placed on the desktop continue to dissipate heat to the desktop, causing significant temperature rise on the desktop. 77% of users drinking hot drinks, dining on the desk are the behavioural factors that cause the desktop temperature to rise. About 45% of users like taking a nap on the desk, and the behaviour of rest will make the respiratory organs close to the formaldehyde (HCHO) and volatile organic compounds (VOCs) release sources, causing people to be seriously affected by dirty air during this behaviour. The above analysis shows that whether the desk is located in a direct sunlight area, the desktop will cause different degrees of warming due to human behaviour, which promotes the volatilization of formaldehyde (HCHO) and volatile organic compounds (VOCs).

3.2.5. Behaviour caused by quiet demand. The behavioural survey shows that 80% of users have a quiet demand for the research office. Referring to the climate characteristics of Taichung City, air conditioning is required indoors for about 35% of the year. The research office is usually in a closed and non-ventilated state. Combined with the premise that formaldehyde (HCHO) and volatile organic compounds (VOCs) act on the human body and have cumulative effects, users who have been engaged in learning activities for a long time in the space have a certain probability of causing physical discomfort. As the number of people in the enclosed space increases, the rise of the concentration of carbon dioxide (CO_2) in the air due to crowding can also cause some physical discomfort. Although dioxide (CO_2) is not the object of this study, it is also one of the air quality evaluation indicators.

3.3. Environmental cognition

While measuring the air quality process, an environmental cognition questionnaire on the comfort of the research office was conducted for the user of the experimental space. The results and related analysis are as follows:

Table 4. The before and after measurement of environmental cognitions

Variable	Mean (Before)	Mean (After)	t-value	Sig.
Freshness of the air	3.45	3.09	1.000	.3410
Comfortableness of the air	3.45	3.45	.0000	1.0000
Stimulating smell of the air	3.84	3.82	.0000	1.0000
Comfortableness of the smell	3.64	3.64	.0000	1.0000
Comfortableness of feelings	3.18	3.36	-.6900	.506
Learning Efficiency	3.55	3.18	.6900	.506

Table 5. The before and after measurement of having bad psychological feeling and having bad physiological feeling

Variable		% (Before)	% (After)
Having bad psychological feeling	No	27.3	27.3
	Yes	72.7	72.7
Having bad physiological feeling	No	18.2	27.3
	Yes	81.8	72.7

According to the statistical analyses, the results (Table 4.) showed that there were no significant differences in environmental cognitions between the before and after measurements. The results also revealed that there were no change of having bad psychological feeling and only a slight change of having bad physiological feeling between the before and after measurements (Table 5.). But as time goes by, the volatilization of formaldehyde (HCHO) and volatile organic compounds (VOCs) will enter a recession period, and the user of the environment will not have any significant difference in the comfort.

4. Conclusions and suggestions

The purpose of this study was to investigate whether people's cognition of the environment and their behaviour in the environment have an impact on air quality in addition to environmental physical factors. After measuring the air quality of the experimental space, investigating the user's environmental awareness and usage behaviour, the conclusions are as follows:

4.1. New perspective on air quality

Air quality acts directly on people. The air quality measurement provided values as a reference is not based on the consideration of users in the environment. This kind of air quality inspection assessment is one-sided. People in different types of space have different behaviours and activities, which requires detailed analysis of the relationship between behaviours and air quality, rather than using the same standards.

4.2. Correlation between behaviour and indoor air quality

In addition to considering the direct correlation between human spatial behaviour interaction and indoor air quality, the indirect correlation between the two is also an important factor leading to air quality, while the temperature is an important indirect factor of human behaviour activity between formaldehyde (HCHO) and volatile organic compounds (VOCs) release.

4.3. Correct cognition and behaviour improve air quality

Timely ventilation is an effective way to improve air quality. The movement of air flow can interfere with the air in the small breathing area in front of the desk and accelerate the dilution of formaldehyde (HCHO) and volatile organic compounds (VOCs) concentration. At the same time, under the condition of no one in the research room, the closed cabinets should be opened while opening the windows for ventilation, and the harmful gas gathered in the cabinets should be discharged in time.

Under the premise of the budget, furniture such as desks and storage cabinets can be replaced with composite panels with high environmental protection level instead of ordinary ones.

Based on the positive correlation between temperature and the release of formaldehyde (HCHO) and volatile organic compounds (VOCs), the blinds can be added to reduce the desktop heating caused by direct sunlight. In order to reduce the long-term contact between the heat source and the desktop, excessive heat transfer to the desktop can be avoided by adding a laptop stand or a heat sink. Hot drinks and food should use trays or heat insulators to reduce heat transfer.

4.4. The disadvantages of this study and further improvement

The purpose of this preliminary study is to raise questions and lead to thinking. Because this research space is exclusive to graduate students in architecture department, the number of users of the experimental space is not big enough, and this experiment adopts the measurement method of before and after comparison, so that the number of sample is limited. It is suggested to try to expand the sample size in future research. The preliminary experiments involved in the process have the disadvantages of single type monitoring equipment and insufficient monitoring time span. The follow-up will be subject to careful experimental design of the research, and the user's behavioural activities need to be reasonably counted and quantified to obtain more accurate experimental results and detailed data analysis.

5. References

[1] D.-M. Shin, C.-N. Kim, C.-H. Park, D.-S. Kim, Y. Nam (2013). On the determination of VOCs emission factors in dry building materials. *Heat Mass Transfer.* **49**:1625-1635

[2] L. Long, G.-R. Li, Y.-C. Zhou (2011). Determination of formaldehyde and other VOC emissions from furniture in the climate chamber . *China Wood Industry.* **25**:13-15

[3] Central Weather Bureau (2017). Monthly weather data. *http://www.cwb.gov.tw.*

[4] D. Chi, L.-Q. Li, Z. Liu, W.-W. Ma, X.-L. Yao (2015). Formaldehyde release parameters and temperature effects of composite panels. *Journal of Central South University (Science and Technology).* **46**:751-758

[5] The International WELL Building Institute (IWBI) (2014). WELL Building Standard™ （WELL）. *http://www.wellcertified.com.*

[6] Environmental Protection Administration Executive Yuan, R.O.C (Taiwan) (2012). Indoor Air Quality. *http://iaq.epa.gov.tw.*

IOP Conf. Series: Earth and Environmental Science **291** (2019) 012006 doi:10.1088/1755-1315/291/1/012006

Access the Change of Water Quality in the Sponge City of LCTIP

N Li[1], P F Du[1], X C Huang[2] and Z P Shu[3]

[1]State Key Joint Laboratory of Environment Simulation and Pollution Control, School of Environment, Tsinghua University, Beijing 100084, China
[2]Beijing WALD Eco-Environment Technology Co. Ltd, Beijing 102200, China
[3] Yunnan WALD Eco-Environment Technology Co. Ltd, Kunming 650000, China

nan-li12@mails.tsinghua.edu.cn

Abstract. The construction of Sponge City (SPC) in China has caused a huge social transformation, and the application of its core low-impact development facilities (LIDs) has changed the lives of residents from the details. Controlling runoff water quantity and its water quality are the two main objectives of low impact development measures. This paper discusses the results of water quality testing, comparing content changes of total phosphorus (TP), total nitrogen (TN), and chemical oxygen demand (COD) and total suspended solids (TSS) measured from water samples collected from the same LID facilities established by LCTIP three times a month during 2018 the local rainy season. The values changes of various pollutants from sampling points to the outlet show that the water quality of after LID facilities is close to the state of natural rain before development. The water purification effect of the LID facility was verified. At the same time, it is necessary to understand that more factors will have an impact on water quality. Therefore, the limitations of current knowledge and recommendations for future research are also discussed.

1. Introduction

According to EPA, the term low impact development (LID) refers to systems and practices that use or mimic natural processes that result in the infiltration, evapotranspiration or use of stormwater in order to protect water quality and associated aquatic habitat. LID is a sustainable concept in land use management that aims to maintain hydrological conditions at a predevelopment level without deteriorating water quality during land development. Based on the sustainable, the Sponge City (SPC) for urban stormwater management was proposed in China. [1] It was first mentioned by the Chinese President in 2013 and proposed in 2014. SPC is a comprehensive concept, focusing on not only the source control (as LID) but also the process control and drainage management. The aim of SPC construction is the detention, onsite storage, infiltration, filtration, reuse, and drainage of stormwater. [2-5] The two main purposes of the series of LID measures are controlling the quantity of storm runoff and improving its quality. Only the removal of pollutants was involved in this study, and the relevant practice measures mainly included rain barrels, bio-filtration facilities, green roofs, permeable pavements, and swale systems, measures that have been commonly applied.

Nutrients, such as Total Nitrogen (TN) and Total Phosphorous (TP) in stormwater can have adverse impacts on receiving water bodies. While these nutrients are necessary for growth by aquatic plants, excessive levels of these nutrients do cause excessive growth these plants, especially algae.

This excessive growth leads to eutrophication of surface waters. Nitrogen is a limiting pollutant in tidal areas, while phosphorous is a limiting pollutant in freshwater systems, particularly ponds and lakes. The common sources of TN and TP are the application of inorganic and organic fertilizers for crop production, lawn fertilization and the disposal of waste residues (bio-solids, animal manure, and compost). As noted above, decomposition of excessive vegetated growth leads to lower dissolved oxygen levels in the water. [6-10]

The chemical oxygen demand (COD) is an indicative measure of the amount of oxygen that can be consumed by reactions in a measured solution. It is commonly expressed in mass of oxygen consumed over the volume of solution which in SI units is milligrams per liter(mg/L). A COD test can be used to easily quantify the number of organics in water. The most common application of COD is in quantifying the number of oxidizable pollutants found in surface water (e.g. lakes and rivers) or wastewater.

Total suspended solids (TSS) is the dry-weight of suspended particles, that are not dissolved, which could block light from reaching aquatic vegetation, while the less light in the water, the less photosynthesis is generated. As photosynthesis is reduced, less oxygen is produced by the vegetation thus affecting many aquatic species. When dead vegetation decomposes, more oxygen is used in the decomposition process thus further reducing oxygen levels in the water. TSS can also clog gravel substrate in breeding streams, affecting insect larvae and fish eggs. In a sample of water that can be trapped by a filter that is analyzed using a filtration apparatus. Meanwhile TSS is a water quality parameter used to assess the quality of a specimen of any type of water or water body, ocean water for example, or wastewater after treatment in a wastewater treatment plant. It is listed as a conventional pollutant in the U.S. Clean Water Act.

These pollutants cannot only kill some forms of aquatic life human beings also could be adversely affected by these pollutants thru the consumption of these higher aquatic species. After they accumulate on impervious surfaces, they are easily washed into the storm drainage with the next rain event. Within this study, water samples tested from two rain barrels and three bio-retentions in LCTIP used for discussion. Through experiments, monitoring the varieties of TP, TN, COD and TSS loads from them

2. Materials and method

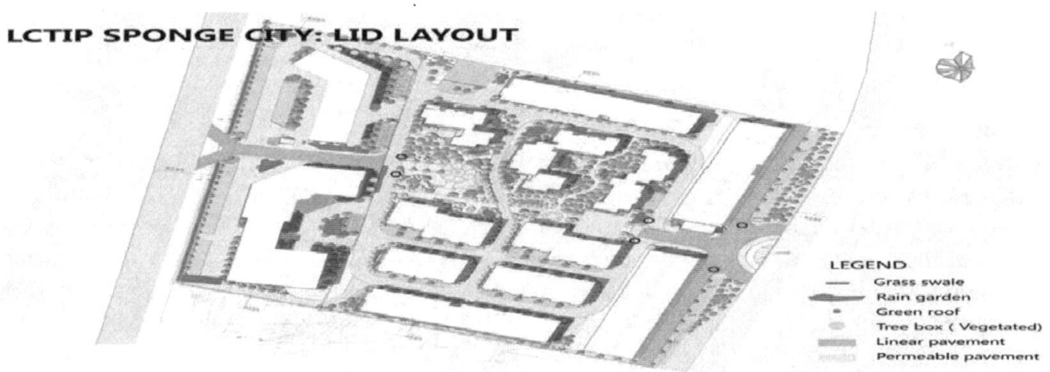

Figure. 1 LID Layout

2.1. Site description

The Lincang Technology Innovation Park (LCTIP) is located in Lincang City industrial park, Yunnan Province, China. Lincang City lies in the southwest of Yunnan Province, approximately between east longitude from 98~40' to100~34' and north latitude from 23~05' to 25~02'. The east adjoins to Pu'er City and the west is adjacent to Baoshan City. Dali Bai Autonomous Prefecture is close to the north. Burma is neighboring the southern borders. And our lab which is just located in Lincang science and Technology Innovation Park(LCTIP). It is an innovation park based on its overall and controls specific planning. It is the first paradigm of SPC in Lincang. Fig.1 illustrates the LID layout. The climate is subtropical with an annual rainfall of 1.158 mm and an average temperature of 17 °C.

The land use type is M1(Department 2012). The area is 3.78 ha, which was occupied by buildings, roads, and green space. The total construction square is 78.500 m2. The land use is a combined use of commercial, residential, and public facilities. The soil belongs to type B and C (clay). The groundwater level is about 1 m–6.8 m. The slope is 4.5 % on the east side higher than the west side. This project was designed in Mar, 2016 by WALD and construction completed at the end of July 2016.

Both design and construction have been approved as qualified SPC project by the local authority. The investment is 1.15 million RMB for LIDs.

2.2. Experiment

2.2.1. Purpose. The data of total phosphorus (TP), total nitrogen (TN), turbidity (TUB) and chemistry oxygen demand(COD) in different water samples from water samples were tested to determine the filtering and purifying effects of plants on water quality and the filtering and purifying effects of sponge facilities on rainwater.

2.2.2. Materials.

. **Table 1**. Sample collection point representation

Sample	
Rainwater → A_0	Rain Barrel 1) → A
Rain Barrel 3) → B	Aquatic plant pool → C
Semi-Aquatic plant pool → D	Terrestrial plant pond → E
Collection pool → F	

As seen in Table 1 respectively collected from roof rainwater (A0) a, rain barrel 1 (A), rain barrel 3 (B), aquatic plants pool (C), semi-aquatic plants pool (D), terrestrial plants pool (E), and sample collection pool (F)b in different rainy days three times a month. Test for changes in total phosphorus, total nitrogen, chemical requirements, and total suspended solids, like Fig.2.

[a] A0 is the rainwater collected on the day using the rain gauge as a rainwater control before entering the LID facilities.

[b] Since the rockery collection pond(F) was connected to the outlet, the collected water samples from the rockery collection pond were considered as the outlet after purification in all LIDs in this study. Using spectrophotometry to obtain the tested data.

Figure 2. LIDs of the test in LCTIP

Total Phosphorus reagent powder, potassium per-sulfate, 1.54N sodium hydroxide solution, deionized water, total nitrogen reagent a, b, c powder pack, per-sulfate powder pack, HR total nitrogen digestion test tube, total phosphorus test tube, COD Reactor, DRB200 Digester, Hach DR6000 UV Spectrophotomet

3. Results

3.1. Rainwater Record

In Fig.3, the rainfall record in 2018 rainy season from May to August reached a cumulative maximum of 538 mm on June 27, with an overall average of 63 mm. Samples were taken on the recorded rainy days, and the corresponding water quality test was carried out.

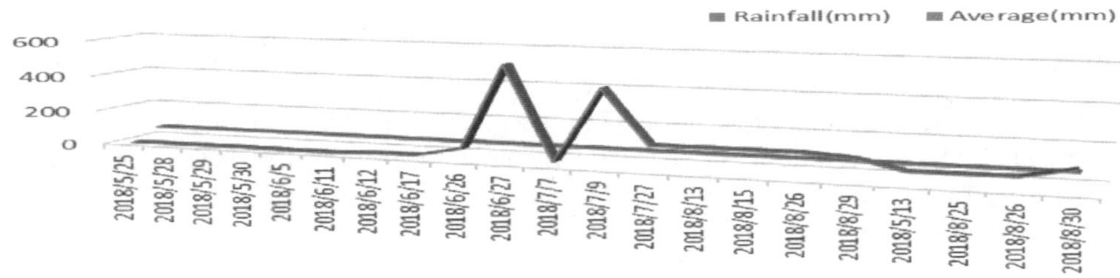

Figure 3. Rainfall record (May-August，2018)

3.2. Total Pollutants

Table 2. Test for total pollutants situation

SAMPLES	Mean				
	TP	TN	COD	TSS	Total
A_0	1.15	2.27	8.75	0.87	3.26
A	2.48	4.17	11.75	2.45	5.21
B	2.29	4.02	10.75	2.18	4.81
C	2.10	3.39	15.75	1.75	5.75
D	2.13	3.55	16.00	1.80	5.87
E	4.86	8.36	28.25	4.63	11.52
F	2.26	3.14	13.00	1.49	4.97

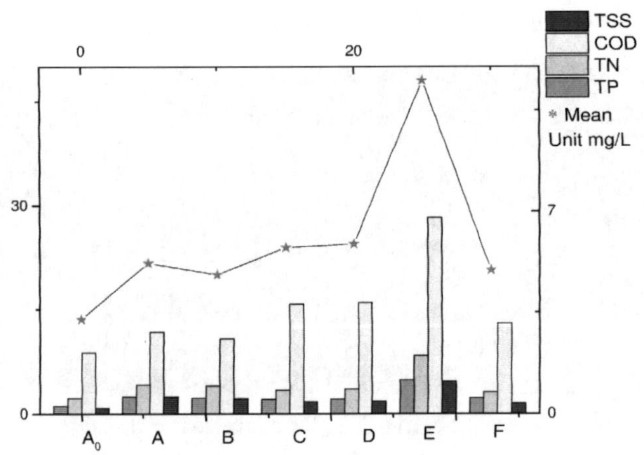

Figure 4. Comparison of total pollutants

As can be seen from Table 2 and Fig.4, that in the different experiment days from rainy season (May to August) in 2018 the mean values of total pollutants in the water collected from sampling points A_0 to F present similar change trends. The same as in LIDs, at sampling point E that terrestrial plants pool got the highest loads, at sampling point F got a drop.

4. Conclusion

In comparison figures could be seen TP, TN, COD and TSS in stormwater would have a drop in values, and the water quality of the collection pond was close to that of undeveloped roof rainwater. Overall, in the collection pool (F) where the discharge (outlet) is located, the pollutants have curtailed apparently. That means that the entire processes from roof rainwater to the outlet collection pond, which will gradually reduce contains of contaminant in the water and improve water quality, through various LID facilities set in LCTIP. This verifies the water purification function of the LID facilities. Meanwhile, compared rain barrels to plants flirting pools, from aquatic plants pool (C) to terrestrial plant pool (E), there was a poof of the filtering and purifying effects of plants especially aquatic plants.

Above all, it can also be seen that the values of various pollutants indicators fluctuate greatly. In addition to the error of the experiment and the improper preservation of the samples, vitalities and metabolic activities of various living beings themselves in the biological filter. Weather and temperature changes and so on, will have certain impacts on the test. Later in the test operation and sample collection should strengthen the specification. Treatment of stormwater for contaminants is necessary in order to sustain ecological and economic benefits related to aquatic resources. For the purpose of further verify the purification effect and effectiveness of the LIDs, the monitoring of other pollutants such as NH3-N, BOD, etc., should be been increased and integration. Follow-up needs to continuously improve for maximize the effectiveness of LIDs.

Acknowledgments

Authors The authors thank Lincang Technology Innovation Park and Shenzhen WALD Urban Design Company for providing original designs and data for this study. Thanks for Freesia Hwang in collecting and analyzing the data. The research was supported by the Major science and Technology Program for Water Pollution Control and Treatment No. 2012ZX07302002.

References

[1] A Neil, F Rolley, and J Walmsley. 2008. "The Sponge City Hypothesis: Does It Hold Water?" Australian Geographer 39(2):109–30.

[2] X N Li, J Q Li, F Xing, Y W Gong, and W L Wang. 2016. "Case Studies of the Sponge City Program in China." Pp. 295–308 in World Environmental and Water Resources Congress 2016.

[3] L P Dai et al. 2017. "Governance of the Sponge City Programme in China with Wuhan as a Case Study." International Journal of Water Resources Development, 1–19.

[4] X Jun et al. 2017. "Opportunities and Challenges of the Sponge City Construction Related to Urban Water Issues in China." Science China Earth Sciences 60(4):652–58.

[5] W W Shao et al. 2016. "Data Integration and Its Application in the Sponge City Construction of CHINA." Pp. 779–86 in Procedia Engineering, vol. 154.

[6] O Reilly, A M., M P Wanielista, N B Chang, Z M Xuan, and W G Harris. 2012. "Nutrient Removal Using Biosorption Activated Media: Preliminary Biogeochemical Assessment of an Innovative Stormwater Infiltration Basin." The science of the Total Environment 432:227–42.

[7] P E G I, T D Fletcher, P L M Cook, A Deletic, and B E Hatt. 2014. "Processes and Drivers of Nitrogen Removal in Stormwater Biofiltration." Critical Reviews in Environmental Science and Technology 44(7):796–846.

[8] S M Wang et al. 2017. "Nitrogen Removal from Urban Stormwater Runoff by Stepped Bioretention Systems." Ecological Engineering 106:340–48.

[9] S D J. et al. 2012. "Assessing Performance of Manufactured Treatment Devices for the Removal of Phosphorus from Urban Stormwater." Journal of Environmental Management 113:279–91.

[10] H Chi-Hsu, A P Davis, and B Needelman. 2007. "Bioretention Column Studies of Phosphorus Removal from Urban Stormwater Runoff." Water Environment Research: A Research Publication of the Water Environment Federation 79(2):177–84.

IOP Conf. Series: Earth and Environmental Science **291** (2019) 012007 doi:10.1088/1755-1315/291/1/012007

Participation patrolling efforts by local people: Case of Nam Tien forest station in Pu Hu Nature Reserve, Vietnam

L. K. Dong[1,2], S. Sinutok[1,3], P. Manop[5] and K. Techato[1,4]

[1]Faculty of Environmental Management, Prince of Songkla University, Hat Yai, 90110, Thailand.
[2]Pu Hu nature reserve, Quan Hoa district 480000, Thanh Hoa province, Vietnam
[3]Coastal Oceanography and Climate Change Center research, Princes of Songkla University, Hat Yai, Thailand.
[4]Environmental Assessment and Technogy for Hazardous Waste Management, Faculty of Environmental Management, Prince of Songkla University Hat Yai, 90110 Songkhla, Thailand.
[5]Faculty of Law, Prince of Songkla University, Hat Yai, 90110 Songkla, Thailand.

E-mail: Ledongthph@gmail.com, uhugua@hotmail.com

Abstract. Protected areas, nature reserves and national parks in Vietnam, has considered foot patrols under traditional law enforcement with local forest guard (LFG). The participant between LFG and rangers in protected areas has been applied as commons way of patrolling on forest protection. However, very few studies have considered the contribution of LFG performance of patrolling activity with GPS-based monitoring system from Nam Tien forest station (FS), Pu Hu nature reserve (NR). Through the contribution of the monthly report was collected and the reliability of patrol data was analyzed in the central office. This study explores that the number of LFG was affected by the distance and patrol hours in the working field ($P < 0.01$). It could also be reducing the rangers' law enforcement if the limitation of LHG's performance was considerably significant. Admittedly, the illegal activity might be declined by ranger's efforts in term of patrolling activity. This study highlights the importance of LFG performance for protected area management was related to the efficacy of conservation targets.

1. Introduction

The main role of tropical forests in biodiversity and climate change has commanded the world to examination for current ways to reduce deforestation [1]. Furthermore, the protected areas, including national parks and nature reserves in Vietnam, are fundamental for conservation [2]. These areas are habitat for the variety of animal and plant species [4] but also to support human well-being [3]. Location of protected area is perceived as a purely protected "sanctuary of nature" without human's activity [5]. Because, strictly protected areas, which proscribe most human activities, are popular in the early days of conservation [1] with keeping far the safe from illegal activities[6].

Typically, conservation policy has been keeping out of people from NRs and operate impingement by law enforcement [7]. Fringe people in protected areas keep on extracting forest resources from PAs by boundary encroachment, including illegal logging, firewood, grazing, hunting, settlements and agricultural farms[8,9]. The continuous struggling between the management board and some immediate inhabitants illustrates some inconsistencies in imposing conservation law because the variety of protected areas are isolated by surrounded settlements and agricultural farms [10].

In order to address threats to conservation, the practiced activity is prohibited illegal activities in tropical forest based on law enforcement efforts that are an essential component of reducing illegal activities [9,11]. Rangers are frontline as the guardians responsible for the management protection [12] in many countries such as Vietnam with often tedious work [13] during patrolling exercise in the huge jungle forest.Forest management recruitslocal people from the fringe villages as local forest guards [14]. That is a part of community forest management that is a widespread conservation approach in thesetopics [1].

This main of this study is to understand how the contribution of local people to participate in patrolling with rangers by monitoring at the Nam Tien forest station (FS) in Pu Hu Nature Reserve (NR). The dual objectives of this study were to (i) to quantify patrol efforts (ii) to analyze the fluctuation of LFG during the time given. This study seeks to concretion of local people who could present their efforts for supporting rangers in conservation aspects in the future.

2. Material and methods

2.1. Location and general description

Pu Hu NR is located in Quan Hoa and Muong Lat districts in Thanh Hoa province with composed mountain areas (latitude 20°30' to 20°40'N and longitude 104°40' to 105°05'E), in northeast Vietnam was established in 1999 as a special-use forest area by Vietnam government. Nam Tien FS is one of the managements of Pu Hu management board. It is home to various endangered and rare animal and plant species [15]. In the protected buffer zone, around 423 households with about 1,512 people in 12 villages of 2 communities are distributed (Figure 1).However, local living areas in and around NR often impact the ability of the PAs to meet conservation objectives[9]. Nam Tien FS has always implemented the patrol activity for the strict protected area. However, the great management issues include the relative unrestricted extraction of bamboo, thatch, medicine plants, and fuelwood by local residents from the surrounding area [9, 16]. With limited the permanent forest officers, Nam Tien FS hasthe number of villagers from the Village Forest Guard Group.

2.2. Data collection

In Pu Hu, Nam Tien FS applied for conventional law enforcement in the method of foot patrols that commonly emanate by ranger from scattered forest stations [15]. The patrol routes were presented on the map (Figure 1) and recorded the using GPS device by rangers or local people. All the patrol tracks were collected from January to December 2017 in central office with collecting monthly reports. The typical patrol track was carried out by one or two village guards with rangers patrolling depend on a head of forest station decision. During the patrolling routes, village guards were supporter rangers to patrol protected areas such as rangers had recorded the patrol day from each member of LFG.

2.3. Participant in patrol operation

Due to using conventional patrol on foot by Global Positing System (GPS)-based monitoring, we analyze patrol data from this area to consider on illegal activity encounters and patrol ranger efforts in the short-term law enforcement intervention [9]. The current patrol tracks and the member of LFG group were decided by the head of the forest station for the specify of sub-area forest [10]. The LFG group could advise the patrol plan due to lesson learn or new information from the village [9]. During the patrolling time, the leader of a team had a decision to make about whether to use an 'arrow head' or a 'single file' formation for his officers, depending on previous information obtained from the earlier patrol [10].Local guards had the discussion the new information with rangers and it could focus on this area. After patrolling each time, the name of patroller in LFG group was recorded by rangers or leader in the forest station. It helped rangers to organize the shifting local guard efficiency.

2.4. Data analysis

Kilometers (km) per Effort index (P/E), the percentage of the distance in kilometers (km) and man-days per unit was patrolled [10]. All the data were analyzed and presented in Microsoft Excel software for descriptive statistics and using the Statistical Package Social Sciences (SPSS, version 20.0)for determine any significant difference among various variables by testing function ($p = 0.05$) and mean values were presented as mean ± S.D (Standard deviation) [9]. Particularly, descriptive statistics were used to summarize the illegal activity found on data set. The data presented not normal distributed, thus, the correlation and compare means were analyzed during non-parametric statistics (Spearman rank correlation and Kruskal-Wallis test if more than two events and Man-Whitney test with two variables, $\alpha = 0.05$).

3. Results and discussion

Generally, the description of possible patrol routes illustrated in Figure 1 and of basic geography of each forest plot presented in Table 1. Most of the patrol routes was established by rangers and local guards based on the available paths and the sensitive biodiversity loss and the density of illegal activity. These tracks are divided into different priorities, like the size of areas, patrol distance, and absolute height.

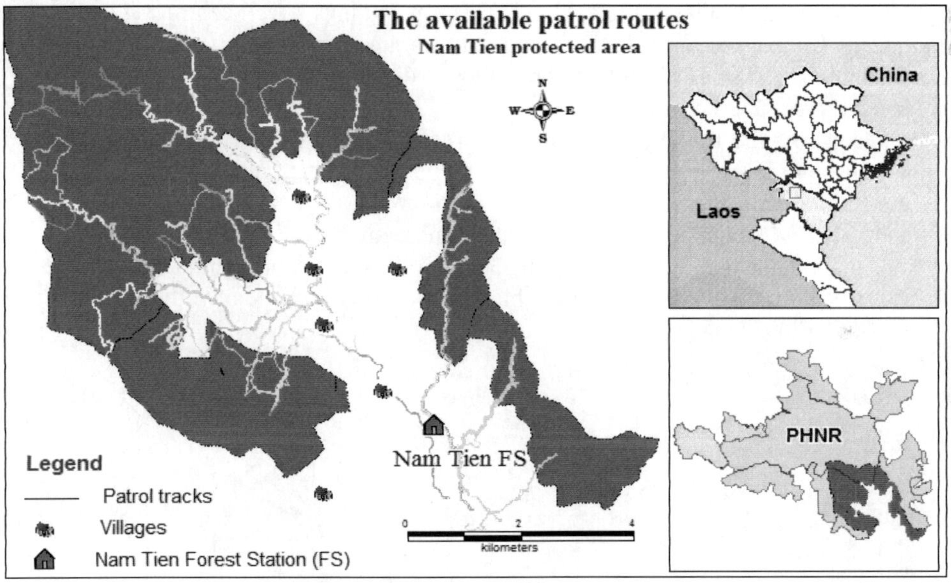

Figure 1. The location of case study area

Table 1. Summary of technical tracks on the protected area

Name of sub-areas	Name of communities	Size of sub-areas (ha)	Patrol distance (km)	Absolute height (m)
102	Nam Tien	1240.21	9 - 10	1,200 - 1,320
120	Nam Tien	1081.03	7 - 9	926 - 1,025
121	Nam Tien	754.46	7 - 8	650 - 730
124	Nam Tien	574.09	5 - 6	541 - 635
132	Nam Tien	658.82	4 - 5	230 - 335

142	Nam Tien	100.35	2 - 4	353 - 430
146	Nam Tien	678.08	1 - 3	454 - 540
147B	Nam Tien	36.06	5 - 6	600 - 752
142	Thien Phu	545.2	7 - 8	720 - 820

Source: [15]

3.1. Fluctuation of distance and patrol-hours

The patrol performance is, regarding patrolling days or hours and distance, the most important part is presented how adaptive management attempt to work [9]. The quarterly average patrol-distances in 2017 were as follows: 10.52 (SD = 5.44), 11.24 (SD = 8.33), 11.00 (SD = 8.08) and 11.73 (SD = 8.87) respectively. And patrol-hours (man-days) from the first quarter to fourth quarter altered in 2017, i.e., from 6.37 (4.48) to 7.23 (SD = 5.65). Particularly, there was no significantly different distances among these quarters ($H = 3.9$, $p = > 0.05$).The patrol efforts might be insufficient due to fitting local guard to the payment of low allowance per patrol day . By contrast, quarterly patrol-hours walked varied significantly across this year ($H = 8.4$, $p < 0.05$). Quite surprisingly, monthly distance and patrol hours by LFG group covered 2017 that did not differ. Different findings have been presented in Ghana [12] who found the significant index of distance and patrol hours, however, another supported to current findings in Vietnam [17, 18]. However, the changing of distance and patrol hour during this year was clearly observed (Figure 2)

On the whole, across the 12 months of patrol between rangers and VFG group, the number of man-day per month was positively correlated with the number of distance in those months ($rs = 0.89$, $p < 0.001$) as the similar consider in Ghana [12]. Furthermore, the number of local patrollers was positively correlated with number of distance and time-patrol ($rs = 0.27$, $p < 0.001$; $rs = 0.3$, $p < 0.001$). It was possible that the higher the number of patrollers on the walk in the larger protected areas explored. Thus, the number of patrollers rises they enabled to cover extended distance and patrol time walked as the similar in Ghana [10] and Vietnam [9].

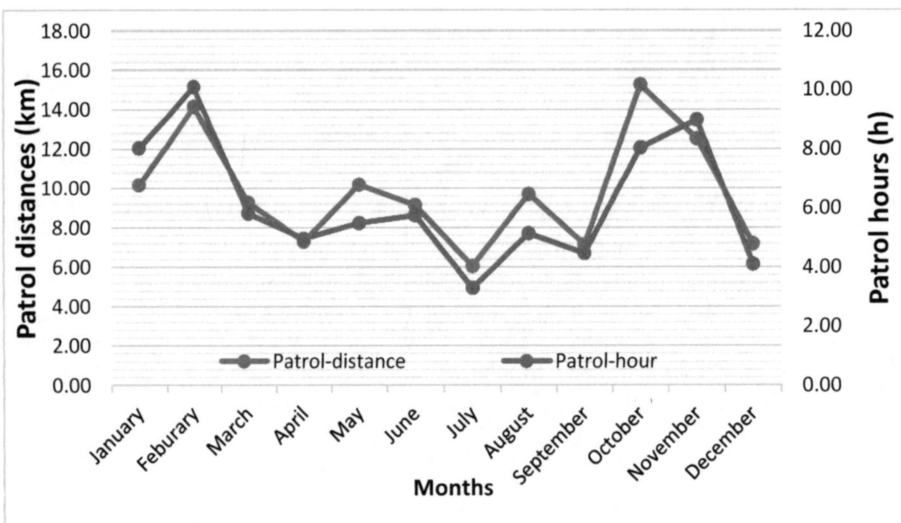

Figure 2. The variation of distance and patrol hour

3.2. Relative altitude and speeds of patrolling

The quarterly relative altitude of patrolling was not significant across all the sub-forest plots ($H = 13.92$, $p > 0.01$). It was no doubt that the patrol routes were significantly stable in topography during

IOP Conf. Series: Earth and Environmental Science **291** (2019) 012007 doi:10.1088/1755-1315/291/1/012007

those months even though patrol movement was unpredictable and subject to adaptation based on lessons learned or natural occurrences[10]. The relative altitude was excessively correlated with the distance and patrol-hours (rs = 0.29, $p < 0.01$ and rs = 0.32, $p < 0.01$, respectively). Admittedly, the most difficult local guards walked, the highest relative altitude local guards was disappointed. As shown in Figure3, the average relative altitude was merely changed.

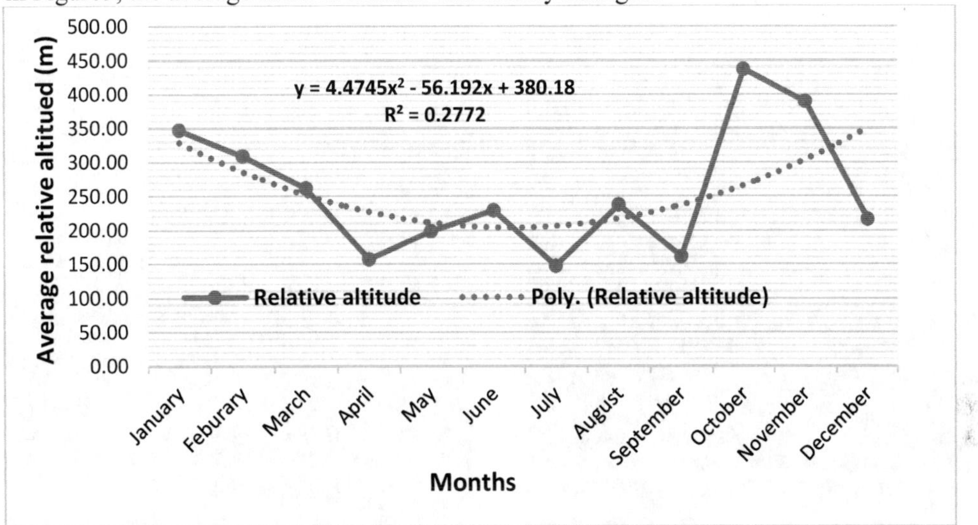

Figure 3. The difference in average relative altitude

Conversely, the quarterly speed of patrolling was not significant (H = 10.12, p <0.01). Depend on the current situation, it required local guards to patrol in the given time with different speeds. Further, the speed of patrolling was the negative correlation with relative altitude (rs = -0.25, $p < 0.01$) that was undoubtedly that level of the topography on conventional patrolling was mainly the factor of discouraging local guard to patrol on the field. However, the increase in patrol speed was significant during the month (Figure4). Interestingly, there was no a signifcant correlation with the speed of patrolling related to the distance and patrol-hours (rs = 0.09, $p > 0.05$ and rs = 0.19, $p > 0.05$, respectively). More importantly, the relationship between the number of local guards given by rangers and relative altitude and speeds of patrolling was not significant (rs = 0.03, $p > 0.05$ rs = 0.27, $p < 0.05$ respectively).

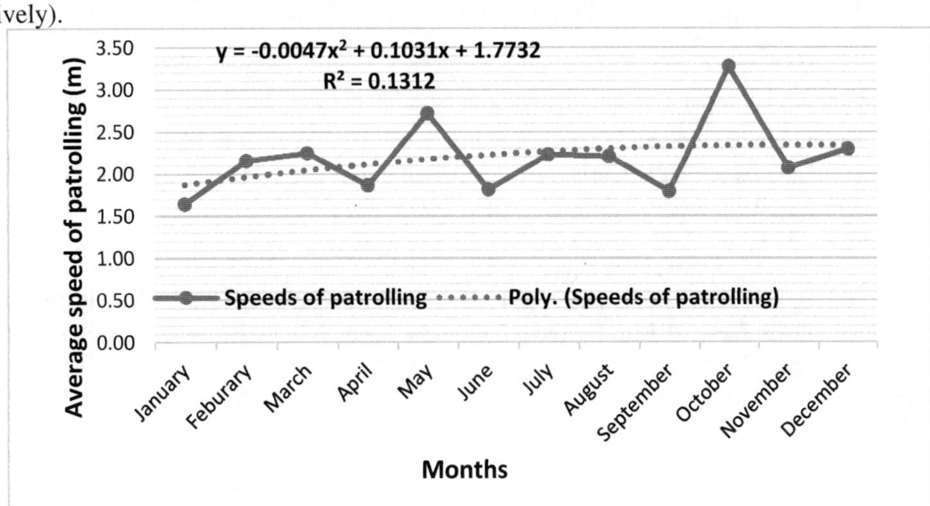

Figure 4. The difference in average speed of patrolling

Figure 5. Picture of the working field

4. Conclusion

This benefit of the paper presents the strategy of a forest activity provides planning being enhanced for the sustainable forest protection at the local level. The objective of community forest management is involved the number of people to participate in patrolling as law enforcement efforts (Figure 5). In spite of the limited staffs, with often no more financial operation flowing to the patrolling fields, the patrol officers is one important factor to continues illegal activities off-take would be reduced drastically. Effective management of LFG group is also dependent on information on the illegal activity found, thus, it is a seemly significant strategy of conservation protection efficiently. Furthermore, research is therefore needed to ascertain the effect of social factors and education can be influenced patrol performance as well as patrol efforts. And, a comparison of patrol efforts of rangers and LFG should consider.

5. References

[1] R. A. Rasolofoson, P. J. Ferraro, C. N. Jenkins, and J. P. Ge. of C. F. M. at reducing deforestation in M. Jones, "Effectiveness of Community Forest Management at reducing deforestation in Madagascar," *Biol. Conserv.*, vol. 184, pp. 271–277, 2015.
[2] A. Erlewein, "Disappearing rivers — The limits of environmental assessment for hydropower in India," *Environ. Impact Assess. Rev.*, vol. 43, pp. 135–143, 2013.
[3] U. C. Berkeley and A. Cole, "Electronic Thesis and Dissertations Wildlife Monitoring and Conservation in a West African Protected Area By Andrew Cole Burton A dissertation submitted in partial satisfaction of the requirements for the degree of Doctor of Philosophy in Environmental Sc," *Univ. Calif.*, 2010.
[4] T. Patarkalashvili, "Some problems of forest management of Georgia," *Ann. Agrar. Sci.*, vol. 14, no. 2, pp. 108–113, 2016.
[5] K. Niedziałkowski, M. Blicharska, G. Mikusiński, and B. Jedrzejewska, "Why is it difficult to enlarge a protected areaα Ecosystem services perspective on the conflict around the extension of the Białowieża National Park in Poland," *Land use policy*, vol. 38, pp. 314–329, 2014.

[6] K. E. Jenks, J. Howard, and P. Leimgruber, "Do Ranger Stations Deter Poaching Activity in National Parks in Thailand?," *Biotropica*, vol. 44, no. 6, pp. 826–833, 2012.

[7] C. No 24/10/2010/ND-CP, *Administration and management of special-use forest.* Hanoi, Vietnam: http://moj.gov.vn/vbpq/lists/vn%20bn%20php%20lut/view_detail.aspx?itemid=26113, 2010.

[8] F. J. Venter, R. J. Naiman, H. C. Biggs, and D. J. Pienaar, "The evolution of conservation management philosophy: Science, environmental change and social adjustments in Kruger National Park," *Ecosytems*, vol. 11, no. 2, pp. 173–192, 2008.

[9] K. T. Dong. K. L, Sutinee. S, Hoa. A. X, Dong. N. P, Ali. A, Mano. P, "A quick comparison of patrol efforts for supportive protection : a case study of two stations in Vietnam," *Appl. Ecol. Environ. Res.*, vol. 16, no. 2, pp. 1767–1781, 2018.

[10] C. Wiafe, "Wildlife laws monitoring as an adaptive management tool in protected area management in Ghana: a case of Kakum Conservation Area," *Springerplus*, vol. 5, no. 1, p. 1440, 2016.

[11] H. Jachmann, "Illegal wildlife use and protected area management in Ghana," *Biol. Conserv.*, vol. 141, no. 7, pp. 1906–1918, 2008.

[12] W. D. Moreto, "Occupational stress among law enforcement rangers: insights from Uganda," *Oryx*, vol. 50, no. 04, pp. 646–654, 2016.

[13] N. J. Dominy and B. Duncan, "GPS and GIS Methods in an African Rain Forest : Applications to Tropical Ecology and Conservation," *Conserv. Ecol.*, vol. 5, no. 2, pp. 1–9, 2001.

[14] S. Humphries, T. Holmes, D. F. C. de Andrade, D. McGrath, and J. B. Dantas, "Searching for win-win forest outcomes: Learning-by-doing, financial viability, and income growth for a community-based forest management cooperative in the Brazilian Amazon," *World Dev.*, 2018.

[15] K. T. Dong. K. L, Sutinee. S, Hoa. A. X, Anh. N. T, Thinh. N.V, Hai. L. V, Mano. P, "Overview of improving patrolling efforts : a case study of forest staction in Pu Hu nature reserve ," *Appl. Ecol. Environ. Res.*, vol. 16, no. 3, pp. 2845–2859, 2018.

[16] H. A. X. Dong Le Khac, Sutinee Sinutok and K. T. Manop Promchana, "Potential of Approached Ecotourism Consideration as Part of Patrolling Efforts Responsibility in Pu Hu Nature Reserve, Vietnam," *EnvironmentAsia*, vol. 7, no. 1, pp. 104–111, 2018.

[17] H. Jachmann, "Monitoring law-enforcement performance in nine protected areas in Ghana," *Biol. Conserv.*, vol. 141, no. 1, pp. 89–99, 2008.

[18] H. Jachmann, "Illegal wildlife use and protected area management in Ghana," *Biol. Conserv.*, vol. 141, no. 7, pp. 1906–1918, 2008.

Acknowledgements

The authors would like to say the thanks to Thailand's Education Hub for Southern Region of ASEAN Countries (TEH-AC), Graduate School of Sustainable Energy Management, faculty of Environmental Management of Princes of Songkla University – Thailand for supporting my research. I would like to appreciate the protected area authorities of the Pu Hu nature reserve and Nam Tien Forest Station where a ranger-based monitoring system have been established. And, I am thankful Department of Law Enforcement-Pu Hu for providing all technical processing of data analysis, their knowledge, and their services as well.

IOP Conf. Series: Earth and Environmental Science **291** (2019) 012008 doi:10.1088/1755-1315/291/1/012008

Estimating the Impact of the MRT Elevated Railway System on Carbon Reduction in Taichung, Taiwan

Kuo-Wei Hsu[1], Jen-Chih Chao[2], Hsiao-Sung Chan[3], Pei-Chen Wu[4]

[1]Department of Landscape and Urban Design, Chaoyang University of Technology, Taiwan
[2]PhD program in Architecture and Urban Design, Department of Architecture, Chaoyang University of Technology, Taiwan
[3]Department of Applied Chemistry, Chaoyang University of Technology, Taiwan
[4]Department of Landscape and Urban Design, Chaoyang University of Technology, Taiwan

E-mail:jenchihchao@gmail.com[2]

Abstract. Amid rapid global economic development and the threat of global warming, green transportation and related issues have risen to prominence in various countries. In Taiwan in_2009, Taichung was selected as a low-carbon demonstration city, in which energy conservation and carbon reduction would be actively promoted, with the wider aim of developing a green economy and lifestyle. This study explores the carbon-reduction benefits of the MRT elevated railway in the Taichung Metropolitan Area. The railway's construction has been associated with intensive redevelopment of existing stations, but this process has been unbalanced due to railway physical. Five new stations will also be added, and belt-like spaces at ground level made available for housing and other non-railway-related purposes. Utilizing literature collection and the Analytical Hierarchy Process (AHP) research method to explore the factors affecting the efficiency of the MRT, this study reveals that the AHP weights of the main criteria for the escalation of MRT development in Taichung are 0.5833 for green transportation and 0.4167 for green environment; while in the overall ranking of the sub-criteria, the top three evaluation factors are mass transportation (0.31337), greenery planting (0.21832), and green building (0.19444).

1. Introduction

Since the industrial revolution, the world's population and economy have grown rapidly, leading to greenhouse gas emissions and thus to global warming. The United Nations adopted its Framework Convention on Climate Change in 1992; this was followed five years later by the Kyoto Protocol, and in 2009 by the Copenhagen Accord. Many countries have embraced energy conservation and carbon reduction as new directions for their economic-development and energy strategies, in pursuit of sustainable low-carbon societies and economies. As such, it can now be said that international norms include reducing energy use and carbon emissions, mitigating climate change, and enabling global sustainability.

The World Wide Fund for Nature defines a low-carbon city as one that maintains low energy consumption and carbon dioxide emissions even under conditions of rapid economic development. The Climate Group's definition, meanwhile, requires that such cities have low-carbon economies. In Taiwan, the goals for low-carbon cities – rooted in wider sustainable-development aims – are to harness economic growth, environmental protection, ecological conservation, social justice and related policies to achieve multiple benefits: i.e., not merely reducing carbon emissions, but also fostering energy security and building resilience to environmental shocks. The development of mass transit can reduce the use of private cars, especially if developed in combination with bicycle networks, pedestrian walkways, and other forms of human-oriented traffic.

As part of its low-carbon strategy, the Taiwanese city of Taichung is actively promoting the development of both mass transit and human-oriented transportation. Specifically, the building of a new elevated railway system there, known as the MRT, will involve the greening and beautification of the ground-level areas previously occupied by the tracks, including but not limited to the installation of bicycle lanes and parks. At the same time, planned improvements to the stations and changes to the urban-planning regulations along these corridors (e.g., to drive the renewal of old buildings and increase the proportion of green buildings) are intended to boost the quality of life of the old metropolitan area; and a green belt along the line will be created as a living garden for residents and a focus for gradual regeneration and economic development, not merely in the city but extending into its rural hinterland. In addition, changes in the existing railway's routes have been devised in part to green parkway.

The present study follows the lead of most railway-related research on carbon reduction, which has focused on low-carbon cities, land-use impacts, and urban-development strategies (e.g., Tu, 2000; Hung, 2010; Kusumaningrum, 2012; Lin, 2013; Wang, 2006) [1-5]. Specifically, it examines the theoretical relationships between low-carbon cities and other low-carbon indicators, and establishes the carbon-reduction benefit index of the MRT elevated railway using the Analytical Hierarchy Process (AHP).

The Taichung MRT Project stretches from Fengyuan Station in the north to Daqing Station in the south, and has a total length of 21.7 kilometers. As well as the two termini, two existing stations – i.e., Tanzi and Taiyuan – will continue in use, and five new commuter stations will be built: at Lilin, Toujiacuo, Songzhu, Jingwu, and Wuquan. Additionally, Taichung Station will be rebuilt on a new site, though its original building will remain as a historical monument, possibly to be repurposed as a railway museum. Songzhu Station and Daqing Station will be served by the MRT's Green Line, and Taichung New Station by the Blue Line and Orange Line, as shown in Figure 1.

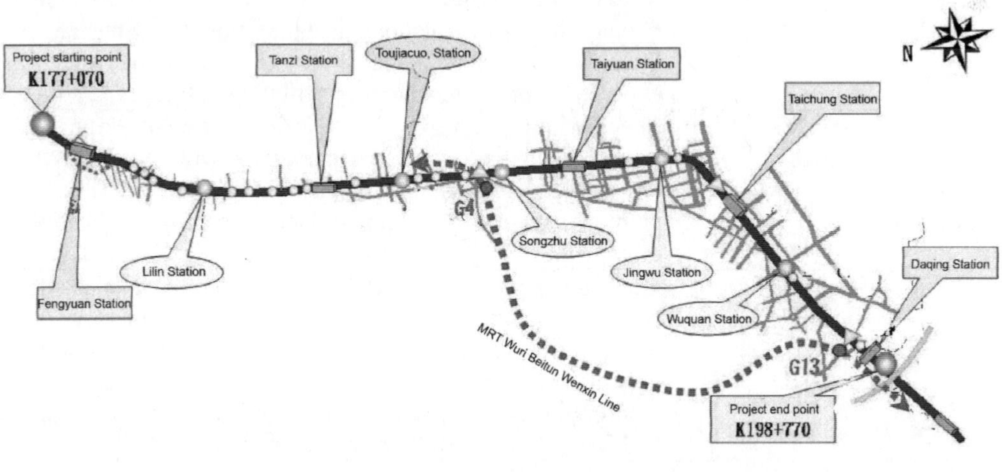

Figure 1. MRT Elevated Railway System in Taichung schematic diagram [6]

2. Research Methods

This study aims to clarify both the definitions and the goals of low-carbon cities and elevated railways, with special reference to the prior literature and actual rail-transit reduction plans in the Taichung

IOP Conf. Series: Earth and Environmental Science **291** (2019) 012008 doi:10.1088/1755-1315/291/1/012008

Metropolitan Area. The literature examined for this purpose included governmental research reports, master's and doctoral theses, online and offline newspaper and magazine articles, the European Green City Index (2009) [7], the Asian Green City Index (2011) [8], and the U.S. Environmental Protection Agency's low-carbon city benefit assessment (2011) [9], among other relevant sources. The structural indicators used in AHP were established by Chih (2009) [10] and Yang (2012) [11] and Wu (2017) [12]. Descriptions of the research dimensions and of these indicators are presented in Tables 1 and 2, respectively.

Table 1. Taichung MRT Research Dimensions

Dimension	Definition
Green Transportation	Development of mass transit can reduce the use of private cars. Bicycle and pedestrian-walkway networks can be designed to complement mass transit, thus achieving the maximum benefit of green traffic. This non-motorized transportation is beneficial to human health and interaction, and relatively safe, as well as an effective means of reducing carbon emissions.
Green Environment	The elevation and re-orientation of the MRT will form a new axis for the city's parks, along which renewal of old buildings, the construction of new green buildings, the creation of new green spaces, and low-carbon economic development will take place. Eventually, it is expected to extend these benefits to surrounding communities that the MRT connects.

Table 2. Taichung MRT Carbon-reduction Indicators

Dimension	Impact Sub-dimension	Definition
Green Transportation	Mass Transit	Five commuter stations will be added to the existing system, of which Songzhu Station and Daqing Station will serve the MRT's Green Line, and Taichung New Station the Blue and Orange lines. In tandem with the bus system, these improvements are expected to promote residents' willingness to use mass transit.
	Bicycle Network	The Taichung Bicycle Road Network, based on the Taichung Heart Green Parkway bicycle road network, complement's the city's established public bicycle-rental system. There are also plans to create a new bicycle lane in the open space along the railway that will connect to the existing bicycle-lane system.
	Pedestrian Trail System	The elevation of the MRT above ground level will release considerable space, which will be devoted primarily to people-oriented functions. It will be connected to the surrounding area's existing pedestrian walkway system to create a friendly and comfortable walking environment and reduce the possibility of collisions between walkers and vehicles.
Green Environment	Green Architecture	The new MRT stations are being built according to green principles, with solar panels and natural ventilation and lighting.
	Greenery Planting	The linear open spaces that will result from the railway's elevation will become a new "garden road" running through Taichung. Most of this space will be used as a park, other green space, and a public plaza. Once it is connected to the city's existing Green Parkway system, it will provide a diverse

	recreational environment that is expected to reduce carbon emissions.

In keeping with the present study's primary interests in low-carbon cities and urban development, the experts surveyed as part of its AHP assessment of the carbon-reduction benefits of the Taichung MRT were recruited from the fields of urban design, ecology, and landscape planning. In this study, a questionnaire sent via e-mail. A total of 17 experts (i.e., either five or six in each of these three fields) were invited to participate, 10 of whom agreed to do so. The eight valid responses these 10 experts provided were then analyzed using Expert Choice.

3. Results and Discussion

3.1. Taichung MRT Carbon-reduction Benefit Measurement

Analysis of the expert questionnaire results indicated that the weight of green traffic (0.5872) is greater than the green environment (0.4128), the weighting difference between the two was only 0.1744, as shown in Table 3.

Table 3. Taichung MRT Carbon-reduction Benefits Dimensions by Rank

Dimension	Relative weight	Absolute weight	Ranking
Green Transportation	0.5833	0.5164	1
Green Environment	0.4167	0.3630	2

As shown in Table 4, the weights of the sub-dimensions of green traffic were mass transit (0.31337), pedestrian trail system (0.17269), and bicycle network (0.10117). For green environment, they were greenery planting (0.21832) and green building (0.19444).

Table 4. Taichung MRT Carbon-reduction Benefits Impact Sub-dimensions by Rank

Impact sub-dimensions	Relative weight	Absolute weight	Ranking
Mass Transit	0.5293	0.3088	1
Bicycle Network	0.1694	0.0988	5
Pedestrian Trail System	0.3013	0.1757	4
Green Building	0.4714	0.1964	3
Greenery Planting	0.5286	0.2203	2

3.2. Analysis of the Weights of Carbon-reduction Indicators for the Taichung Metropolitan Area's Overhead Rapid Transit Plan

3.2.1.Green Transportation. Our review of the prior literature on green transportation suggests that, while the development of mass transit is critically important, people-oriented transportation should never be overlooked, as it can dramatically increase the convenience of living in urban areas by widening access to all areas (mass transit facilities included). Moreover, the convenience of mass transit frequently leads to increases in the overall number of trips people make; and if there is no effective plan for moving people from their homes to the trains, and from the trains to their eventual destinations, the net effect of rail improvements might actually be increased congestion. From the perspective of long-term urban development, therefore, the completely integrated planning of mass transit and human-traffic routes will increase the efficiency of the city, reduce waste, and effectively achieve the goal of carbon reduction.

3.2.2.Green Environment. Although the weighting that the experts assigned to the green-environment dimension is lower than the one they assigned to green transportation, the importance of its contribution to the Taichung MRT program's carbon-emission reduction effort was still strong. Guided by the overarching goal of reinventing itself as a low-carbon city, Taichung has built a ring road and incentivized the construction of green buildings. Parkways connect small local green spaces to larger parks, forming a green network that mitigates the air pollution and carbon reduction while providing a space that is sociable, environmentally friendly, energy-saving, healthy and in harmony with the area's natural ecology.

3.2.3.Mass Transportation. The prior literature has often advanced arguments that public transport is the key factor in urban carbon reduction, and this study's panel of experts indeed gave mass transit a much higher weighting (0.31337) than either the pedestrian-walkway system (0.17269) or the bicycle network (0.10117).

Proceeding from the concept of sustainable urban development, it is reasonably clear that high-efficiency mass transportation systems are the backbone of high-quality, sustainable urban development, and important drivers of the use of human-powered transportation and reductions in the use of private motor vehicles (Du, 2000). After the Taichung MRT project is completed, the new station and the MRT Green Line will be shared, which as well as improving the metropolitan area's mass-transit network will effectively encourage its people to reduce their carbon emissions.

3.2.4.Greenery Planting. The category of greenery planting includes green spaces and parks. As briefly noted above, the area beneath and beside the 21.7-kilometer route of the Taichung MRT will become the new Green Parkway in Taichung City, in tandem with the original Green Parkway in Taichung Metropolitan Area. Green spaces and gardens in cities have become important tools for regulating climate and reducing carbon. The canopy of trees in the city can intercept, reflect and absorb solar radiation; and evaporation from trees and other plants can also absorb heat, mitigating the urban heat-island effect. In addition to the linear open space on the ground that will be made available due specifically to the elevation of the railway, open space will be cleared on both sides of each station to reduce the system's overall impact via greening. Planning restrictions can ensure that it will be used as a greenbelt in combination with the space once occupied by the original railway, and thus connect surrounding areas into a new low-carbon living circle.

3.2.5.Green Building. All building work associated with the MRT project – including repair and renewal, as well as new construction – will be conducted on green-building principles. It is anticipated that the completion of each new or refurbished station will revitalize the surrounding area, with existing buildings being given the opportunity to be renewed, repaired or rebuilt using green methods and green building materials, and guided by the design principles of energy saving and carbon-emissions reduction. The plan's inclusion of historical preservation of some station architecture will bring about further opportunities for development and redevelopment of the adjacent area.

3.2.6.Pedestrian Trail System. Fairly small changes to the human environment can have large impacts on the use of mass transit, and thus on the greening of transportation in general. In recent years, Taichung's planning officals have actively promoted human-oriented road spaces, amid a shift in the overall emphasis in the development of urban space and transportation from efficiency to liveability. In addition to their basic function of moving people and goods from place to place, urban transportation services are increasingly expected to meet the requirements of humanization and environmental protection, and attract a greater share of the public to use mass transit. In recent years, amid increases in air pollution caused by railway engines and the negative impact of global warming, people are paying ever more attention to the concept of perseverance and re-engaging with the concept of green transportation. In the development of low-carbon cities, people-oriented transportation is assigned a high value, due to its environment-friendliness and sustainability. Pedestrian walkway systems emphasize freedom, convenience, accessibility, aesthetics, safety, comfort and sustainability. Among such systems' sustainability and comfort indicators, factors such as green coverage, plant diversity, and planting green beautification are frequently mentioned; and because – if done right – walkways can both enhance interaction between communities and connect people to mass transit, thus enhancing their willingness to use it, perfecting the pedestrian system is one of the most important tasks of mass- transit design and construction.

3.2.7.Bicycle Network. Taichung's bicycle lanes and its plans to construct others are discontinuous, and there is currently no mechanism for creating a single, connected bicycle road network there. Nonetheless, recent years have seen active promotion of the bicycle network's development as a green corridor connecting the Taichung Central Life Circle to commuter transportation and recreation areas, as an intrinsic part of the emerging green corridor network. However, it is clear that despite people's extensive use of bicycles for leisure purposes, they are not used for commuting in Taichung, so it is perhaps unsurprising that the nascent bicycle network's carbon-reduction benefits were assigned the lowest AHP weighting among all the factors this study examined. If the overall bicycle network is eventually perfected to the point that Taichung's residents can use it as a continuous commuting route, it will become a viable alternative to both mass transit and personal motor vehicles, and thus make an important contribution to carbon reduction in the city. The public bicycle-rental system was implemented specifically to encourage short-distance use of this low- mode of transportation, not only to reduce environmental pollution and energy use, but to mitigate urban road-traffic congestion by reducing demand for the ownership and use of private motor vehicles.

4. Conclusion

Based on a literature review and an AHP panel study, the present work has explored the carbon-reduction impact of Taichung's MRT elevated-railway project. The results suggest that the impact structure of the MRT on carbon-reduction benefits consists of two major facets: green transportation and green environment. These in turn can usefully be divided into a total of five evaluation factors: i.e., green transportation into mass transit, bicycle networks and pedestrian-walkway systems; and green environment into green buildings and greenery planting.

Although the AHP weighting assigned to green transportation (0.5872) was greater than that of green environment (0.4128), the weighting difference between the two was relatively small. This implies that the development of green transportation must proceed in tandem with green-environment initiatives if the city and metropolitan area are to effectively achieve their carbon-reduction goals.

The carbon-reduction benefit index was highest for mass transportation, with an absolute weight of 0.31337. The next highest was for greenery planting, at 0.21832. The third- and fourth-ranked positions went to green building (0.19444) and the pedestrian walkway system (0.17269), while by far the lowest absolute weight was assigned to the bicycle network, at 0.10117.

5. References:

[1] Tu, Yun-Lung (2000) Transit-Oriented Development-Applications for Tanhai New Town, Graduate Institute of Urban Planning, National Taipei University, Thesis for the Degree of Master.
[2] Hung, Min-Chen (2010) Simulation Modeling of Taiwan Railway Administration's Rapid Transit Operations by Using Stochastic Petri Nets, Department of Transportation Management, Tam Kang University, Thesis for the Degree of Master.
[3] Rahayu Kusumaningrum (2012) Impact of Road Improvement Projects on Greenhouse Gas Emission: A Case Study of the Railway Elevation Project in Taoyuan, Taiwan, Department of Civil Engineering, National Central University, Thesis for the Degree of Master.
[4] Lin, Chih-Chun (2013) The Study for Low Carbon City Indicators, Department of Urban Planning and Spatial Information, Feng Chia University, Thesis for the Degree of Master.

[5] Wang, I-Fan (2006) The Analysis of Influence Factors in the Land Use Changes along Metro line—An Empirical Study of the Metro Blue Line, Institute of Traffic and Transportation College o f Management, National Chiao Tung University, Thesis for the Degree of Master.
[6] Taichung Metropolitan Area Elevated Railway Project ONLINE. Available online: https://www.rrb.gov.tw/04100.aspx?id=5&lan=ch (accessed on 2 October 2018).
[7] Economist Intelligence Unit (London) (2009) European Green City Index – assessing the environmental impact of Europe's greatest cities, Munich: Siemens AG.
[8] Economist Intelligence Unit (London) (2011) Asian Green City Index – assessing the environmental impact of Europe's greatest cities, Munich: Siemens AG.
[9] U.S. Environmental Protection Agency's (2011) Low Carbon Sustainable Information System. Available online: https://lcss.epa.gov.tw/Default.aspx. (accessed on 2 October 2018).
[10] Cheng-Hsien Chih, 2009, Hierarchical Analysis of the Social Education for Promoting Carbon Reduction, Department of Environmental Engineering and Management, Chaoyang University of Technology, Thesis for the Degree of Master.
[11] Yang, Chia-Hui (2012) Assessment of the Priorities of Green Low-Carbon Campus Application in Elementary and Junior High School by AHP, Graduate Institute of Architecture and Urban Design, National Taipei University of Technology, Thesis for the Degree of Master.
[12] Wu, Yi-Shiun (2017) Exploring impact factors of MRT joint development using AHP: Case study of Taoyuan MRT green line, Graduate Institute of Construction Engineering and Management, National Central University, Thesis for the Degree of Master.

The Road Ahead for Municipal Solid Waste Management in the 21st Century: A Novel-standardized Simulated Paradigm

M S S Danish[1,2], H Zaheb[2], N R Sabory[2], H Karimy[2], A B Faiq[3], H Fedayi[3] and T Senjyu[1]

[1] Electrical and Electronics Department, University of the Ryukyus, 1 Senbaru 9030213, Japan
[2] Energy Engineering Department, Kabul University, Jamal Mina, 1006, Afghanistan
[3] Electrical and Electronics Department, Kabul University, Jamal Mina, 1006, Afghanistan

Email:mirsayedshah@yahoo.com

Abstract. The recent global trend in lifestyle change, alter in economic and human civilization growth has a dramatic increase in municipal solid waste (MSW) generation that led to a renewed interest in researches and innovations in this context. A primary objective of the study is to identify influencing factors in term of MSW from different standpoints hiring tangible and non-tangible cause and effect theory to foster sustainability through an exhaustive decision-making model and strategy. Explicitly, this research effort breakdown a business model into manageable chunks to address how to systematically manage MSW; and which tools and techniques should be efficaciously applied to adapt each step of the process. Decidedly, an inclusive strategy and model can be demonstrated with a real-world application with some significant implications for future practice as a multilateral planning and design reference for students, researchers, scholars, and practitioners in the context of MSW management. Therefore, in a general term, the outlook of this study is based on the evolution of the state of the art of a novel model for MSW management based on multi-oriented approach, comprises technical, social, environmental, institutional, business, management, policy, and cultural factors.

1. Introduction

A rapid change in economic growth and human civilization have led to a dramatic increase in MSW generation in the urban and rural levels, especially in the least developed countries. That faces these nations with the challenge to maintain a clean air and an effective framework for MSW disposal and treatment. According to [1], the top ten pollutant cities in the world with poor air quality due to many factors; the increasing population growth and MSW generation are parts of this problem, shown in Table 1.

Table 1. The most top ten pollutant cities in the world.

Rank	City	Pollution Index	Exp Pollution Index
1	Accra, Ghana	97.88	177.34
2	Tetovo, Macedonia	96.85	177.01

3	Faridabad, India	96.24	174.46
4	Ghaziabad, India	96.10	174.32
5	Kathmandu, Nepal	95.54	173.46
6	Ulaanbaatar, Mongolia	94.93	175.33
7	Kabul, Afghanistan	94.84	171.67
8	Cairo, Egypt	93.79	169.84
9	Dhaka, Bangladesh	92.96	167.41
10	Ho Chi Minh City, Vietnam	92.79	168.02

Numerous studies have attempted to explain the term of sustainability, in general consensus [2-5] the concept of sustainably can be referred that how to use the energy resources in a way to be sufficient for now, and do not compromise the ability of future generations to meet their needs. The most important criteria for sustainable energy production are accessibility, affordability, disparity, safety, use efficiency, supply and production efficiency, cost effectiveness, and environmental impacts on air, water, and soil quality [6, 7]. Mismanagement of MSW can lead to unexpected environmental and social disasters. According to the [8, 9], 97% of cities in low and middle-income countries with more than 100,000 inhabitants do not meet the World Health Organization (WHO) air quality guidelines that can cause the risk of stroke, heart disease, lung cancer, and chronic and acute respiratory diseases, including asthma.

2. Methodology

The methodology of this paper is concerned with an amalgamated approach, which targeted various domains and stakeholders (residence, scholars, researcher, technology developers, enterprise, the government body, policy-makers, and so on). A systematic process of the model development is given in Fig 1, which composes qualitative and quantitative elements within logical consequence attitudes and analysis. Each step is demonstrated with applicable tools and techniques, which are parameters-sensitive that can be described based on the adequate inputs (objectives, factors, identifiers, constrains, hypothesis, characteristics, etc) in order to model a proper solution for MSW.

3. Standardized paradigm

Increasing globalization enforces standardizations to align interchangeability and converge diversities within defined global paradigms for qualitative and quantitative values optimization and economic integration. The first-ever efforts of standardization were proposed in the early twentieth century [10], now it applies globally beyond its initial objective as a body of certification and certainty.

The ISO 14001 standard is promulgated in 1996, in compliance with British BS 7750 standard, which aims to strengthen toward a green paradigm for production and consumption in term of Environmental Management System (EMS) [10]. The adaption of ISO 14001 in the main global economic zone countries is shown in Figure 1.

The IC-GDP ratio is calculated by ICI that shows the share between the number of worldwide issued certificates and percentage share of total world GDP in 2015 (in USD at current exchange rate of the time) Figure 2 [10, 11].

4. The proposed framework-scheme

Since, this research is a multi-dimensional study that involves various potential partners with different roles and expected contributions, the major objective of this research is to provide equal opportunities to all stockholders involved in MSW management and governance process. Reference to our finding, there is need for an interconnected scheme of collaboration among the stockholders to address a close coordination.

The proposed scheme is projected to simulate scenario-based methodology. In which, scenarios' optimum outcomes are separately obtained in term of suitability criteria to align the optimum solutions in form of an integrated moderate model of implementation. The cause and effect technique can be hired to explore the root cause and propose a reasonable solution. There are various approaches for weighing and scoring according to different codes and methodologies. The weighing approach can be applied to score and reach a decision. The purpose of a dynamic-flexible scenario is to adapt to the real situation with optionality of wizard methodology to select different feasible option. These scenarios can be door-to-door, pay-for-use and many more options.

While, delineating indicators for each influential factor is a prerequisite for weighting approach. A detailed discussion on weighing approach and indicators is not the scope of this study, the authors will deal with these topics extensively future. The driven scheme in Figure 3 is flexible roadmap to adjust desired parameters and millstones for Triple-R control. These parameters and objectives can vary from city to city due to local resiliency and demand. For instance, to reduce solid waste disposal production and foster collection can set incentives and penalties, engage private and public sectors, promote awareness programs, suggest alternatives, and so on. An overall optimization of current MSW management system through upgrading and adding new futures (technically and technologically) to simply MSW collection and disposal flow can enhance this trend.

A complete instruction of this scheme will be provided in the future studies along with some case studies to evaluate the viability of the proposed mechanism, and to validate the effectiveness of the model within the given hypothesizes and parameters/characteristics.

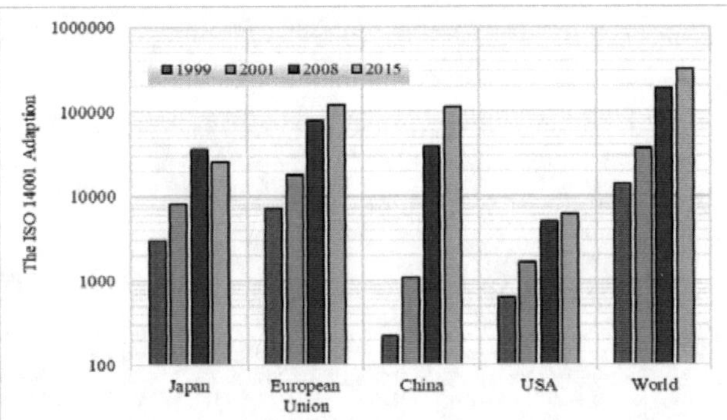

Figure 1. The global economic zone countries ISO 14001 statistics [9]).

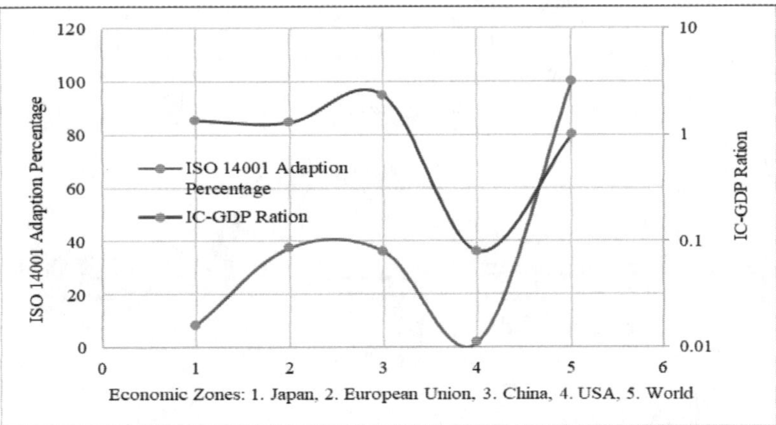

Figure 2. The global economic zone countries ISO 14001 adaption percentage.

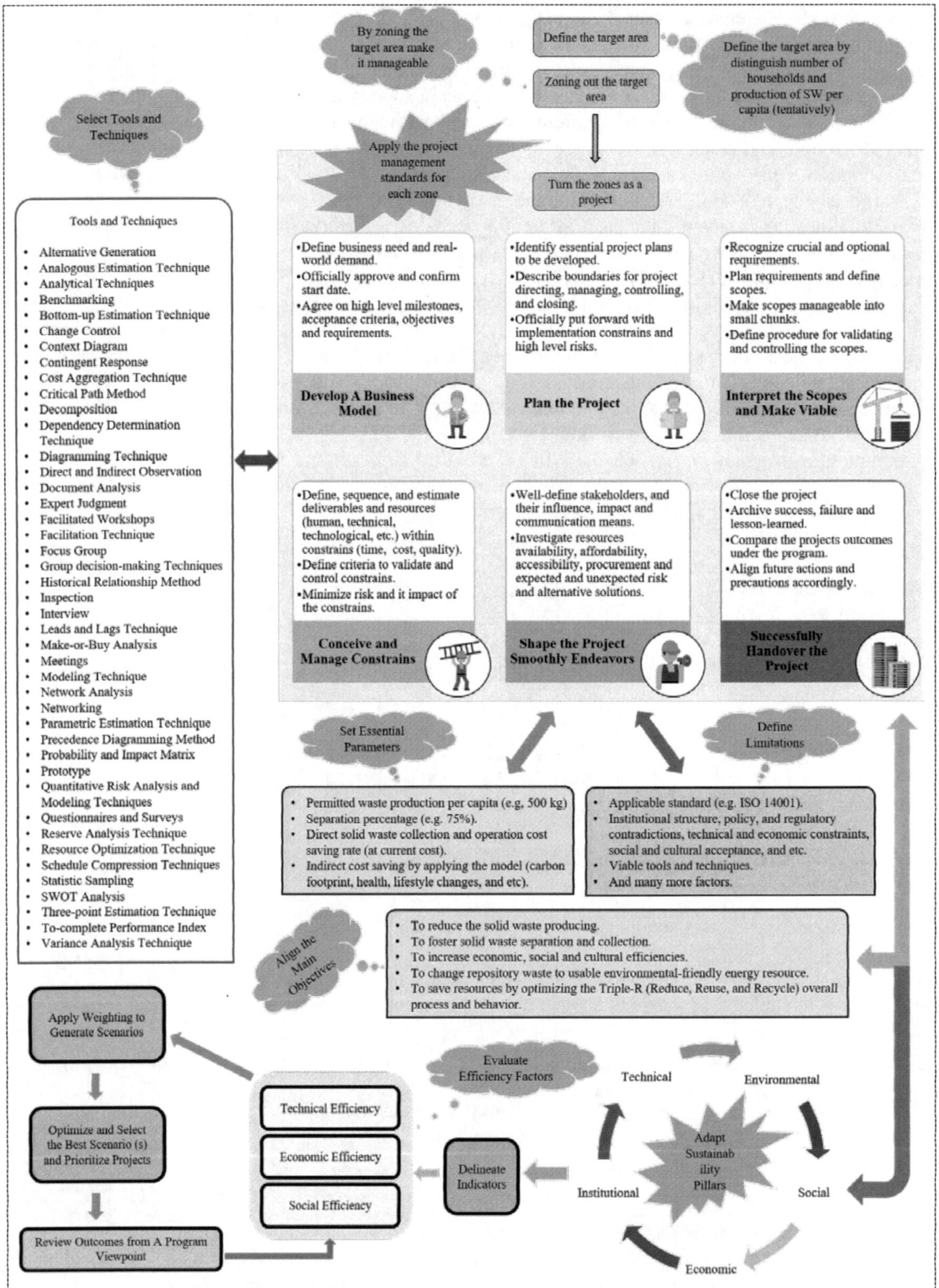

Figure 3. The proposed framework-scheme for MSW management.

5. Conclusion

The overall objective of this study is to shape a sustainable-cost-effective model that enhances Triple-R (Reduce, Reuse, and Recycle) efficiency and effectiveness considering technical and technological efficiencies in term of MSW management. That the proposed scheme relies on the most crucial factors to meet desired objectives. Unlike recently, this research deals with first-ever effort to conceptualize a viable roadmap/Scheme for a proper management of municipal solid waste along with it implementation systematic phases. The special measures are hired in this study to draw an exhaustive framework in compliance of resiliency and sustainability criteria throughout its life-cycle. This study can have some significant implications for future practice as a multilateral planning and design reference for students, researchers, scholars, and practitioners related to a municipal solid waste management and operation. The authors are willing to put forward a series of procedures paradigm in details within management, engineering, and business domains based on the world-wide accepted approach in a synergistic manner in the future.

6. References

[1] Pollution Index Rate 2018. https://www.numbeo.com/pollution/rankings_current.jsp, *Numbeo* (accessed on: 12 November 2018)

[2] Danish M S S, Sabory N R, Ershad A M, Danish S M S, Ohta R, Sediqi M, Ahmadi M and Senjyu T 2017. The least developed countries need for changing the passive trend of renewable energy exploitation to a proactive trend, *International Journal of Energy and Power Engineering*. **5** 215-21.

[3] O'Neill-Carrillo E, Irizarry-Rivera A A, Colucci-Rios J A, Perez-Lugo M and Ortiz-Garcia C 2008. Sustainable energy: Balancing the economic, environmental and social dimensions of Energy. *In Energy 2030 Conference, 2008*. Nov 17 Atlanta 1-7

[4] Brundtland G, Khalid M, Agnelli S, Al-Athel S, Chidzero B, Fadika L, Hauff V, Lang I, Shijun M, de Botero M M and Singh M 1987. Our common future (Brundtland report). *Medicine and War* **4** 17-25

[5] Danish M S S, Senjyu T, Sabory N R, Danish S M S, Ludin G A, Noorzad A S and Yona A 2017. Afghanistan's aspirations for energy independence: Water resources and hydropower energy. *Renewable Energy* **113** 1276-87

[6] Danish M S S, Senjyu T, Ibrahimi A M, Ahmadi M and Howlader A M 2019. A managed framework for energy-efficient building, *Journal of Building Engineering* **21** 120-8

[7] Danish M S S, Sabory N R, Ershad A M, Danish S M S, Yona A and Senjyu T 2017. Sustainable architecture and urban planning trough exploitation of renewable energy, *International Journal of Sustainable and Green Energy* **6** 1-7

[8] Scafà M, Carbonari S, Papetti A, Rossi M and Germani M 2018. A new method for product service system: the case of urban waste management, *Procedia CIRP,* Linköping **73** 67-72

[9] WHO Global Ambient Air Quality Database 2018. http://www.who.int/airpollution/data/cities/en (accessed on: 20 November 2018)

[10] Heras-Saizarbitoria I, Boiral O and Allur E 2018. Three decades of dissemination of ISO 9001 and two of ISO 14001: Looking back and ahead. *In ISO 9001, ISO 14001, and New Management Standards*, Cham, Springer 1-15

[11] Heras-Saizarbitoria I 2018 *ISO 9001, ISO 14001, and new management standards* (Ferrara: University of Ferrara) p 218

7. Acknowledgment

This study is a part of International Exchange Agreement research efforts between Engineering Faculties of University of the Ryukyus and Kabul University signed on 19th April 2017, followed by a sub-agreement (Exchange Collaboration Agreement, 7th May 2018), by the support of Center for Strategic Research Project, University of the Ryukyus (Senbaru 1, Nishihara, Okinawa, 903 -0213, Japan) under the research budget of JFY 2018 (98394640).

IOP Conf. Series: Earth and Environmental Science **291** (2019) 012010 doi:10.1088/1755-1315/291/1/012010

Feasibility study of municipal solid waste for producing refuse-derived fuel (RDF): A case study of Bangkok, Thailand

N Eaktasang[1,*], Y Suma[1] and S Kingkeaw[1]

[1]Faculty of Public Health, Thammasat University, 99 M.18, Klong 1, KlongLuang, Pathumthani, 12120, Thailand

Email: numfon.e@fph.tu.ac.th

Abstract. The objective of this study was investigated the potential of municipal solid waste (MSW) to produce refuse-derived fuel from Bangkok municipality. The composition, heating value and proximately content were analyzed. The amount of MSW in Bangkok generated was approximately 4.86 million tons or 13,327 tons/day in 2017. About 79.5% was disposed by landfill and 20.5% was recycled. The compositions of MSW were biodegradable (49.58%), combustible (36.32%) and non-combustible (14.10%) wastes. Moisture content was 59.5±8.6% (as received), volatile matter, fixed carbon and ash were 72.4±3.2%, 8.2±0.5% and 19.4±0.5% (as dry basis), respectively. Heating value was 20.03±0.52 MJ/kg. These results indicated that MSW from Bangkok municipality is potential to produce the refuse-derived fuel. To reduce the environmental impacts as well as energy recovery, the MSW might be alternative sources of renewable energy in the future.

1. Introduction

In last decade, solid waste is an important issue in various countries. The amount of solid waste is increasing with the number of population. Currently, Thailand is facing solid waste management crisis. Solid waste was increased because the traditional and life style has been changed. Most of waste comes from human activities. Municipal solid waste (MSW) generation in Thailand was 27.37 million tons in 2017 or 74,998 tons/day, generation rate of 1.13 kg/capita/day [1]. The proper disposed was approximately 43% and recycled of 31%. Although, most of MSW was properly disposed and recycled but the amount of 7.17 million tons (26%) or 19,656 tons/day was improperly disposed by open dumping, open burning and dumping into the rivers.

Bangkok is a capital of Thailand as well as one of the destination of tourists that made Bangkok became too high population density. Nowadays, MSW generation in Bangkok was 4.86 million tons or 13,327 tons/day. After segregation, the amount of 0.99 million tons of solid waste was recycled whereas the remaining of 3.87 million tons was disposed by landfilling and incineration methods. Although, the number of waste recycled was increased but the total waste generation was trend to increase. Lack of a good management of MSW, is possible effected on human health and environment such as odor, lose of aesthetics, source of vector diseases, water pollution, air pollution, etc. In addition, improper disposal method such as open burning, which is the rapidly method and low cost but enhance environmental problem such as smoke, shoot, particulate matter, CO_2 which contributed to greenhouse gases (GHGs) and cause of respiratory system. The MSW is a critical environmental and health issue that needs more and better management [2].

The majority (90%) of world's energy comes from non-renewable energy sources such as coal, crude oil and natural gases which have limited availability [3]. Therefore, search for alternative energy resources as renewable energy is an important to solve the energy crisis. To increase the value of MSW and reduce environmental pollution as well as alternative energy resources, the concept of waste to energy (WTE) is one of the most attractive option in waste management [4]. WTE technology is the generation of energy as heat or power from waste by transformation of non-recyclable material through different procedures, including thermal and non-thermal technologies. Thermal technologies include incineration, pyrolysis, gasification, plasma arc and refuse derived fuel (RDF), and non-thermal technology includes anaerobic digestion. WTE technologies can usually reduce the volume of the original waste by 90%, depending on the composition and use of outputs.

The mixed solid waste without segregation and pre-treatment as a feedstock to power production is enhance the efficiency approach to WTE recovery. RDF is the mechanical processing to recovery recyclable materials to produce a combustible product. The main advantages of using RDF as a fuel are an important reduction in the volume of waste and the possibility of energy recovery. According to ASTM E-75, RDF can be classified into 7 types depending on process. RDF has high calorific value comparing with collected solid waste. RDF is simple for storage, transportation, operation. The aim of this research was investigated the potential of MSW in Bangkok municipality to produce RDF. Determination of composition, heating value and proximately content was necessary to analyze its potential as an energy sources.

2. Material and methods

This study was conducted by taking samples of MSW on six occasions during three months of research operation. The study site was transfer station at On-Nuch station that collected MSW from Bangkok areas. Samples were collected from dumping pit for two times per day (morning and afternoon). Each sample was collected about 200 liters of waste [2]. Waste composition and waste characterization were conducted through the established sampling techniques and lab analyses of physical and chemical parameters. Waste composition analysis was performed through established sampling techniques following ASTM standards D-5231-92 [5] and manual sorting of waste samples with standard protocols, followed by lab analyses of the representative samples for its physical and chemical parameters. The proximate analysis was accomplished by the analysis of the moisture content, volatile matter content, fixed carbon and ash content according to ASTM standards D7582-12 [5]. Heating value was determinate by bomb calorimeter [4].

3. Results and discussion

3.1 MSW composition analysis

The samples of MSW that obtained from On-Nuch transfer station in Bangkok was analyzed the composition following ASTM standards D-5231-92. The average percentage of MSW composition representative of Bangkok municipality is presented in table 1. The large fraction of MSW was found to be of food waste (42.68%), followed by plastic (14.88%) and paper (12.09%). Further, these individual waste components were classified into three categories: biodegradable (food waste and yard waste), combustibles (plastic, paper, rubber/leather and textile), and non-combustibles (glass, metal, ceramic and others), their composition are as shown in the figure 1. The per capita generation rate in Bangkok (1.54 kg/capita/day) was more than generation rate of Thailand (1.13 kg/capita/day). This is definitely due to the high influx of commercial activities in Bangkok [6].

Table 1. MSW composition of Bangkok municipality.

Content	Waste composition (%)
Food waste	42.68
Paper	12.09

Plastic	14.88
Glass	6.63
Metal	3.54
Rubber/Leather	2.57
Yard waste	6.9
Ceramic	3.93
Textile	4.68
Others	2.10
Total	**100**

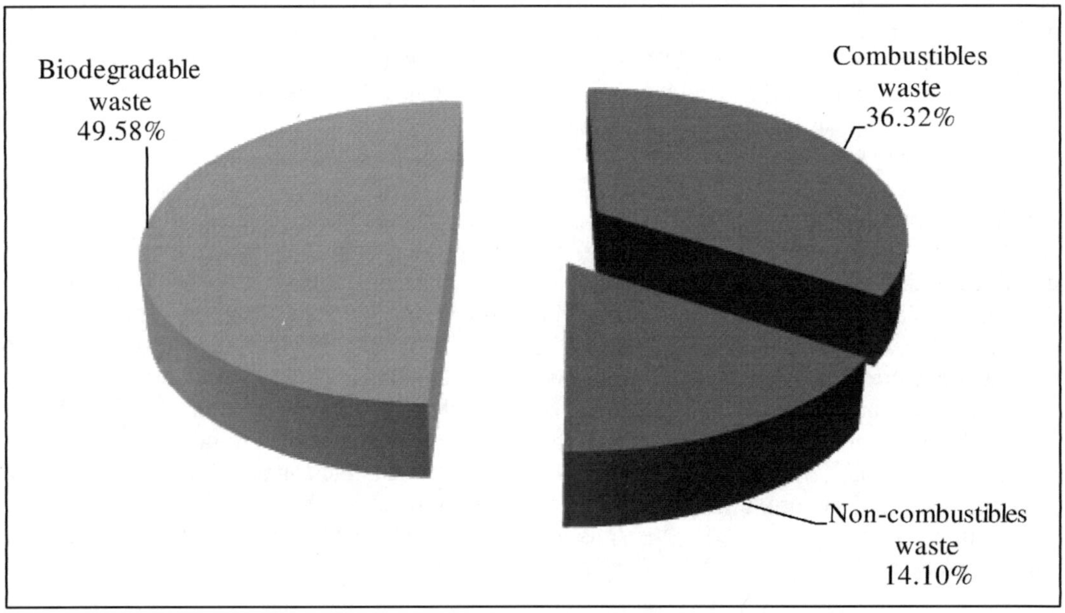

Figure 1. Classification of MSW in Bangkok municipality.

3.2 Chemical characteristics of MSW analysis

The proximate analysis was conducted according to the ASTM standards D7582-12 including determination the percentage of moisture content, volatile matter, fixed carbon and ash content. The percentage of moisture content, volatile matter, fixed carbon and ash was 59.5±8.6 % as received, 72.4±3.2 %, 8.2±0.5 and 19.4±0.5 % as dry basis, respectively. Moisture content and ash both related to the heating value of MSW. Since the moisture content measured in this study exceeded the maximum value for the RDF standard (25%, wet basis), so the pre-treatment of MSW of Bangkok is recommended to remove excess moisture content prior to produce the RDF. Whereas, the ash content in this study was found to be below the RDF standard of 20% as dry basis [2]. The result of calorific value or heating value of MSW was found to be of 4,785±125 kcal/kg as dry basis or 20.03±0.52 MJ/kg, which is higher than the RDF standard [7]. This result could be recommended the MSW of Bangkok municipality is appropriated to produce RDF. Table 2 shows the comparison of heating value that obtained from this study and previous study. The moisture content was a key factor affected the heating value of municipal solid wastes in variety sources.

Table 2. The comparison of heating value of this study and previous study.

RDF material	Heating value (MJ/kg)	References
Solid waste from open dump site	29±1.62	[2]
Municipal solid waste	21.28-27.10	[8]
Municipal solid waste	13.51	[9]
Wastewater sludge	17.97	[10]
Municipal solid waste	14.40	[11]
Municipal solid waste	17.40	[12]
Mechanical-biological treatment	15.53	[4]
Compost rejected	22.88	[4]
Municipal solid waste	20.03±0.52	This study

4. Conclusion

Results obtained in the present study show that the characteristics of MSW in Bangkok municipality are expected to produce a high energy content of RDF. It is also important to extend the landfill lifetime as well as reduction of CO_2 and other greenhouse gases emissions from landfill site. In addition, MSW is one of an alternative energy source for the future.

5. References

[1] Pollution Control Department 2018 Thailand state of pollution report 2017 *Ministry of Natural Resources and Environment*. p 23-24

[2] Weerasak T and Sanongraj S 2015 Potential of producing refuse derived fuel (RDF) from municipal solid waste at Rajamangala university of technology isan surin campus *App Env Res* **37** (2) p 85-91

[3] Bandara W and Kowshayini P 2017 Evaluation of the Performances of biomass briquettes produced with invasive *Eichornia crassipes* (water hyacinth), wood residues and cow dung for small and medium scale industries *J Fundam Renewable Energy Appl* **8** (1) p 247

[4] Brás I, Silva M E, Lobo G, Cordeiro A, Faria M and Lemos L T 2017 Refuse derived fuel from municipal solid waste rejected fraction-a case study *Energy Procedia* **120** p 349-356

[5] AbdAlqader A and Hamad J 2012 Municipal solid waste composition determination supporting the integrated solid waste management in Gaza Strip *International Journal of Environmental Science and Development* **3** (2) p 172-176

[6] Chiemchaisri C, Juanga J P and Visvanathan C 2007 Municipal solid waste management in Thailand and disposal emission inventory *Environ Monit Assess* **135** (1-3) p 13-20

[7] Gendebien A, Leavens A, Blackmore K, Godley A, Lewin K and Whiting K J et al. 2003 Refuse derived fuel, current practice and perspective: B4-3040/2003/306517/ MA R/E3, July 2003. *European Commission-Directorate General Environment.*

[8] Kimambo O N and Subramanina P 2014 Energy efficient refuse derived fuel (RDF) from municipal solid waste rejects: a case for Coimbatore *International Journal of Environment* **3** (2) p 205-215

[9] Zhou H, Meng A,Long Y Q, Li Q H and Zhang Y G 2014 An overview of characteristics of municipal solid waste fuel in China: Physical, chemical composition and heating value *Renewable and Sustainable Energy Reviews* **36** p 107–122

[10] Alsheyab M A T, Schingnitz D, Al-Shawabkeh A F and Kusch S 2013 Analysis of the potential use of major refuse-derived fuels in Jordan as supplementary fuel *Journal of the Air & waste Management Association* **63** (8) p 902-908

[11] Chang Y M, Liu C C, Hung C Y, Hu A and Chen S S 2008 Change in MSW characteristics under recent management strategies in Taiwan *Waste Management* **28** p 2433-2455

[12] Dong C, Jin B and Li D 2003 Predicting the heating value of MSW with a feed forward neural network Waste *Management* **23** p 103–106

Acknowledgments
This study was supported by the Thammasat University Research Fund and Faculty of Public Health, Thammasat University.

In Situ Aerobic Biodegradation In Soil Contact Using The Standard Test Method of ASTM D 5988-12

S.M. Al-Salem, A. Al-Nasser, H. Sultan and H. Karam

Environment and Life Sciences Research Centre, Kuwait Institute for Scientific Research, P.O. Box: 24885, Safat 1319, Kuwait.

Abstract. With an over growing problem of plastic waste accumulation, Kuwait faces a serious threat to its natural resources. Latest estimates show that plastic solid waste (PSW) is generated at an alarming rate of over 200 ktpa, making up over 16% of the total municipal waste load. In this paper, the standards test method of ASTM D 5988-12 for determining the degree and rate of biodegradation of plastics in contact with soil was applied to three commercial plastic products sourced locally. The plastic samples are claimed to be of biodegradable nature by the manufacturer and the converter. The test method was applied in situ using analytical desiccators and have proven to be a good indicator to biodegradable measurements. In addition to technical control and reference tests, the test method was also applied on the natural soil secured from the field simulating the largest operating landfill site. The test has proven to be reliable and has produced repeatable results. The biodegradation extent was noted to be very rapid and can be described by comparison to other works as accelerated. This was attributed to both what is described by this work as external and internal factors of the polymer-soil matrix. The former can be noted as environment and soil characteristics, pH level and presence of contaminants in the biota. On the other hand, the latter can be described as type of soil (mixture) ratio, polymer surface area and carbon dioxide (CO_2) evolution kinetics which were noted to be of a first order with respect to biodegradation time and dissolution.

1. Introduction

The rapid development in plastic articles and their properties has made their management post disposal an alarming threat to the natural environment. Polymeric articles constitute a number of chemicals and additives that are used to enhance their properties [1]. However, biodegradation has made it possible to alter these hydrocarbon (HC) chains and make them susceptible to chemical bonds' cleavage [2-3]. Therefore, it is deemed necessary to start having test procedures and standards that can govern the use and assessment of biodegradable polymers. Therefore, devising a standard test methodology for the determination of biodegradability extent is noted to be of essential need nowadays. The American Society for Testing & Materials (ASTM) has developed and reported such a test method (D 5988-12). The test determines the degree and rate of biodegradation under controlled laboratory conditions in contact with soil. This method was chosen in our work due to the fact that it reports the biodegradation extent and rate, in addition to, simulating the actual microorganism attack by the local environment.

The purpose of this communication is to report the study results of three commercially produced and converted (claimed to be) biodegradable plastic film products. The reliability of the test method is discussed, in addition to the results obtained from the analytical work, which is the first of its kind in the Middle-East region. The test is also noted to provide very essential information on the biodegradation of local soil secured from Kuwait's arid land (Al-Maqwa area) which also simulates the activities of the largest operating landfill site in the country (seventh ring road - south).

2. Experimental

Three types of plastic bags used in the Kuwaiti market were secured from local dealers with their main characteristics and sample dimensions depicted elsewhere [5,6]. The samples acquired were presumed to be environmentally friendly and biodegradable, which were secured from various sources within the state. The samples were cut into standard testing dimension specimens accepted by international protocols for thin plastic film characterization as described previously [1,5]. The determination of the biodegradability extent was conducted in this work under the presumption that the samples were biodegradable in nature with suitable content of pro-degradants (or oxo-biodegradable) as declared by the company's marketing these polymeric products. The biodegradation extent was investigated via aerobic bacteria in contact with soil under controlled conditions [6]. The standard testing procedures of ASTM D 5998 [7] were followed to determine the rate of aerobic biodegradation of plastic materials in soil. Natural soil (particle size < 2 mm) was used after reclaiming it from a local dealer in Kuwait in the amount of 50 kg. The soil originates from the Maqwa'a area which is the closest to the largest operating landfill site in Kuwait, dug up from 1 to 3 m (top soil) stored in a cold room storage with a maintained temperature between 1 to 3°C. Compost (sheep manure) sourced from local dealers was also used in this work. Potato starch (MSDS CAS no. 009005-84-9) purchased from Loba Chemie Company was used for positive control experimental runs. The soil was aerated with the compost in laboratory conditions and mixed in a 1:25 ratio (e.g. 18:450 g compost/g soil) with a moisture content of 80-100% maintained throughout the full duration of the experiment by adding distilled water. The moisture holding capacity (MHC) of the soil was determined initially after drying it in an oven for 24 h at 105°C and was estimated to be 0.6%. Each experimental run consisted of six air tight 2 litre desiccators. The first was denoted as blank (control) which contained the soil. A technical control experimental run to make sure that the experimental setup was free from leakage, was also executed. A positive control experiment was conducted in conjunction with the first two containing the starch and soil. The remaining three desiccators contained the samples studied. Plastic bags sample size were used as described in the experimental protocol of ASTM D 5998 [7] and Rudeekit et al.[8] which were 2x2 cm^2. The amount of 200 mg for each type and positive control (starch) sample were used. The carbon (C) to nitrogen (N) ratio was maintained in the experimental setups after the addition of Ammonium Phosphate ($(NH_4)_2HPO_4$) (4.72 g/l, Sigma Aldrich Company reagent >98%) and maintaining the pH 8.2 [6]. The C:N ratio was maintained at 10:1 after addition to the soil.

The test method described aims at the determination of aerobic biodegradability, obtaining soil as a matrix and source of inoculum, as well as, estimating the degree of biodegradability [7]. To this end, both barium hydroxide ($Ba(OH)_2$) and deionized water (DI) were used within each experimental setup, placed inside each of the desiccators used. The $Ba(OH)_2$ solution was prepared by dissolving 4.0 g anhydrous $Ba(OH)_2$ (Aldrich Company, technical grade 95%) in distilled water and confirmed by manual titration. The solution was sealed and stored to prevent absorption of CO_2 from the air. The amount of 100 ml of 0.025 N $Ba(OH)_2$ and 50 ml of distilled water were kept constantly throughout the experiments within each of the desiccators used on the perforated plate inside the vessel. All tested were conducted in dark room in laboratory conditions. Manual titration was conducted in a frequency of 3 days for the first three weeks and once every three weeks thereafter as recommended previously by Gomez and Michael [9] and ASTM D5998 [7]. The titration aims at determining the accumulative carbon dioxide (CO_2) evolution, conducted using hydrochloric acid (HCl, 0.05 N) and $Ba(OH)_2$ to a phenolphthalein end-point. The theoretical total carbon (TC) was determined for each type of plastic bags and starch using dynamic flash combustion using modified Dumas method [10] in a Thermo-Fisher Scientific (Flash 2000 Model) CHNS-O analyser. Samples were conducted in triplicates for quality assurance. The theoretical CO_2 quantity was calculated using reaction stoichiometry thus

$$CO_{2(th.)} = \frac{44 \times Y}{12} \quad (1)$$

IOP Conf. Series: Earth and Environmental Science **291** (2019) 012011 doi:10.1088/1755-1315/291/1/012011

where Y is the net total theoretical carbon (C) charged to each vessel (mg), and $CO_{2(th.)}$ is the theoretical CO_2 available (mg). The amount of net CO_2 produced is calculated by subtracting the titration readings from the test material titration thus

$$Z_n = Z_b - Z_t \tag{2}$$

where Z_n is the calculated amount of HCl needed to titrate the CO_2 generated solely from the test material (ml), Z_b is the amount of HCl used to titrate the blank vessels with soil only (ml) which is termed as technical control vessel in this work, and Z_t is amount of HCl used to titrate the vessels containing the soil plus test material (ml). The amount of CO_2 produced is calculated as per the formula given below after accounting for the Normality of HCl [7]

$$CO_{2(\text{Produced})} = \frac{1.1\left(0.05\,\text{N} \times Z_n \times 44\right)}{2} \tag{3}$$

The percentage of CO_2 evolution (E) is determined as per the following equations, which on the other hand is used to estimate the standard deviation (s_e) of the experimental runs and biodegradation extent thereafter.

$$E\,(\%) = \frac{CO_{2(\text{Produced})}}{CO_{2(\text{th.})}} \times 100 \tag{4}$$

$$s_e = \sqrt{\left(s_{test}^2/n_1\right) + \left(s_{blank}^2/n_2\right)} \times 100/C_i \tag{5}$$

where n_1 and n_2 are the number of replicate test material and blank vessels, respectively, s is the average standard deviation of the total gaseous carbon produced (mg), and C_i is the amount of carbon (mg) initially present in the vessel.

3. Results and discussion

As an initial trail and in order to assure that the soil material used will accommodate the ASTM D 5988 test method undertaken in this study, trails on the soil without the presence of samples was conducted to assure microbial activity within the degrading environment. The method and experimental design is similar in nature to the blank (control) experiment. The difference was in the soil material preparation procedure and conditioning. For these initial trails, the soil was aerated for the duration of three days and prepared in similar fashion to the previous description. The soil was stimulated under 35°C for maximum microbial activity for the period of nine days. Throughout the experimental efforts, including the ones dedicated for the plastic specimens, no substantial change in the amount of distilled water was noticed. However, droplets creating a thin film was noticed around the desiccators walls. This points towards the formation of barium carbonate ($BaCO_3$) which precipitates due to the reaction of the formed CO_2 with the barium hydroxide ($Ba(OH)_2$). This was continuously avoided by gently shaking the vessels as the test method suggests. This will also neutralize CO_2 to OH^- ion which is not dependent on the position of the $CO_3^{2-} \leftrightarrow HCO_3^-$ equilibrium at the pH phenolphthalein endpoint [11].

Fig.1. indicates the CO_2 evolution rate with time for the soil stimulated at 35°C for the nine days monitoring time of the blank test. This verifies that the soil is compatible for conducting the test method described in the previous section. By association, the kinetics of evolution can be described by Michaelis-Menten mechanism in a homogenous-phase solution [11]. By comparison to the rate of CO_2 evolution estimated for the actual test, it can be noted that actual test conditions facilitated more CO2 production as time has progressed and passed the 10 days mark. Therefore, the soil stimulation with temperature can be noted to be less of a factor in the increased CO_2 production

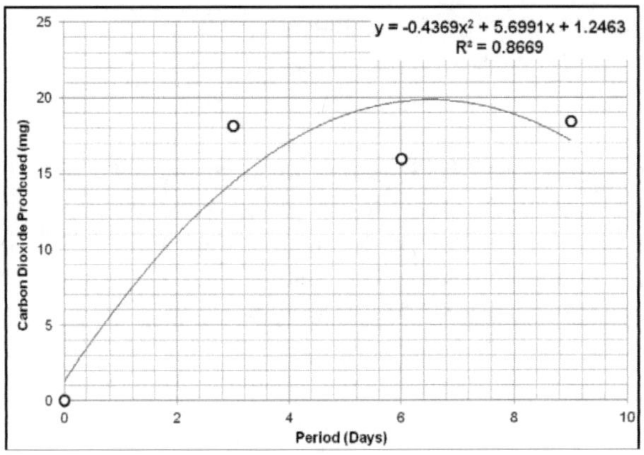

Figure 1. Production of carbon dioxide (CO2) as a function of time (○) measured in the initial trail test without addition of sample with soil aeration period of three days based on the titration volume of 5 ml.

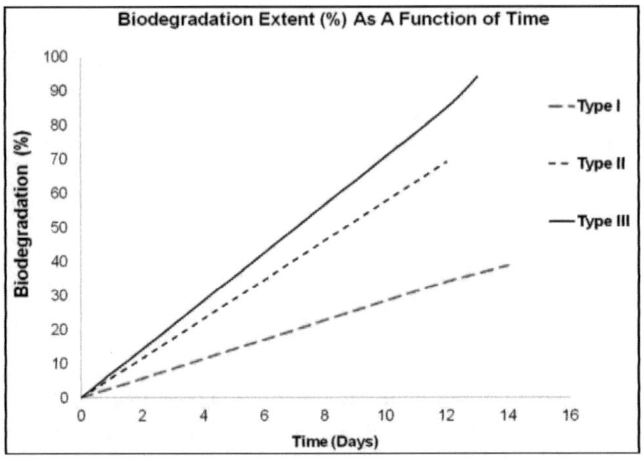

Figure 2. Biodegradation extent (%) determined for the three plastic samples tested in this work (raw data).

The biodegradation rate was also estimated for the types of the samples studied. It was noted that the sample thickness and possible starch content, can be major factors in the rate of biodegradation. Fig.2 shows the biodegradation extent (%) of the samples considered. It can be noticed that the samples of type I sustained the longest period of time with biodegradation extent reaching 14 days. However, samples of type II and III have deteriorated at higher rates. The plateau level of CO_2 production evolution in this test method and, optionally, the quantified residual test material will suggest the degree of biodegradability of the plastic material.

4. Conclusion

The biodegradation extent using the standard test methods of ASTM D5988, was conducted on three different types of plastic samples. The test have started using blank soil tests to investigate the biota activity. It was noted that the test method supported soil and polymer matrix influential factors noted as major influencers on the biodegradation rate. Two sets of factors have been noted to affect the bio-

IOP Conf. Series: Earth and Environmental Science **291** (2019) 012011 doi:10.1088/1755-1315/291/1/012011

testing, external and internal ones. The kinetics, with biodegradable substrate and CO_2 production rate, has also followed the Michaelis-Menten mechanism. The former can be noted as environment and soil characteristics, pH level and presence of contaminants in the biota. On the other hand, the latter can be described as type of soil (mixture) ratio, polymer surface area and carbon dioxide (CO2) evolution kinetics which were noted to be of a first order with respect to biodegradation time and dissolution.

5. References

[1] Al-Salem SM 2009 Influence of natural and accelerated weathering on various formulations of linear low density polyethylene (LLDPE) films Mater. Des. vol. 30, no. 5, pp. 1729-1736.

[2] Al-Salem SM, Bumajdad A, Khan AR, Sharma BK, Chandrasekaran SR, Al-Turki FA, Jassem FH, Al-Dhafeeri AT 2018 Non-isothermal Degradation Kinetics of Virgin Linear Low Density Polyethylene (LLDPE) and Biodegradable Polymer Blends J. Polym. Res. (Accepted).

[3] Rosa, D. S., M. A. G., Bardi, L.D.B. Machado, D.B. Dias, L.G.A. Silva and Y. Kodama 2010 Starch plasticized with glycerol from biodiesel and polypropylene blends mechanical and thermal properties 2010 J. Anal. Therm. Calor. vol. 102, pp. 181–186.

[4] Müller R 2005 Biodegradability of Polymers: Regulations and Methods for Testing General Aspects and Special Applications Wiley first edition.

[5] Al-Salem SM, Abraham G, Al-Qabandi OA, Dashti AM 2015 Investigating the effect of accelerated weathering on the mechanical and physical properties of high content plastic solid waste (PSW) blends with virgin linear low density polyethylene (LLDPE) Polym. Test. vol. 46, pp. 116-121.

[6] Al-Salem SM, Al-Nasser A, Al-Wadi M, Sultan H, Al-Foudaree M, Karam H, Al-Dhafeeri A, 2018 Biodegradable Films Physical Properties Change With Respect To Degradation Exposure To Various Media International Proceedings of the 7th International Conference on Engineering for Waste and Biomass Valorisation, Prague, Czech Republic, 2nd-5th July.

[7] American Society for Testing and Materials (ASTM) 2012 Standard Test Method for Determining Aerobic Biodegradation of Plastic Materials in Soil ASTM D 1435-12, Philadelphia, Penn.

[8] Rudeekit Y, Leejakpai T, Euaphantasate N, Kongsuwan K 2004 Comparative degradation of biodegradable plastics aerobic microorganim using ASTM D 5338-98 (03) and ASTM D 5988-03 as standard methods. International Proceedings of International Conference on Powder, Thailand, 24-25 November.

[9] Gómez EF, Michel Jr. FC 2013 Biodegradability of conventional and bio-based plastics and natural fiber composites during composting, anaerobic digestion and long-term soil incubation Polym. Degrad. Stab. Vol. 98, no. 12, pp. 2583-2591.

[10] Krotz L, Leone F, Giazzi G 2016 Nitrogen/Protein Determination in Food and Animal Feed by Combustion Method (Dumas) using the Thermo Scientific Flash Smart Elemental Analyzer, Application Note, Themro Fisher.

[11] Modelli, A., Calcagno B., Scandola M., 1999 Kinetics of Aerobic Polymer Degradation in Soil by Means of the ASTM D 5988-96 Standard Method, J. Polym. Environ. Vol. 7, No. 2.

6. Acknowledgments

This work is funded by the Kuwait Institute for Scientific Research and Kuwait Municipality under the grant for Project EM097C.

IOP Conf. Series: Earth and Environmental Science **291** (2019) 012012 doi:10.1088/1755-1315/291/1/012012

The application of organic solvents and thermal process for eliminating EVA resin layer from waste photovoltaic modules

Wei Sheng Chen[1], Yen Jung Chen[1], and Yu An Chen[1]

[1]National Cheng-Kung University, Resource Engineering

qaz5932201@gmail.com

Abstract. With rapidly increasing production and installation, recycling of PV modules has become the main issue. In this study, we developed the application to recover the tempered glass from panels and remove Ethylene-vinyl acetate (EVA) from PV cells. The processes divided into two parts, organic solvents soaking and thermal treatment. In the organic solvents process, this study soaked PV modules into organic solvents to dissolve EVA resin and recover tempered glass. In the thermal treatment, this study removed EVA resin remaining on the PV cell. This study can comprehensively recover tempered glass and remove 99.97% EVA resin from PV cell. Moreover, it can contribute to not only subsequent leaching processes but also the environment and humankind health.

1. Introduction

The PV technology is considered to be one of the greenest and most promoting green energy-generating technologies as it generates electricity directly from the solar energy and therefore avoid fossil energy consumption and global warming gases emissions during energy production process. Over this decade, PV technology has shown the attempt to become the majority of green power generation for the world. According to 2018's report, in 2017, 94.6GW PV modules were installed worldwide which represents 55% of newly installed renewable energy capacity. The Cumulative installation of PV panels achieved to 402GW in 2017. [1] However, the dilemma of PV system end-of-life (EoL) is now a big problem because of the lack of relevant PV modules recycling research. PV module average life, 20-25 years, reflects a large number of PV modules waste after 2030. According to the statistics, the mass of waste will reach 8.6k tons in 2049. [2] Additionally, the PV module contains hazardous materials such as Lead, Cadmium, Tellurium, Selenium, Silver which have caused serious health problems. [3] Therefore, the end-of-life PV modules should be recycled for the environment.

PV modules are mainly divided, Polycrystalline Silicon Solar PV (c-Si) and Thin-Film Solar PV (CdTe, CIGS, a-Si). Polycrystalline Silicon Solar PV (c-Si) account for 93% in PV modules, and the others account for 7%. The main issue of recovering PV modules is to eliminate the EVA resin layer which is used to become the adhesive layer on PV modules. This study mainly provides the process for the separation of the tempered glass and PV cells to recycle c-Si PV. Several solvents such as sulfuric acid, nitric acid, organic solvents (Alkane, Alcohol) were employed to separate tempered glass in other researches. [4][5] According to the literature review, the EVA resin can be easily removed by heating. [6][7]

In this study, the tempered glass was recovered by two alternative types of solvent, one for swelled EVA resin and one for dissolved EVA resin. The residual EVA was also removed by the thermal process.

2. Materials and experiment

2.1. Materials

PV modules are composed of aluminum frame (6.98%), tempered glass (74.1%), Ethylene-vinyl acetate (11.3%), Crystalline silicon batteries (6.81%), and Junction Box (1.3%). Fig.1 shows the schematic representation of the layers of the waste c-Si PV module of 1636x983mm2/18.5kg/250W which was used in this essay.

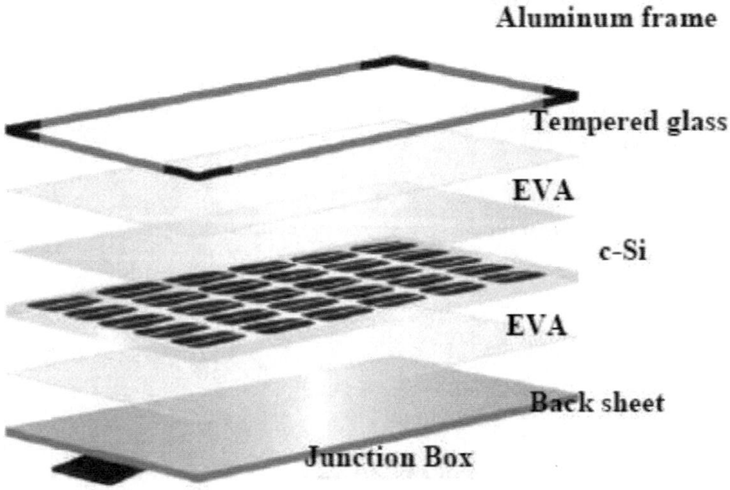

Figure 1 Schematic representation of the layers of a crystalline silicon photovoltaic panel.

2.2. Organic solvent processes for c-Si recycling

The tempered glass and PV cells were separated organic solvents. There were three status performed, non-separated, swelled, and separated. After then, this study compared effective organic solvents for investigating the separation efficiency. This study screened two ways to recover tempered glass: separate directly organic solvents and swelled but need hand sorting organic solvents.

2.3. Thermal process

This study provided the thermogravimetric analysis graphs of pure EVA resin and PV modules. This process compared the main mass loss of PV modules and EVA resin to set the condition temperature for removing EVA resin from the PV cell. After the aforementioned process, the shattered PV cell with high purity of Silicon and valuable metal can make subsequent wet extraction techniques more efficient.

3. Result and discussion

3.1. Separation of the tempered glass and the PV cell using organic solvents

According to the experiment, the PV modules at the swelled status and separated status can easily remove the tempered glass from the PV cell. As the result, according to Table 1, Tetrahydrofuran, Toluene, Chloroform, and D-Limonene had a well-sorting effect on the separation between the tempered glass and PV cell, while Acetone and Ethyl alcohol had no change. After the organic solvents process, the tempered glass can be easily removed and recycled. The recovery of tempered glass can reach 100%. According to Table 1 and Fig2, D-Limonene and Toluene have the higher solubility of EVA resin. They spent less time completing the process of expansion and decomposition. Tetrahydrofuran and Chloroform cannot comprehensively dissolve EVA resin but can effectively swell it. In this case, this study screened two ways to recover tempered glass: separate directly organic solvents (D-Limonene, Toluene) and swelled but need hand sorting organic solvents (Tetrahydrofuran, Chloroform). Solvents can be selected for different requirements.

Table1. Status of PV modules immersed by the organic solvents in 90℃

solvent	1 min	30 min	1hr	2hr	24hr	48hr
D-Limonene	Δ	○	○	○	○	○
Tetrahydrofuran	X	Δ	○	○	○	○
Toluene	○	○	○	○	○	○
Chloroform	X	Δ	Δ	Δ	Δ	Δ
Acetone	X	X	X	X	X	X
Ethyl alcohol	X	X	X	X	X	X

X: non-separated **Δ: swelled** **○: separated**

Table 2. 6M, 40°C organic solvents for EVA solubility. Stir an hour.

Solvent	Weight change
D-Limonene	21.14%
Tetrahydrofuran	13%
Toluene	30.60%
Chloroform	18.2%

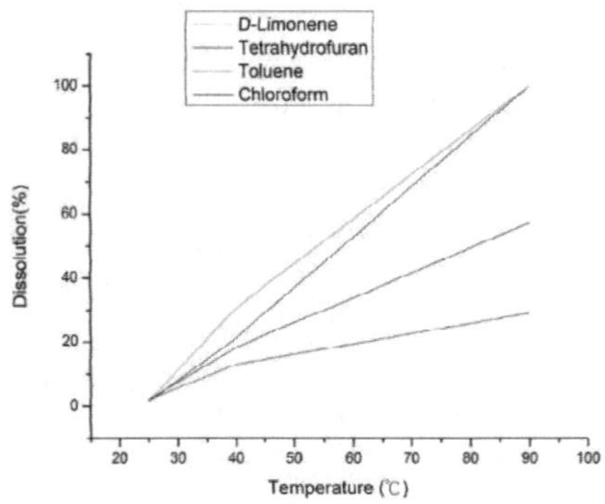

Figure 2. Solubility increases with temperature.

3.2. Removal of the EVA resin by heat treatment

As Fig. 3 shown, after the separation process, EVA resin still remained on the PV cell. Heat treatment process was employed to remove EVA resin from the PV cell. Fig. 4(a) and (b) showed that EVA resin and PV modules both have a huge mass decline at 380°C [8]. When the temperature reaches 500°C, the weight of EVA resin approached zero. Therefore, it can be known that the EVA resin on the surface of the solar cell can effectively reduce the weight of the EVA resin after heating to 500°C.

IOP Conf. Series: Earth and Environmental Science **291** (2019) 012012 doi:10.1088/1755-1315/291/1/012012

The temperature of thermal treatment dropped to 500°C for 2 hours because of the power generation efficiency of Polycrystalline Silicon decline when the temperature heat over 500°C. [9] The temperature was raised to 900°C for 4 hours to eliminate all organic matter. After heating to 900°C for 4 hours [10], the weight change of the PV cell accounted for 0.028%. The consequent showed that this study removed 99.97% of EVA resin.

3.3. Characterization of the PV cell after recovering
The PV cell in our experiment mainly contains pure silicon, Aluminum Oxide, and Copper Oxide. Fig. 5 showed the peak of the PV cell, Si accounted for the most part of the PV cell. Table 3 showed the concentration of the PV cell. In cell recycling, interconnected cells are first immersed in a leaching solution to dissolve four metals, Al, Ag, Pb, and Cu. These metals are then recovered from the leaching solution one by one through electrowinning [11].

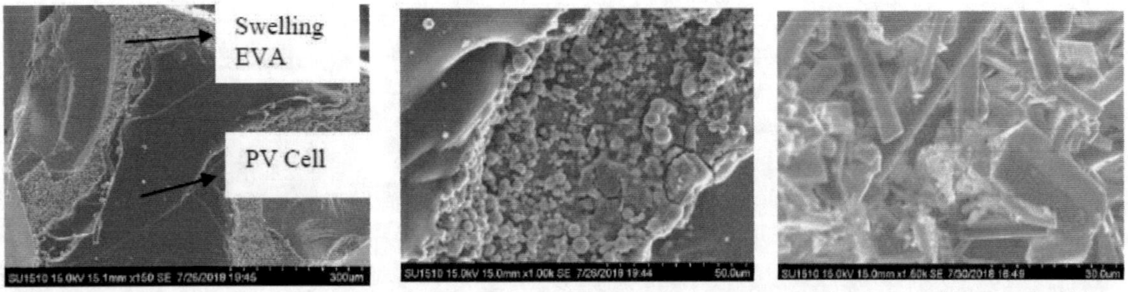

Figure 3(a) the PV cell surface after the swelling of EVA resin in microscopic view. **3(b)** EVA bubble resin remained on the PV cell. **3(c)** PV cell after thermal treatment. Nearly no bubble resin remained on the PV cell.

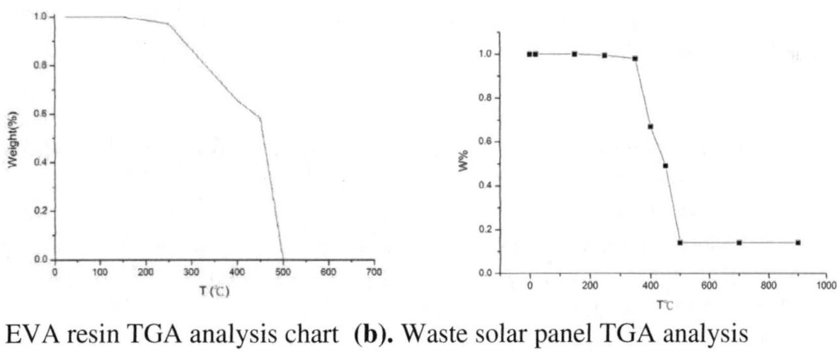

Figure 4(a). EVA resin TGA analysis chart **(b).** Waste solar panel TGA analysis

Table 3 Weight percentage of valuable metal in the PV cell

metal	percentage
Si	**36.19%**
Al	**0.15%**
Cu	**0.8%**
Ag	**0.01%**
Pb	**0.06%**

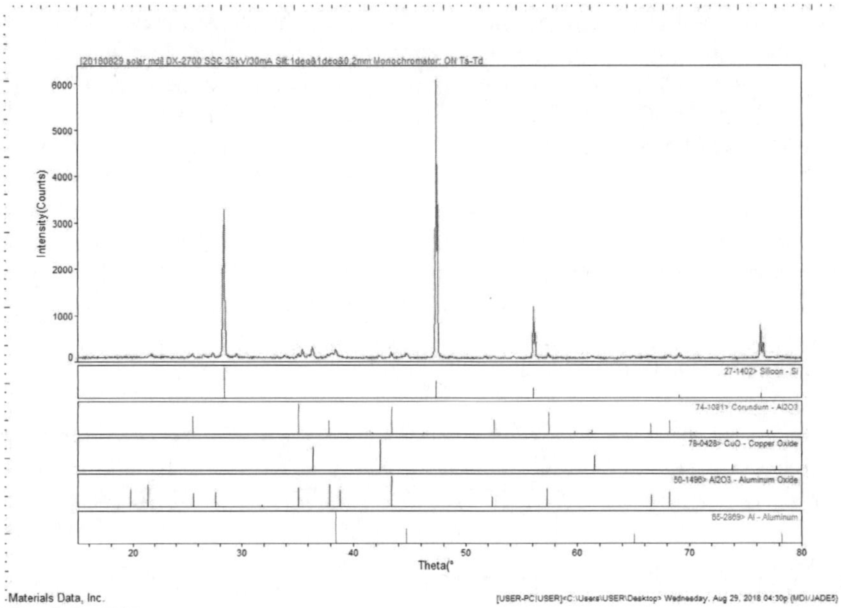

Figure 5 XRD of the PV

4. Conclusion

Organic solvents and heat treatment can effectively recover PV modules. This study gave two types of organic solvents for recovering tempered glass, separated and swelled. As the result, D-Limonene and Toluene made more effect on dissolved EVA resin and comprehensively recovered the tempered glass. This study set 500°C 2 hours for effectively eliminating EVA resin. The thermal process removed 99.97% EVA resin. This study can provide effective processes to treat the waste of solar panels containing tempered glass or EVA resin.

5. References

[1] T. Doi, I. Tsuda, H.Unagida, A.Murata, K.Sakuta, and K.Kurokawa, "Experimental study on PV module recycling with organic solvent method," Sol. Energy Mater. Sol. Cells, vol. 67, no. 1–4, pp. 397–403, 2001.

[2] S.Weckend, A.Wade, and G.Heath, End-of-life management: Solar photovoltaic panels. 2016.

[3] M.Marwede and A.Reller, "Future recycling flows of tellurium from cadmium telluride photovoltaic waste," Resour. Conserv. Recycl., vol. 69, pp. 35–49, 2012.

[4] V.Savvilotidou, A.Antoniou, and E.Gidarakos, "Toxicity assessment and feasible recycling process for amorphous silicon and CIS waste photovoltaic panels," Waste Manag., vol. 59, no. August, pp. 394–402, 2017.

[5] S.Kang, S.Yoo, J.Lee, B.Boo, and H.Ryu, "Experimental investigations for recycling of silicon and glass from waste photovoltaic modules," Renew. Energy, vol. 47, pp. 152–159, 2012.

[6] T.Doi, I.Tsuda, H.Unagida, A.Murata, K.Sakuta, and K.Kurokawa, "Experimental study on PV module recycling with organic solvent method," Sol. Energy Mater. Sol. Cells, vol. 67, no. 1–4, pp. 397–403, 2001.

[7] A.Doni and F.Dughiero, "Electrothermal heating process applied to c-Si PV recycling," Conf. Rec. IEEE Photovolt. Spec. Conf., pp. 757–762, 2012.

[8] A.Badiee, I. A.Ashcroft, and R. D.Wildman, "The thermo-mechanical degradation of ethylene vinyl acetate used as a solar panel adhesive and encapsulant," Int. J. Adhes. Adhes., vol. 68, pp. 212–218, 2016.

[9] E.Technology, Y.Yuan, C.Materials, and Z. N.Road, "Silicon Resource Recovery of Spent Monocrystalline Silicon Solar Batteries," pp. 9–13, 2012.

[10] A.Marcilla, A.Gómez, and S.Menargues, "TG/FTIR study of the thermal pyrolysis of EVA copolymers," J. Anal. Appl. Pyrolysis, vol. 74, no. 1–2, pp. 224–230, 2005.
[11] W.Huang, W.Jung, L.Wang, W.Sun, and M.Tao, "Strategy and technology to recycle wafer-silicon solar modules," Sol. Energy, vol. 144, pp. 22–31, 2017.

Acknowledgments
This study was supported by National Cheng-Kung University Dept. of Resource Engineering, R.O.C.

IOP Conf. Series: Earth and Environmental Science **291** (2019) 012013 doi:10.1088/1755-1315/291/1/012013

Preliminary study of dye removal from aqueous solution using elephant dung activated carbon

Y Suma[1], N Pasukphun[2], N Eaktasang[2] and P Laor[3]

[1] Faculty of Public Health, Thammasat University (Lampang Campus)
248 Moo 2 Lampang-Chiang Mai Road, Pongyankok, Hangchat, Lampang 52190, Thailand
[2] Faculty of Public Health, Thammasat University (Rangsit Campus)
99 Moo 18 Paholyothin Road, Klong Nueng, Klong Luang, Pathumthani 12121, Thailand
[3]Environmental Health Program, School of Health Science, Mae Fah Luang University
333 Moo 1 Thasud Muang, Chiang Rai 57100 Thailand

Corresponding author:[1] yanasinee.s@fph.tu.ac.th,yanasinees@gmail.com

Abstract. Dyes are widely used in various industries. It is significant to remove dye from wastewater before discharge. Adsorption onto activated carbon is more than beneficial than the other traditional methods as a result of its high adsorption capacity for coloured particle. The main objective of this study is to prepare activated carbon from elephant dung (EDAC). The adsorption performance of EDAC was studied by conducting batch adsorption experiments. The adsorption data are correlated with Langmuir and Freundlich under various conditions methylene blue (MB) concentration and contact time. The optimum dosage of EDAC was found to be 100 mg. EDAC adsorption was fitted with Langmuir model. The calculated maximum adsorption capacity was 18.98 mg/g. A dimensionless separation factor (R_L) indicated a favourable adsorption of MB onto EDAC. The study indicates that EDAC could be employed as a cost-effective alternative adsorbent for removing MB from wastewater streams. EDAC also was cheap, more eco-friendly and effective waste utilization.

1. Introduction

Dyes are widely used in various industries, for example, textile, paper and leather to produce its colour properties [1]. The increasing existence of dyes in the aqueous bodies is one of the most significant environmental issues [2]. The textile industries discharged the highest amount of dye effluent (54%), which found more than half of dye effluents being in the environment around the world [1]. Dyes also affect photosynthetic activity owing to reduced light penetration, leading to disturb the ecosystem [2]. Dye effluents existing in water sources are undesirable as vital water for animals and humans' daily activities [3]. Therefore, it is important to apply some colour removal treatment methods to decrease the harmful effects before releasing the dye effluents into water bodies.

There are the dye treatment from wastewater methods including anaerobic decolourisation, chemical oxidation, reverse osmosis, ion exchange, adsorption and biological treatment methods [4]. Among these techniques, adsorption onto activated carbon has been demonstrated to be the most effective process to remove colour because of its initial cost, uncomplicatedness of design, easy operation and insensitivity to toxic constituents [5]. Adsorption onto activated carbon is more than beneficial than the other traditional methods as a result of its high adsorption capacity for both organic and inorganic substrates, especially, for coloured particle [4]. Nevertheless, application of commercial activated carbon for colour adsorption has some limitations due to its high cost and regeneration problem. The challenges have been explore some effective, cheaper and easily available alternatives adsorbents. Previous studies demonstrated alternatives low cost activated carbons prepared from natural products such as sun flower stalks, corn cobs and barley husk, rice husk, modified saw dust, peanut hulls, peanut shells, pineapple stem and orange peel [6]. These activated carbons derived from various natural materials are broadly used as adsorbents because of their high adsorption capacity, large surface area and micro-porous structure, which a result of their chemical and thermal stability.

Elephant dung is available in large quantities in the Thai Elephant Conservation Center (Lampang province, Thailand). Elephant dung is used to produce biogas, paper, paper products and fertilizer. However, it is encounter elephant dung to become extremely valuable products. Therefore, use of elephant dung as a bio-sorbent will undertake to verify to be the potential utility of low-cost activated carbon from elephant dung waste for dye treatment from aqueous solution.

In this study, elephant dung waste was chosen as the raw material to manufacture activated carbon by carbonization and chemical activation processes. The adsorption performance of elephant dung activated carbon was studied by synthesized wastewater. The adsorption data were associated with the Langmuir and Freundlich models.

2. Method

2.1 *Preparation of elephant dung activated carbon*

Elephant (*Elephas maximus*) dung was collected from Thai Elephant Conservation Center, Lampang province, Thailand. The dung was dried by sunlight for a week and dried in a hot air oven at 103 °C for 24 hours. The preparation method was modified from the previous study [4]. The dried material was as activated by concentrated sulfuric acid (36 N) (1:2) for 12 hours. The acid was then detached by repeatedly washing with a large quantity of deionized water. It was then dried in a hot air oven at 103 °C for overnight. The elephant dung activated carbon (EDAC) was ground and sieved with apertures of 250 μm and stored in an air tight container for adsorption studies.

2.2 *Preparation of dye solution*

The adsorption studies were obtained with color of methylene blue (MB), supplied by Fisher Scientific and used without further purification. This dye has an empirical formula of $C_{16}H_{24}ClN_3O_3S$, molecular weight of 373.896 g/mol. The dye solution of 1000 mg/L was prepared from methylene blue for single solution which was dissolved in deionized water and adjusted to 1000 mL for the concentrated stock solution. The concentration of MB was analysed by Hitachi U-2900 spectrophotometer using wavelength of 665 nm for measuring its colour concentrations.

2.3 *Equilibrium studies and effect of dye concentration, contact time and EDAC dosages on dye removal*

Batch adsorption experiments were carried out at room temperature by varying the different initial dye concentrations (10, 20, 30 and 40 mg/L). The dye solutions of 50 mL were obtained to a series of 250 mL Erlenmeyer flasks with a desired dose of adsorbent (EDAC). The batch experiment was conducted without any pH adjustment. The Erlenmeyer flasks were then shaken by incubating shaker at a constant agitation speed (100 rpm). The samples were withdrawn from the shaker at determined time intervals, and the supernatant liquid portions were centrifuged by centrifuge machine at 8000 rpm for 20 minutes and analysed for remaining dye concentration using spectrophotometer. For the adsorbent dosage study, 0.02-0.10 g of EDAC was introduced into a series of 250 mL Erlenmeyer flasks containing 50 mL of different concentrations (10, 20, 30 and 40 mg/L) of MB solution. The samples were withdrawn, centrifuged and estimated for remaining dye concentration spectrophotometrically.

The extent of MB removal by EDAC was calculated to fit with the Langmuir and Freundlich model as shown in equation (1) and in equation (2), respectively [7, 8].

$$1/Q_e = (1/K_L q_m) 1/C_e + 1/q_m \quad (1)$$

Where; Q_e is the uptake of dye, C_e is the concentration of dye in wastewater, q_m represents the mass of dye adsorbed over mass if adsorbent for a complete monolayer, K_L is the Langmuir constant, and C_e is the equilibrium concentration

$$Q_e = K_F C_e^{\ 1/n} \quad (2)$$

Where; Q_e is the uptake of dye, Ce is the concentration of dye in wastewater, K_F is represents the Freundlich constant related to adsorption capacity and n is the Freundlich constant related to adsorption intensity (n>1). The linearized form of Freundlich isotherm can be written as in equation (3)

$$\log Q_e = \log K_F + (1/n)\log C_e \quad (3)$$

3. Result and discussion

Batch equilibrium studies

Considering on the R^2 values achieved, the goodness-of-fit for the batch equilibrium studies to the isotherm models found that the R^2 of Langmuir model is more than that of Freundlich model according to R^2 at 0.9894 and 0.9411, respectively, as shown in Figure 1. For Langmuir isotherm, the essential feature of Langmuir isotherm can be expressed in terms of a dimensionless separation factor (R_L), given by the following equation [9]

$$R_L = 1/(1+b_L C_0) \quad (4)$$

where, C_0 is the initial dye concentration (mg/L). R_L indicates the type of isotherm to be either unfavourable (R_L>1), linear (R_L=1), favourable (0<R_L<1) or irreversible (R_L=0). In this study, the R_L values were found in the range of 0.019-0.072, which lie between 0 and 1 for all dyes, hence suggesting a favourable adsorption of dyes onto EDAC. The Freundlich isotherm constant (mg/g) (K_F) and adsorption intensity (n) were 9.028 and 2.490, respectively. Since the 1/n value, it was showed that 1/n being greater than 1 one indicates cooperative adsorption [10] and 1/n is a heterogeneity parameter, if n lies between one and ten, this mark a favourable sorption process [11] as shown in Table 1.

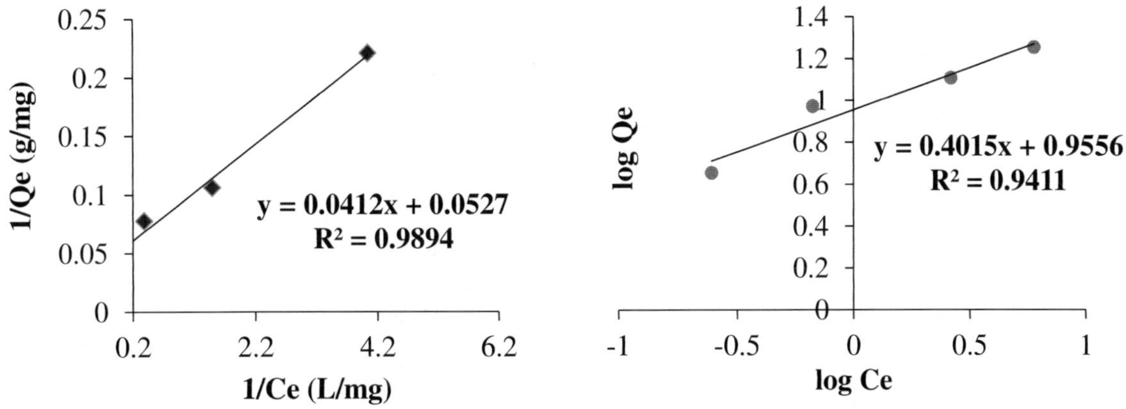

Figure 1. (a) Langmuir adsorption isotherm and (b) Freundlich adsorption isotherm of MB dye removal using EDAC

Table 1. Langmuir and Freundlich adsorption isotherm constants and regression correlation coefficients (R^2) for MB dye

Isotherm model		R^2
Langmuir	K_L = 1.279 q_m =18.975	0.9894
Freundlich	K_F=9.028 n =2.490	0.9411

Effect of initial MB concentration, EDAC dosage and contact time

The ability of EDAC to remove MB at different initial concentrations (10-40 mg/L) is presented in Figure 2. After 120 minutes contact time, no difference was recorded in the amount of MB uptake; to the contrary, a decrease was observed in removal, i.e. desorption was occurring at longer contact times. As higher concentrations resulted in increased removal efficiency, this suggests an association with higher availability of active binding sites present on the surface of EDAC [7].

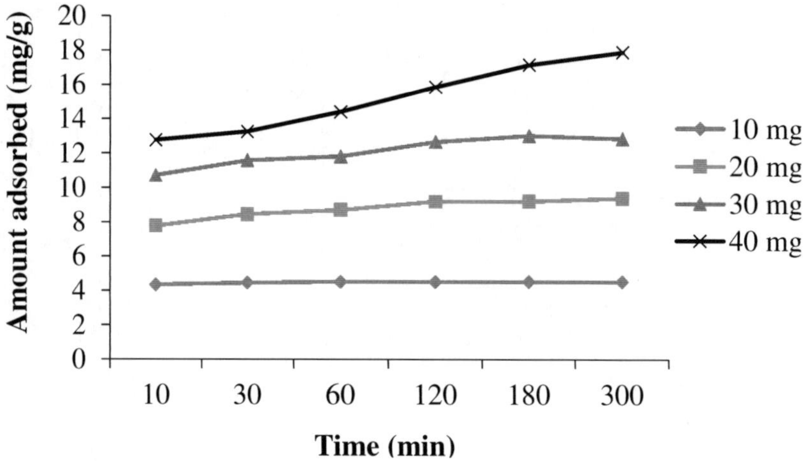

Figure 2. Effect of initial MB dye concentration on dye removal by EDAC

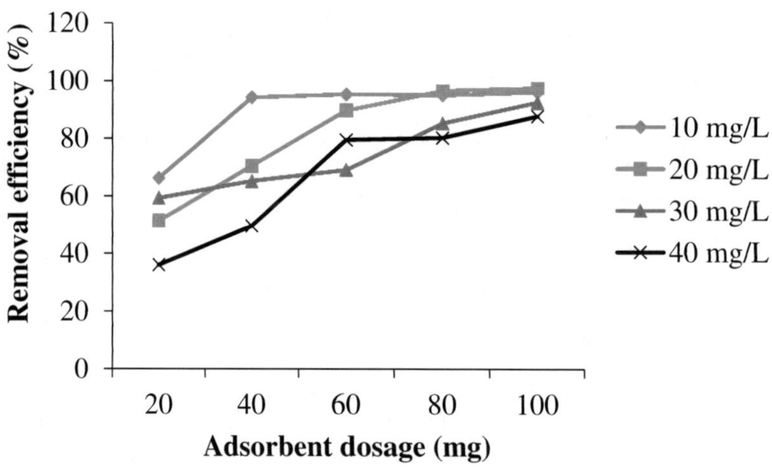

Figure 3. Effect of EDAC dosage on MB removal

The effect of EDAC dosage (20 -100 mg), it is clear from the results that removal efficiencies of dyes increases with increase in the dosage of EDAC and reaches saturation after a particular amount of EDAC for all the dyes studied as shown in Figure 3. This is caused by the fact that, increase in adsorbent dose, increases the number of adsorption sites on the surface of the adsorbent [12]. It is noticeable that adsorbent dose increasing, the removal efficiency of EDAC increases; however, the removal efficiency was low with increase in initial dye concentrations from 10 mg/L to 40 mg/L. This is due to the reason that adsorption sites remain unsaturated during the adsorption process [13-14]. The optimum dosage of EDAC was found to be 100 mg.

4. Conclusion

In this study, a preliminary study of the utility of elephant dung activated carbon (EDAC) for methylene dye found that EDAC could be effectively used as a low cost adsorbent for the removal of colour and dye molecules from aqueous solution. The optimum dosage of EDAC was found to be 100 mg. Furthermore, EDAC adsorption was fitted with Langmuir model. The calculated maximum adsorption capacity was 18.98 mg/g. A dimensionless separation factor (R_L) indicated a favourable adsorption of MB onto EDAC. Further studies should focus on effect of pH and temperature on dye removal. Economic analysis should also be completed in order to evaluate the feasibility of its use in the large-scale community.

5. References

[1] Katheresan V 2018 J. Environ. Chem. Eng. 6 4676–97
[2] Kulkarni M R, Revanth T, Acharya A and Bhat P 2017 *Resource-Efficient Technologies* **3** 71– 7
[3] Wang Z, Xue M and Liu Z 2011 *Advances in Treating Textile Effluent* ed Prof. Peter Hauser (Rijeka, Croatia) chapter 5 p 92- 116
[4] Theivarasu C and Chandra S 2011 *Orient j. chem.* **27** 573-85
[5] Hamdaoui O 2006 *J Hazard Mater.* **135** 264-73
[6] Soni M, Sharma A K, Srivastava J K and Yadav J S 2012 *International Journal of Chemical Sciences and Application*, **3** 338-45
[7] Mahamad M N, Zaini M A A and Zakaria Z A 2015 *Int. Biodeterior. Biodegradation* **102** 274-80
[8] Itodo A U, Itodo H U and Gafar M K 2010 *J. Appl. Sci. Environ. Manage.* **14** 141-5
[9] Özacar M and Şengil A 2004 *Colloids Surf. A Physicochem. Eng. Asp.* **242** 105-13
[10] Foo KY and Hameed B H 2010 *Chem. Eng. J.* **156** 2-10
[11] Goldberg S 2005 *Equations and Models Describing Adsorption Processes in Soils*. (Madison: Soil Science Society of America)
[12] Mane V S, Mall I D and Srivastava V C 2007 *Dyes Pigm.* **73** 269-78
[13] Aljeboree A M, Alshirifi A N and Alkaim A F 2017 *Arab J. Chem.* **10** S3381-93
[14] Ata S, Din M I, Rasool A, Qasim I and Mohsin I U 2012 *J. Anal. Methods Chem.* **2012** 1-8

Acknowledgments

The authors gratefully acknowledge the financial support provided by Thammasat University under the TU New Research Scholar, Contract No.8/2561. We are also grateful to Thai Elephant Conservation Center in Lampang province, Thailand, who provided elephant dung for the study. We are thanks to Environmental Health Laboratory, School of Health Science, Mae Fah Luang University, Thailand to permit us to use laboratory facilities. In addition, we are grateful for TU Research Fund and Faculty of Public Health, Thammasat University that provided the financial support for our research presentation.

IOP Conf. Series: Earth and Environmental Science **291** (2019) 012014 doi:10.1088/1755-1315/291/1/012014

Effect of salinity, single and binary ionic compounds' low salinity water on wettability alteration in carbonate rocks

C Ho[1] and M A Ayoub[1,2]

[1] Petroleum Engineering Department, Universiti Teknologi PETRONAS, Seri Iskandar 32610, Perak Malaysia.

Email: hoching1258@gmail.com, abdalla.ayoub@utp.edu.my

Abstract. Low salinity water flooding is one of the emerging enhanced oil recovery technologies as it has been proven economical and environmentally friendly. However, the recovery mechanism of low salinity water (LSW) is still under debatable due to the complex effect of low salinity water and its ionic compositions. Therefore, this study aims to discover the optimum seawater dilution salinity and influence of single and binary ionic compounds low salinity water on wettability alteration of carbonate core slices at optimum salinity. To achieve that, a modified Design of Experiments (DOE) has been implemented. Contact angle measurement was carried out to characterize the wettability of core slices at 0 hour, after 24 hours and after 48 hours. The results revealed that dilution of seawater reduced the contact angle of carbonate core slices towards more water wet until the optimum salinity of 1750ppm. Further dilution to 700ppm only shown a slight impact in shifting the wettability of the carbonate slices towards more water wet. In single ionic compound LSW, $MgCl_2$ showed the greatest ability in altering wettability. In binary ionic compounds LSW, it was found out that $MgCl_2$ mixed with ionic compounds containing monovalent ion is more effective in altering contact angle than $MgCl_2$ mixed with ionic compounds containing divalent ion which serve as a new finding in current low salinity water study on carbonate rocks.

1. Introduction

Most of the oil and gas reserves are held in carbonate reservoirs. Its complex nature of characters such as oil wet to intermediate wet, heterogeneous characteristic and highly fractures resulted in large amount of hydrocarbon left untapped in the reservoir [1]. Until last decade, one green popular enhanced oil recovery (EOR) method known as low salinity water flooding (LSWF) is being explored extensively on carbonate reservoirs. Low salinity water flooding is not only eco-friendly and inexpensive; the simplicity of injection into oil bearing formation and capability of sweeping light to medium gravity crude oil leading it more attractive over the other EOR methods [2], [3]. There are different conclusions that have been drawn on the performance of low salinity water flooding in carbonate rocks. In the studies of [3-5], lowering the salinity of water could have lowered the contact angle towards more water wet and thus increasing oil recovery. However, Fathi et al. [6] claimed that low salinity water has negative impact on oil recovery of carbonate rock. Sharifi and Shaikh [7] concluded that there is no direct relationship between contact angle and salinity and Su et al. [8] has revealed that there is an optimum salinity to achieve maximum wettability alteration. On the other hand, many laboratory investigations showed favourable oil recovery by tuning the ionic compositions like types and concentration of ions in low salinity water. Hognesen et al. [9] reported that the increase in SO_4^{2-} concentration in seawater could increase the oil recovery through their spontaneous

IOP Conf. Series: Earth and Environmental Science **291** (2019) 012014 doi:10.1088/1755-1315/291/1/012014

imbibition experiment. In the work of Zhang and Austad [10], the authors mentioned that both SO_4^{2-} and Ca^{2+} are the potential determining ions. Zhang et al. [11] were then proved that SO_4^{2-} is required to interact with either Ca^{2+} or Mg^{2+} to improve oil recovery. Contradicting result was found by Al-Attar et al. [5] work who claimed that the increase in Ca^{2+} concentration of water injection has an adverse impact on oil recovery. Interestingly, contact angle measurement on an Iranian carbonate rock carried out by Lashkarbolooki et al. [12] reported that the monovalent ions (K^+ and Na^+) are able to improve water wetness of carbonate rock compared to divalent ions (Ca^{2+} and Mg^{2+}) when tested individually. Conversely, Sodium chloride is known to be a non-active ion by [6], [13]. Therefore, to improve understanding on the effect of low salinity water on carbonate rock, this study aims to investigate further into the optimum salinity and the effect of ions in single ionic compound and the relative effect of ions in binary ionic compounds low salinity water on wettability alteration of Indiana limestone.

2. Experiment Materials and methodology

2.1. Core

Indiana limestone outcrop core was provided by Kocurek Industries Inc. The petrophysical properties of Indiana limestone were measured by Poroperm using Helium gas and are tabulated in table 1.

Table 1. Petrophysical properties of Indiana limestone.

Air permeability (mD)	Porosity (%)	Pore volume (cc)	Grain volume (cc)	Bulk Volume (cc)	Grain density (g/cc)	Bulk density (g/cc)
66.03	11.571	9.418	71.976	81.394	2.540	2.246

2.2. Oil

Real crude oil has been used in this study. Oil was filtered by using 11mirons pore size Whatman filter paper before conducting experiment to eliminate the presence of coarse particles in the oil. The properties of crude oil are as shown in table 2.

Table 2. Properties of crude oil.

Density at 80°C (g/cm³)	API @ 60°F	Viscosity at 80°C (cP)
0.7937	36.87	1.6065

2.3. Brines

Synthetic seawater, low salinity waters and formation water were prepared by dissolving Merck reagent grade salts with purity of more than 99.0% into distilled water such as Sodium Chloride, Calcium Chloride, Magnesium Chloride, Potassium Chloride, Sodium Sulphate and Sodium Bicarbonate. Seawater was synthesized closed to the compositions of typical seawater [14]. Formation water was synthesized based on the formation water of a reservoir limestone at which the air permeability is roughly the same as the air permeability of the core used in this study [15]. Through employing the modified design of experiment (DOE), the brine formulations in this study have been made possible as in table 3. Seawater (SW) was diluted to 5, 10, 20 and 50 times and are represented as SW/5, SW/10, SW/20 and SW/50 respectively. Binary ionic compounds are made up of 50%

IOP Conf. Series: Earth and Environmental Science **291** (2019) 012014 doi:10.1088/1755-1315/291/1/012014

$MgCl_2$ and 50% other ionic compounds (KCl, NaCl, $CaCl_2$ and Na_2SO_4) at a fixed salinity of 3500ppm. Therefore, there are a total of 14 samples that were used to investigate the low salinity water effect on wettability alteration of Indiana limestone. Density of brines were measured at both 25°C and 80°C by using Mettler Toledo density meter (DM40).

Table 3. Formulation of brines.

Composition	Na^+ (mg/L)	Ca^{2+} (mg/L)	Mg^{2+} (mg/L)	K^+ (mg/L)	SO_4^{2-} (mg/L)	HCO_3^- (mg/L)	Cl^- (mg/L)	TDS (mg/L)	Ionic Strength (mol/L)	Density at 25°C (g/cm^3)	Density at 80°C (g/cm^3)
Formation Water	49933.0	14501.0	3248.0	0.0	234	0.0	111947.0	179863	3.6605	1.1228	1.0949
Seawater, SW	10669.0	420.0	1330.0	385.0	2730.0	142.0	19324.0	35000	0.6979	1.0241	0.9988
SW/5	2133.8	84.0	266.0	77.0	546.0	28.4	3864.8	7000	0.1396	1.0024	0.9777
SW/10	1066.9	42.0	133.0	38.5	273.0	14.2	1932.4	3500	0.0698	1.0003	0.9752
SW/20	533.5	21.0	66.5	19.3	136.5	7.1	966.2	1750	0.0349	0.9988	0.9738
SW/50	213.4	8.4	26.6	7.7	54.6	2.8	386.5	700	0.0140	0.9979	0.9729
KCl	0.0	0.0	0.0	1835.7	0.0	0.0	1664.3	3500	0.0469	0.9997	0.9747
NaCl	1376.9	0.0	0.0	0.0	0.0	0.0	2123.1	3500	0.0599	0.9999	0.9749
$MgCl_2$	0.0	0.0	893.7	0.0	0.0	0.0	2606.3	3500	0.1103	1.0002	0.9755
$CaCl_2$	0.0	1264.0	0.0	0.0	0.0	0.0	2236.0	3500	0.0946	1.0002	0.9754
Na_2SO_4	1133.0	0.0	0.0	0.0	2367.0	0.0	0.0	3500	0.0739	1.0006	0.9761
$MgCl_2$ + KCl	0.0	0.0	446.8	917.8	0.0	0.0	2135.3	3500	0.0786	1.0003	0.9752
$MgCl_2$ + NaCl	688.4	0.0	446.8	0.0	0.0	0.0	2364.7	3500	0.0851	1.0004	0.9752
$MgCl_2$ + $CaCl_2$	0.0	632.0	446.8	0.0	0.0	0.0	2421.2	3500	0.1024	1.0006	0.9755
$MgCl_2$ +Na_2SO_4	566.5	0.0	446.8	0.0	1183.5	0.0	1303.2	3500	0.0921	1.0008	0.9756

2.4 Experimental Procedures

Core sample was cut into 14 core slices (length of 2cm, width of 2cm and thickness of 3mm-5mm) by using Geological cutting machine. To produce a smooth surface of core slices for contact angle measurement, core slices were polished by using Forcipol 300-1V Grinder and Polisher before they were cleaned with methanol by using Soxhlet Extraction Apparatus. Core slices were then dried in oven for 24 hours at 90°C. To establish the initial water condition of core, core slices were saturated with formation water using Vacuum pump connected to Desiccator until no gas bubbles were released from the core slices. The core slices immersed in formation water were then placed into oven for 3 days at 80°C before it was aged in crude oil for 5 days at 80°C to create initial oil wet condition of core slices. The initial contact angle on aged core was then determined by dropping a water droplet on the cleaned and polished rock surface and the contact angle between the solid surface and tangent of wetting fluid from the surface of solid was then recorded at atmospheric condition. Finally, each aged core slice was immersed in each low salinity waters at 80°C before subsequent contact angle measurements were conducted after core slices being immersed for 24 hours and 48 hours. Contact angle measurement was performed by using Vinci Technologies IFT 700. The method of contact angle measurement is similar to what has been described in literature [16], [17].

3. Results and Discussions

3.1. Seawater and its dilution low salinity water (LSW)

Based on figure 1, the contact angles decrease with time, which indicating that wettability is shifted towards more water wet; hence improving oil recovery. This result is in line with literature as illustrated by [3], [4], [5]. Seawater however could not alter the wettability of core slice. Interestingly, it is found out that diluting seawater to 1750ppm is the minimum concentration of low salinity water to have the greatest change in contact angle for Indiana limestone as depicted in figure 2. Further dilution to 700ppm only shows a slight impact in shifting the wettability, which also means additional cost is needed to desalinate seawater with minimum impact on recovery. The reason is deemed to be lack of potential determining ions concentration [6]. Therefore, there is a certain dilution extent for maximum wettability alteration to occur. A similar trend of the result was also obtained by Su et al. [8]. Due to significant change in contact angle has been observed when 7000ppm low salinity water is diluted to 3500ppm and only marginal changes when 3500ppm is diluted to 1750ppm, therefore 3500ppm is marked as the model concentration of low salinity water for subsequent contact angle measurements.

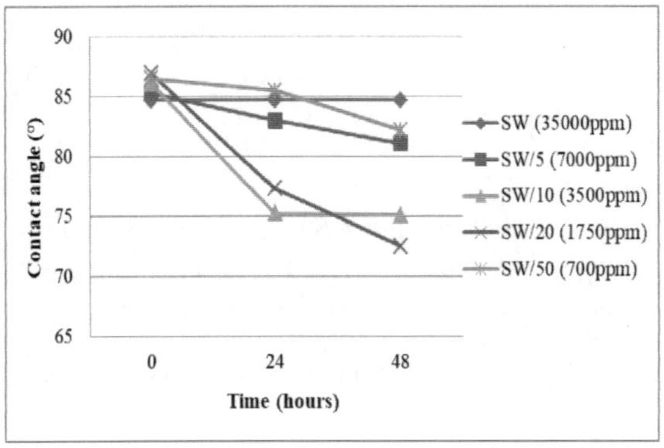

Figure 1. Contact angle measurement for seawater and its dilution with time.

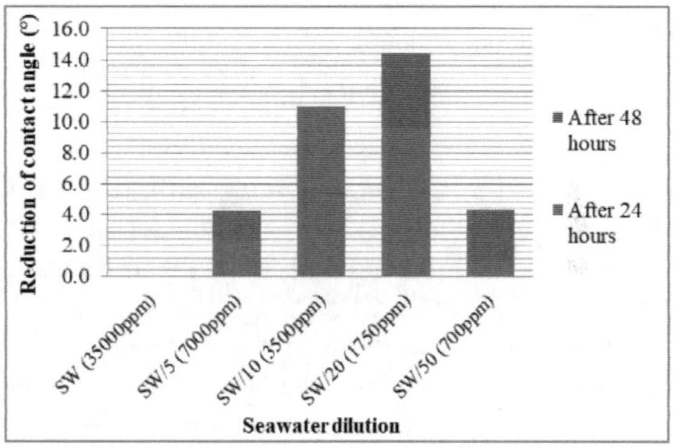

Figure 2. Total cumulative reduction of contact angle for seawater and its dilution after 24 hours and 48 hours.

3.2. Single ionic compound low salinity water (LSW)

Figure 3 presented the contact angle reducing with time for single ionic compound low salinity waters at a salinity of 3500ppm. Relatively appreciable change in contact angle is notable for $MgCl_2$ composition as illustrated in figure 4. The contact angle has changed from 84.9° to 78.2° with total of 6.7° change in contact angle. Comparable results can be seen for Na_2SO_4 and $CaCl_2$ composition which has the total change in angles of 4.7° and 3.5° respectively. The results are analogous with the low salinity water contact angle measurements conducted by other researchers [18], [19] who found out that Magnesium ions can shift the wettability towards more water wet than Sulphate ions. Ionic compounds like NaCl and KCl showed the least impact on contact angle changes in carbonate rock. Contradiction was found in the work of Lashkarbolooki et al [12] who mentioned NaCl and KCl have higher capability in wettability alteration than $MgCl_2$ and $CaCl_2$. The effect and mechanism of low salinity water are not well understood yet.

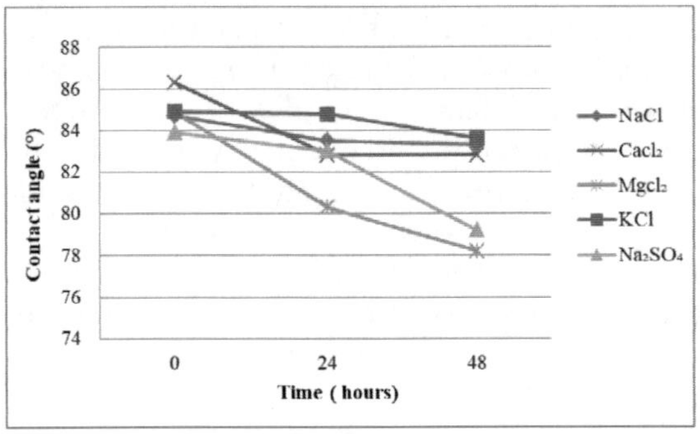

Figure 3. Contact angle measurement for single ionic compound low salinity waters with time

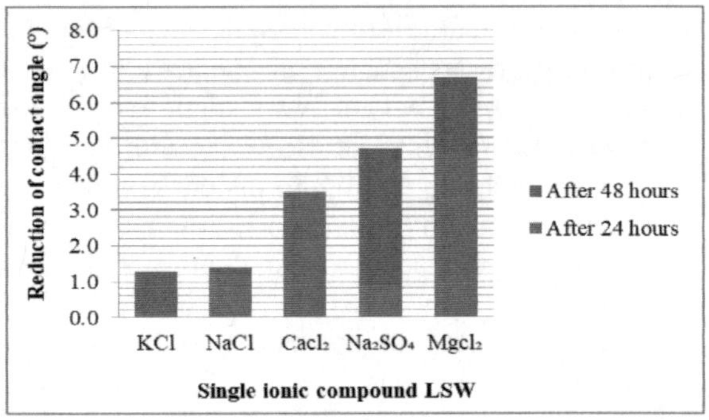

Figure 4. Total cumulative reduction of contact angle for single ionic compound low salinity waters after 24 hours and 48 hours.

3.3 Binary ionic compounds low salinity water (LSW)

Like figure 3, figure 5 shows the contact angle of binary ionic compounds LSW fixed at 3500ppm is observed to be reducing with time. From figure 6, it is apparent that the effect of the combination of 50% $MgCl_2$ + 50% KCl on wettability is tremendous amongst others. The results revealed that binary ionic compounds of 50% $MgCl_2$ + 50% KCl or 50% NaCl which has monovalent ion resulted in greater contact angle modification compared to binary ionic compounds of 50% $MgCl_2$ + 50% Na_2SO_4 or 50% $CaCl_2$ which has divalent ion. In other words, replacing 50% of single ionic compound such as KCl and NaCl with $MgCl_2$ increases the reduction of contact angle while replacing 50% of single ionic compound such as Na_2SO_4 and $CaCl_2$ with $MgCl_2$ decreases the reduction of contact angle.

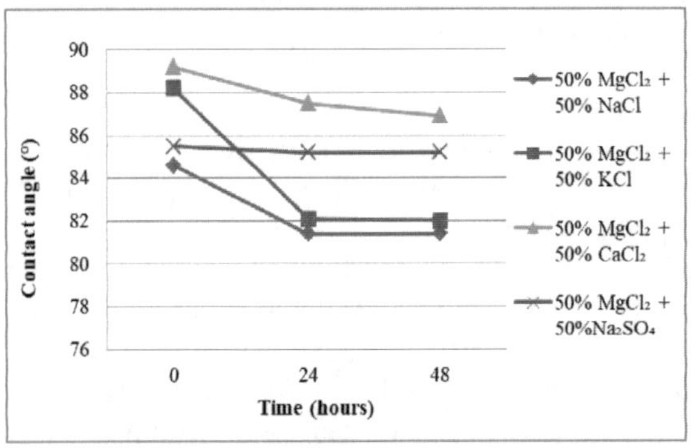

Figure 5. Contact angle measurement for binary ionic compounds low salinity water with time

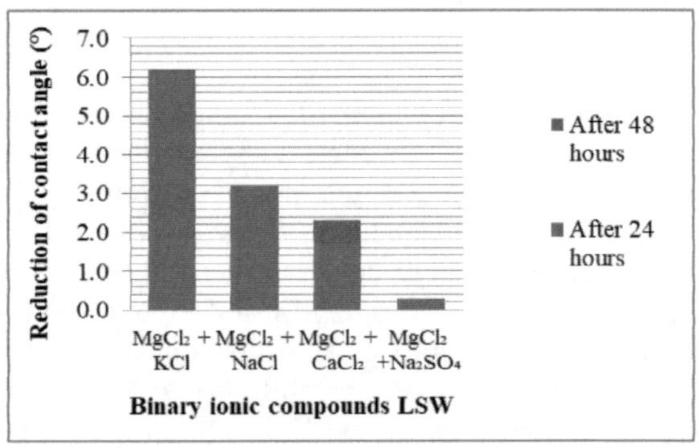

Figure 6. Total cumulative reduction of contact angle for binary ionic compounds low salinity water after 24 hours and 48 hours.

4. Conclusions

In this study, it has been demonstrated that the dilution of seawater as well as tuning ionic composition low salinity water could affect the wettability of carbonate rocks. The more the salinity is reduced, the greater the alteration of contact angle and thus improving the wettability of carbonate rock towards water wet. However, there is an ideal salinity below which diluting the water reduces the change in contact angle. Overly dilution leads to a total change in contact angle reduces. In single ionic compound LSW, $MgCl_2$ has the greatest ability in shifting wettability towards more water wet followed by ionic compounds containing divalent ion (Na_2SO_4 and $CaCl_2$) which are more capable to alter contact angle compared to ionic compounds containing monovalent ion (NaCl and KCl).

However, in binary ionic compounds LSW, the final wettability for 50% $MgCl_2$ + 50% ionic compounds containing monovalent ion (NaCl or KCl) tends to be more water-wet than 50% $MgCl_2$ + 50% ionic compounds containing divalent ion (Na_2SO_4 or $CaCl_2$). Overall, from these 14 samples, reducing seawater salinity to 1750ppm has the greatest change in contact angle (from 86.9° to 72.5°), therefore for future work further tuning of types and concentration of low salinity water ions is vital to obtain higher wettability alteration to improve this eco-friendly enhanced oil recovery method.

5. References

[1] Sheng J J 2013 Review of surfactant enhanced oil recovery in carbonate reservoirs *Advances in Petroleum Exploration and Development* **6** 1-10

[2] Al-Shalabi E W and Sepehrnoori K 2016 A comprehensive review of low salinity/engineered water injections and their applications in sandstone and carbonate rocks *J. Pet.Sci. Eng.* **139** 137-161

[3] Yousef A A, Al-Saleh S H, Al-Kaabi A and Al-Jawfi M S 2011 Laboratory investigation of the impact of injection-water salinity and ionic content on oil recovery from carbonate reservoirs *SPE Reserv. Eval. Eng.* **14** 578-593

[4] Alameri W, Teklu T W, Graves R M, Kazemi H and AlSumaiti A M 2014 Wettability alteration during low salinity-waterflooding in carbonate reservoir cores *SPE Asia Pacific Oil & Gas Conference and Exhibition* (Adelaide, Australia: Society of Petroleum Engineers)

[5] Al-Attar H H, Mahmoud M Y, Zekri A Y, Almehaideb R and Ghannam M 2013 Low-salinity flooding in a selected carbonate reservoir: experimental approach *J. Petrol. Explor. Prod. Technol.* **3** 139-149

[6] Fathi S J, Austad T and Strand S 2010 "Smart water" as a wettability modifier in chalk : the effect of salinity and ionic composition *Energy & Fuels* **24** 2514-2519

[7] Sharifi M and Shaikh M 2013 Investigation of optimum salinity of injected water in carbonate reservoirs using wettability measurement and core flooding *SPE Reservoir Characterisation Simulation Conf. and Exhibition* (Abu Dhabi, UAE: Society of Petroleum Engineers)

[8] Su W, Liu Y, Pi J, Chai R, Li C and Wang Y 2018 Effect of water salinity and rock components on wettability alteration during low-salinity water flooding in carbonate rocks *Arab J Geosci* **11** 260

[9] Hognesen E J, Strand S and Austad T 2005 Waterflooding of preferential oil-wet carbonates: oil recovery related to reservoir temperature and brine composition *SPE Europec/EAGE Annual Conf.* (Madrid, Spain: Society of Petroleum Engineers)

[10] Zhang P and Austad T 2006 Wettability and oil recovery from carbonates: Effects of temperature and potential determining ions *Colloids Surf. A Physicochem. Eng. Asp.* **279** 179-187

[11] Zhang P, Tweheyo M T and Austad T 2007 Wettability alteration and improved oil recovery by spontaneous imbibition of seawater into chalk: Impact of the potential determining ions Ca^{2+}, Mg^{2+}, and SO_4^{2-} *Colloids Surf. A Physicochem. Eng. Asp.* **301** 199-208

[12] Lashkarbolooki M, Ayatollahi S and Riazi M 2017 Mechanistical study of effect of ions in smart water injection into carbonate oil reservoir *Process Saf. Environ.* **105** 361-372

[13] Rashid S, Mousapour M S, Ayatollahi S, Vossoughi M and Beigy A H 2015 Wettability alteration in carbonates during "Smart Waterflood": Underlying mechanisms and the effect of individual ions *Colloids Surf. A Physicochem. Eng. Asp* **487** 142-153

[14] Castro P and Huber M E 2008 *Marine Biology* ed Reidy P E and Henricks D A (New York: McGraw-Hill) p 44

[15] Chandrasekhar S and Mohanty K K 2013 Wettaability alteration with brine composition in high temperature carbonate reservoirs *SPE Annual Technical Conf. and Exhibition* (Louisiana, USA: Society of Petroleum Engineers)

[16] Zhang P and Austad T 2005 Waterflooding in chalk: Relationship between oil recovery, new wettability index, brine composition and cationic wettability modifier *SPE Europec Annual Conf.* (Madrid, Spain: Society of Petroleum Engineers)
[17] Shabib-Asl A, Ayoub M A and Elraies K A 2015 Laboratory investigation into wettability alteration by different low salinity water compositions in sandstone rock *SPE/IATMI Asia Pacific Oil & Gas Conf. and Exhibition* (Bali, Indonesia: Society of Petroleum Engineers)
[18] Gomari K A R, Karoussi O and Hamouda A A 2006 Mechanistic study of interaction between water and carbonate rocks for enhancing oil recovery. *SPE Europec/EAGE Annual Conf. and Exhibition* (Vienna, Australia: Society of Petroleum Engineers)
[19] Karimi M, Al-Maamari R S, Ayatollahi S and Mehranbod N 2016 Wettability alteration and oil recovery by spontaneous imbibition of low salinity brine into carbonates: impact of Mg^{2+}, SO_4^{2-} and cationic surfactant *J.Pet Sci. Eng.* **147** 560-569

Acknowledgments
The authors wish to acknowledge Centre of Research in Enhanced Oil Recovery (COREOR), Institute of Hydrocarbon Recovery, Universiti Teknologi PETRONAS for providing an opportunity to conduct this study. This study is supported via YUTP grant under the cost centre 0153AA-E65.

IOP Conf. Series: Earth and Environmental Science **291** (2019) 012015 doi:10.1088/1755-1315/291/1/012015

Evaluation of Antifouling Coating performance With Commercial and Lemon Natural Biocide on Composite Fiberglass Surface.

Asmalina Mohamed Saat[1], Muhammad Irfan Sabri[1], Hanisah Johor[1], Fatin Zawani Zainal Azaim[1], Mohd Iqbal Lokman[2]

[1]Universiti Kuala Lumpur, Malaysian Institute of Marine Engineering Technology, Lumut Perak,

[2]Bina Integrated Industries Sdn Bhd, Bandar Baru Bangi, Selangor, Malaysia.

E-Mail: asmalina@unikl.edu.my

Abstract. Marine growth is a common problem for a near coastal marine vessel including composite fiberglass boat. Natural antifouling coating for composite fiberglass hulls towards efficient, cleanliness, sustainability and greener oceans are crucial especially in near coastal area. Antifouling performance of commercial and natural lemon biocide were analyzed by referring to ASTM 3623 – 78a for 30 days in seawater. Both commercial and natural biocide coating shows no marine growth occur within 30days of immersion due to good interaction between biocide and gelcoat. Natural biocide with 6% concentration observed similar performance as 4% commercial biocide, thus it shows lemon has a great potential of lemon as new natural biocide in marine antifouling paint.

1. Introduction

Ships have been used as a mode of transportation for hundreds of years. Woods, fiberglass and steel are the most common materials that were used to build a ship. All types of hull vessel are detrimental to the marine growth from thousands of years. Marine growth is a natural phenomenon that occurs on ship propeller and hull that affects performance of a ship such as increase in ship resistant (86%), reduce in speed loss approximately 2% and increase in fuel consumption up to 40-50% [1]. Free marine growth hull leads to increase of fuel efficiency thus reduce the percentage of CO_2 global greenhouse gas emission produced from shipping industry [2]. The marine growth on steel hull will increase maintenance cost due to frequent dry docking needed [3]. However, in the cruise ship, hull scrubbing is not an option, thus effective antifouling material is a critical challenge. Cruise ships also face important challenge for fuel saving and credential of good environment to sustain a spotless waterline [2]. Biocides are chemical substances commonly added in antifouling paint to deter or kill the microorganisms from marine growth. Historically, lead and copper were primarily used as antifouling to protect wooden vessel. Then, the usage of heavy metal continues until TBT was introduced as biocides in 1960. TBT reported to have efficient performance inhibiting the biofouling and minimize the frequency of dry docking for vessels. However, an abnormal of oyster shell growth reported in 1974 due to TBT usage, later this leads to banned of TBT by International Maritime Organization in 2008. The banned of TBT has generated a lot of interest to search efficient antifouling, from development of fouling release antifouling (slippery, nontoxic coat) [2], self-polishing antifouling (biocides prevent establish of fouling) and combination of both systems [3]. Current challenges on the development of antifouling are to produce future antifouling that contributes to the great concern in cleanliness, sustainability of oceans, less toxic and effective (fuel efficiency and less emission CO_2) especially to near coastal ship. Other challenges face in development of antifouling that need to be considered are the various type of vessels, operation profiles and locations [2]. Development of effective nontoxic antifouling also requires understanding of interaction between marine growth and coated surface which is still lack behind [4]. Most common commercial biocides used in antifouling paints founded are chlorothalonil, dichlofluanid, DCOIT (4,5-dichloro-2-n-octyl-4-isothiazolin-3-one, Sea-nine 211®), Diuron, Irgarol 1051, TCMS pyridine (2,3,3,6-tetrachloro-4-methylsulfonyl pyridine), zinc pyrithione and Zineb [5]. Meanwhile common natural biocides usually extracted from marine sources such as soft corals, seaweed, sponges, mangroves and etcetera. They are reported to have antibacterial such as carotenoids, furanones, alkaloids, peptides, lactones, steroid and terpenoids [6]. The alternative of natural biocide earlier reported on the extract of fishes used in (vinyl chloride–vinyl acetate) copolymer [7] algae embedded in epoxy matrix[8] [9]. Lemons are also reported to have biocide properties due to the content of saponins, alkaloid, phenolic, terpenoid and flavonoid which is similar to the one found in marine sources biocides [10]. Thus, it is interesting to explore the potential of lemon as biocides in antifouling paint. This research is focused on analyzing the performance of newly developed natural antifouling paint apply on fiberglass surface immerse in a seawater.

2. Methodology

Composite fiberglass panels with two types of surface, gelcoat and polyester resin were prepared as uncoated (reference) and coated by rosin modified antifouling paint contained commercial and natural biocide. Antifouling coating was prepared using rosin modified base and it was obtained from BINA Integrated Sdn Bhd. Meanwhile natural biocide fresh lemon juice was obtained from cleaned squeezed lime after filtration process. Five mixtures of antifouling were prepared with 2%, 4% and 6% of natural biocide and 2% and 4% of commercial biocide and then stirred thoroughly. The specimen was coated with two layers of antifouling coating materials and left to dry for three days as recommended by the manufacturer before second layer of antifouling was applied. The panels were then immersed in seawater for marine growth exposure at shallow submerge according to ASTM D3623 – 78a[11]. Visual inspection for percentage marine growth and weight change conducted for 30 days and recorded every 5 days. Regression method of statistical analysis was conducted for all samples in correlation to the change occur every 5 days.

3. Results and Discussion

Figure 1 shows the pH of water and percentage of marine growth on composite fiberglass polyester resin surface in uncoated and coated with commercial (2% and 4% CA) and natural biocide (2%, 4% and 6% NA). Percentage of marine growth on resin surface after 30 days of immersion for uncoated specimen observed was 29.61%. Coated specimen with natural biocide 2%, 4% and 6%, percentage of marine growth 2.96%, 1.97% and 0.99% respectively. Higher concentration of natural biocides observed better inhibition properties with less percentage of marine growth. Meanwhile commercial biocide observed slightly less percentage than 6% natural antifouling with 0.49% of marine growth. Natural lemon biocides 6% observed similar performance as 4% commercial biocides in both observations. This shows that lemon natural antifouling produces similar rate of inhibition from marine growth as commercial biocide. Thus, proves good interaction between natural biocides with polyester resin surface. Natural and commercial biocides in antifouling paint observed better inhibition properties coated on gelcoat surface compared to polyester surface. Coated specimen on fiberglass gelcoat surface observed better inhibition behavior with no marine growth occur within 30 days of immersion as summarized in Table 1. The natural biocides coated on gelcoat surface observed a slight of slime adhered compared to uncoated gelcoat surface occurred 4.48% of marine growth. This agrees with other studies that microfouling biofilm formed at initial stage before animal type fouling occurred within 7 days of immersion[12]. The percentage of marine growth was slow with the combination of gelcoat and antifouling, thus delaying the occurrence of marine growth compared to uncoated gelcoat. Both natural and commercial biocides antifouling paint coated on both polyester resin and gelcoat surface on fiberglass also observed stable condition in variable of pH from 7.6 to 8.4 with no significant effect in the percentage of marine growth. This finding similar reported earlier that biofouling are stable during fluctuating environmental conditions despite changes in salinity and in dissolved oxygen within 7 days of immersion [4]. Figure 2 shows the water temperature and performance percentage of weight change on composite fiberglass uncoated and coated with commercial (2% and 4% CA) and natural biocide (2%, 4% and 6% NA). The seawater temperature varies between 30 to 37 ℃ throughout 30 days of immersion. Uncoated fiberglass sample observed highest weight changes with 1.95% increased. The weight of specimen increases in linear relation as days of immersion increased. Total of weight changed after 30 days of immersion for natural biocide specimen with 2%, 4% and 6% ratio observed value of 0.48%, 0.49% and 0.51% weight increased respectively. Meanwhile commercial biocides observed similar weight change for 2% and 4% ratio with 0.50% and 0.51% of weight increased. The percentage of weight increased due to growth of animal types marine growth that contributes to increased weight of the panels. Natural biocides antifouling with 6% ratio shows similar inhibition performance as commercial biocides 4% in both percentage of mariner growth and weight change. Both percentage marine growth and weight change observed increase in percentage as number of days increased. Regression analysis observed p-value less than 0.05 for uncoated specimen in percentage marine growth and weight change shows its relationship with number of days are highly significant. Coated specimen shows significant relation with similar p-value less than 0.05 in percentage marine growth; however, in percentage of weight change the p-value shows less significant. Coefficient correlation observed positive values due to increase in both percentages of marine growth and weight change as number of days increased. Meanwhile, observation on percentage marine growth shows high percentage (>80%) of R square fit the linear equation for natural biocides antifouling and uncoated specimen. The non-significant value observed in coated specimen at polyester resin shows that the marine growth and performance of both antifouling coating affected by other factors such as location of immersion, seawater conditions and types of vessels [2]. Other than that, the performance of antifouling coating was also affected by the organic content, current flow, water pH, salinity and coating surface [5].Table 1 summarizes the percentage growth and weight change in composite fiberglass uncoated and coated with commercial (2% and 4% CA) and natural biocide (2%, 4% and 6% NA). This finding shows that both commercial and natural antifouling coating have good interaction with gelcoat surface and give better inhibition properties to the composite fiberglass hull.

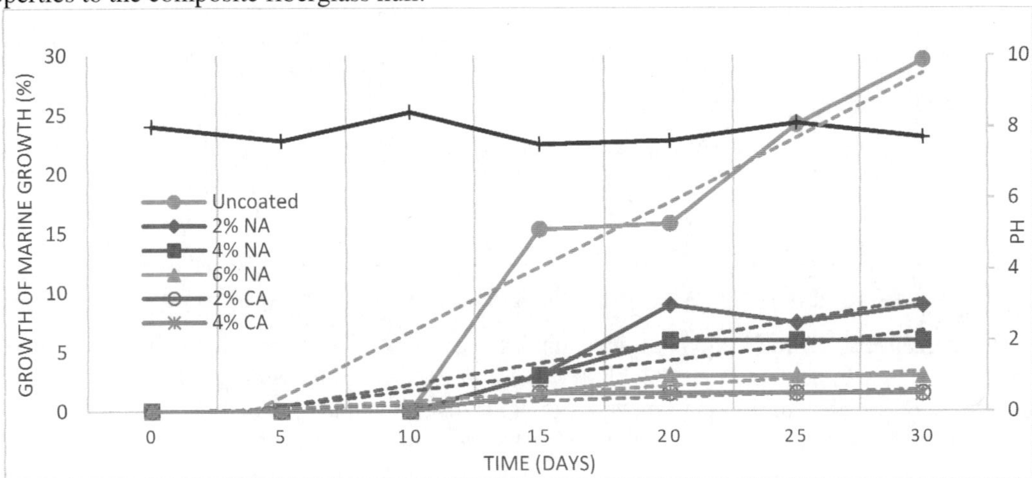

Figure 1. Percentage of marine growth on composite fiberglass polyester resin surface in uncoated and coated with commercial (2% and 4% CA) and natural biocide (2%, 4% and 6% NA)

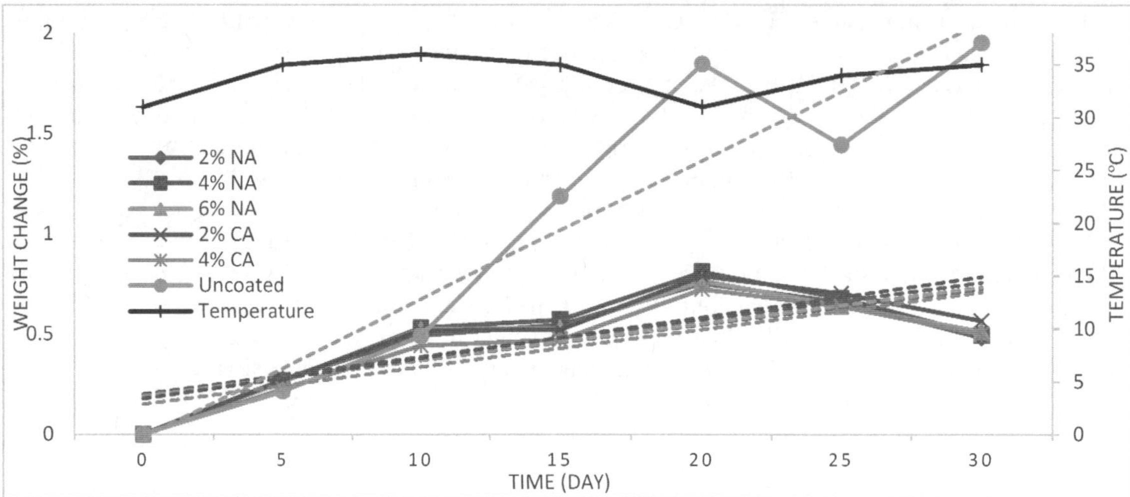

Figure 2. Percentage of weight change on composite fiberglass uncoated and coated with commercial (2% and 4% CA) and natural biocide (2%, 4% and 6% NA)

Table 1. Summary of percentage growth and weight change in composite fiberglass uncoated and coated with commercial (2% and 4% CA) and natural biocide (2%, 4% and 6% NA)

Descriptions	Types	p-value	R-square	Coefficient Correlation	Linear Equation
Percentage marine growth on polyester resin surface	2% NA	0.0040	84%	0.1198	y = 0.5992x - 1.0573
	4% NA	0.0028	86%	0.0423	y = 0.4229x - 0.7049
	6% NA	0.0028	86%	0.0423	y = 0.2115x - 0.3524
	2% CA	0.0117	75%	0.0211	y = 0.1057x - 0.1410
	4% CA	0.0117	75%	0.0211	y = 0.1057x - 0.1410
	Uncoated	0.0008	91%	1.0926	y = 5.4629x - 9.7274
Percentage marine growth on gelcoat surface	2% NA	NIL			
	4% NA				
	6% NA				
	2% CA				
	4% CA				
	Uncoated	0.0070	80%	0.1797	y = 0.8986x - 1.3134
Percentage of Weight change	2% NA	0.2409	57%	0.0277	y = 0.2457x + 278.55
	4% NA	0.2770	53%	0.0288	y = 0.2557x + 277.30
	6% NA	0.2176	59%	0.0293	y = 0.2521x + 278.95
	2% CA	0.1327	69%	0.0356	y = 0.275x + 272.740
	4% CA	0.1332	69%	0.0355	y = 0.2621x + 282.43
	Uncoated	0.0095	92%	0.1879	y = 0.9307x + 268.80

4. Conclusion

Both commercial and natural lemon biocides antifouling coating show great interaction with gelcoat fiberglass surface with no marine growth observed within 30 days of immersion. Higher concentration of biocides shows better antifouling performance with low percentage of marine growth and weight change observed. Natural lemon biocides 6% observed similar performance as 4% commercial biocides in both observations. This concludes that natural lemon biocides antifouling has good inhibition with both gelcoat and polyester resin in composite fiberglass. Thus, shows that it has great potential as new biocide in marine antifouling coating for application in composite fiberglass surface.

However, the performance of this new natural lemon biocides still need physical and chemical analysis as well as further observation of antifouling performance evaluation factor as discussed earlier. It is also interesting to explore the new natural lemon biocides antifouling hydrodynamic performance.

5. Conflict of interests.

The authors declare that there is no conflict of interests regarding the publication of this paper.

6. Acknowledgement

This work was financially supported by UniKL Short Term Grant (UniKL/Cori/STRG/16135). Author also appreciates of the collaboration with Bina Integrated Industries Sdn Bhd, Bandar Baru Bangi, Selangor, Malaysia. The author gratefully acknowledges on both of support.

References

[1] M. P. Schultz, J. A. Bendick, E. R. Holm, and W. M. Hertel, "Economic impact of biofouling on a naval surface ship," *Biofouling*, vol. 27, no. 1, pp. 87–98, 2011.

[2] J. Kavanagh, "The future of antifouling," *The marine Professional*, 2018.

[3] Y. K. Demirel, M. Khorasanchi, O. Turan, and A. Incecik, "On the importance of antifouling coatings regarding ship resistance and powering," *Low Carbon Shipp. Conf.*, vol. 44, no. 0, pp. 1–13, 2013.

[4] F. Regan, A. Barrett, C. Briciu-Burghina, and T. Sullivan, "Antifouling studies and coating strategies for marine deployed structures," in *OCEANS 2017 - Aberdeen*, 2017, vol. 2017–Octob, pp. 1–6.

[5] F. A. Guardiola, A. Cuesta, J. Meseguer, and M. A. Esteban, "Risks of using antifouling biocides in aquaculture," *Int. J. Mol. Sci.*, vol. 13, no. 2, pp. 1541–1560, 2012.

[6] V. Gopikrishnan, M. Radhakrishnan, R. Pazhanimurugan, T. Shanmugasundaram, and R. Balagurunathan, "Natural products : Potential and less explored source for antifouling compounds," *J. Chem. Pharm. Res.*, vol. 7, no. 7, pp. 1144–1153, 2015.

[7] Y. A. Soliman, A. S. Mohamed, and M. NaserGomaa, "Antifouling activity of crude extracts isolated from two Red Sea puffer fishes," *Egypt. J. Aquat. Res.*, vol. 40, no. 1, pp. 1–7, 2014.

[8] J. Chapman *et al.*, "Bioinspired synthetic macroalgae : Examples from nature for antifouling applications," *Int. Biodeterior. Biodegradation*, vol. 86, pp. 6–13, 2014.

[9] J. Chapman *et al.*, "Bioinspired synthetic macroalgae: Examples from nature for antifouling applications," *Int. Biodeterior. Biodegrad.*, vol. 86, pp. 6–13, 2014.

[10] Hardoko and F. Yuliana, "Stability Study of Antibacterial Activity of Mixed Lime Juice and Honey of Heating Temperature on Staphylococcusaureus and Streptococcuspyogenes," *Int. J. Pure Appl. Sci. Technol.*, vol. 21, no. 2, pp. 1–7, 2014.

[11] ASTM, "Standard D3623 Standard test method for testing antifouling panels in shallow submergence," *ASTM*, vol. 06.02, no. Reapproved, pp. 1–8, 2015.

[12] J. Chapman, "Antifouling performances of macro- to micro- to nano-copper materials for the inhibition of biofouling in its early stages," *J. Mater. Chem. B*, vol. 1, no. November, pp. 6194–6200, 2013.

IOP Conf. Series: Earth and Environmental Science **291** (2019) 012016 doi:10.1088/1755-1315/291/1/012016

A new predictive model for Plants Photosynthesis Influenced by Major Climatic Conditions

ZS Liu[1], WY Yang[2], XL Yu[3]

[1] College of Agriculture, Shandong Agricultural University, Taian, Shandong, China
[2] College of automation, Shenyang Aerospace University, Shenyang, Liaoning, China
[3] School of Science, Beijing University of Posts and Telecommunications, Beijing, China

Email: zzc1890@hotmail.com

Abstract: Climate change, particularly global warming, is significantly affected by atmospheric CO_2 dynamics. Plant photosynthesis is capable of fixing a large amount of airborne CO_2 and converts it into vegetation biomass and thus alleviates the greenhouse effect from atmospheric CO_2. However, how climate change and climate condition impact the dynamics of plant photosynthesis is still highly uncertain. Here we combined high frequency land surface measurements of photosynthetic CO_2 fixation data and information theory to understand the casual relationship from climate drivers on the photosynthesis rate. We found that temperature and shortwave radiation dominated photosynthesis more at forest site, while precipitation dominated photosynthesis more at grass land site. More importantly, linear regression based analysis failed to identify such relationships, which confirmed the important role of information theory in identifying nonlinear relationship within a complex system.

1. Introduction

Since preindustrial era, human activities have significantly increased atmospheric greenhouse gas (e.g., carbon dioxide) concentrations, the sources of which were mainly from fossil fuel emissions, biomass burning, land use and land cover changes [1]. Positive radiative forcings induced by airborne greenhouse gases largely warmed up the earth system and led to a substantial and observable increase in surface temperature [2]. For example, the atmospheric CO2 concentration increased from 280 ppm (ppm = parts per million) before industrial evolution to 410 ppm at present day. Consequently, global mean surface temperature increased more than half a degree Celsius and will further increase in the future [3].

Although anthropogenic activities significantly enhanced atmospheric CO2 concentration, natural ecosystems are capable of absorbing a large fraction (about a half) of human induced CO2 emissions, thus effectively mitigate the increasing trends of both greenhouse gas and surface temperature [1]. Terrestrial ecosystem (mainly forest and grass) sequestrates atmospheric CO2 molecules through photosynthesis, a biochemical reaction that combines CO2 and water and store as carbohydrate in the vegetation biomass [4]. The photosynthesis reaction is fueled by light, and significantly controlled by the activity of RuBisCo enzyme, which is highly temperature sensitive. Therefore, theatrically, terrestrial ecosystem photosynthetic capacity is co-controlled by substrates (CO2 and water), temperature (enzyme activity), and solar radiation (energy supply). However, which one is the dominant controlling factor of photosynthesis reaction is still highly uncertain for different vegetation types.

Understanding the temporal dynamics of photosynthesis carbon uptake as well as its relationships with the changing climate is challenging. Previous efforts used observations (either in situ [5] or remotely sensed [6]) to establish predictive relationship between photosynthesis rate and climate drivers, and also improved process-based model predictability based on the empirical relationships mined from observations [7]. Machine learning algorithms have been widely used to infer the empirical relationship. For example, Beer 2010 [6] used tree-based regression model to first obtain the predictive relationship between photosynthesis rate and climate drivers, then upscaled FLUXNET observed gross carbon uptake to generate global terrestrial ecosystem carbon sequestration rate maps. In this study, we used an advanced information theory based machine learning technique to infer the robust non-linear relationship between photosynthesis rate and the climate, which will generate significant insights into predictive modeling of photosynthesis rate.

2. Methodology

2.1 FLUXNET data

Eddy Covariance (EC) technique has been widely used in observing CO2 and water fluxes between land surface and the atmosphere. FLUXNET is a globally distributed observational network based on EC technique. Nowadays, hundreds of FLUXNET sites are established and are well maintained to generate reliable fluxes data covering the major vegetation types [5]. We used the observed photosynthesis rates as well as relevant climate drivers from four representative and high-quality FLUXNET sites: 1) FLX_AT-Neu grass site; 2) FLX_DE-Tha evergreen forest site; 3) FLX_NL-Loo evergreen forest site; and 4) FLX_US-Var grass site. Figure 1 showed the density distribution of observed GPP (Gross Primary Productivity, hereafter we will use GPP in stead of photosynthesis rate), solar radiation (FSDS), precipitation (Prcp), air temperature (T), and longwave radiation (FLDS). Black, blue, green, and red shaded area denoted the previously mentioned four sites, respectively.

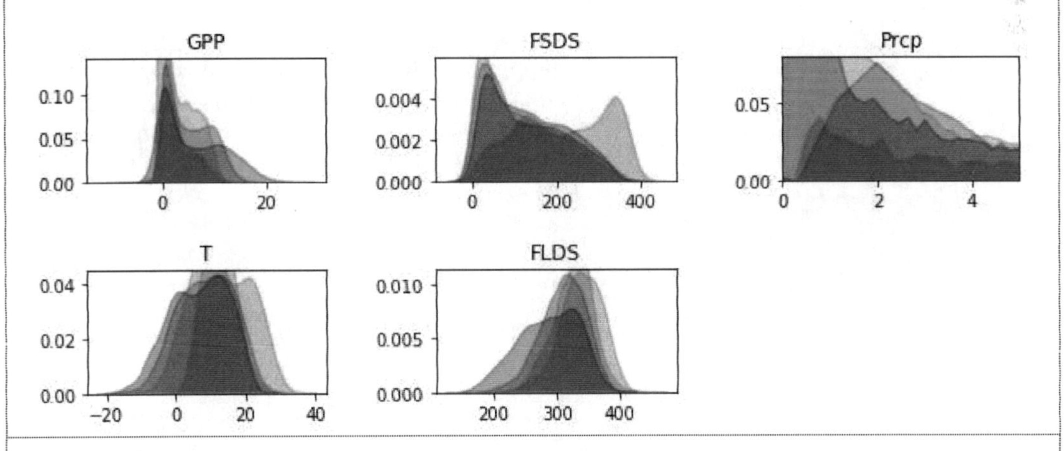

Figure 1. Density distribution of the observed Gross Primary Productivity (GPP), short wave radiation (FSDS), precipitation (Prcp), air temperature (T), and long wave radiation (FLDS). Black, blue, green, and red colors are FLX_AT-Neu, FLX_DE-Tha, FLX_NL-Loo, and FLX_US-Var sites, respecitively

2.2 Mutual information

The responses of Gross Primary Productivity (GPP) to different climate drivers are presumably highly nonlinear, therefore, traditional linear regression based analysis may generally fail. In order to robustly infer non-linear relationship between GPP and the climate drivers, we employed the mutual information concept form information theory. First the Shannon information entropy is defined as [8]:

$$H(X) = -\sum_{i=1}^{n} p(x_i)\log_2 p(x_i) \quad (1)$$

where X is a random variable (in this study could be GPP, temperature and so on), p(x) is the probability density distribution of X. Based on Eqn. 1, we could further define conditional information entropy as:

$$\begin{aligned} &H(X|Y) \\ &= \sum_y p(y) H(X|Y=y) \\ &= -\sum_y \sum_x p(x,y)\log_2 \frac{p(x,y)}{p(y)} \end{aligned} \quad (2)$$

where X, Y are two random variables, p(x), p(y) are marginal distribution of X or Y, p(x,y) is the joint distribution of X and Y. Combine Eqn. 1 and 2, we further derive mutual information as:

$$\begin{aligned} &I(X,Y) \\ &= H(X) - H(X|Y) \\ &= \sum_y \sum_x p(x,y)\log_2 \frac{p(x,y)}{p(x)p(y)} \end{aligned} \quad (3)$$

where H(X) is information entropy of random variable X, H(X|Y) is conditional information entropy of X to Y (Figure 2). The mutual information represents the level relevance between X and Y.

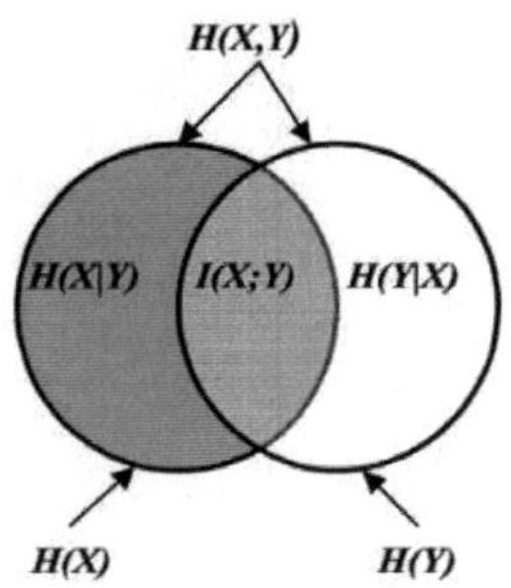

Figure 2. Illustration of mutual information I(X,Y) and its relationship with Shannon information entropy and conditional Shannon information entropy

3. Results and discussion

3.1 Linear coupling

We first analysed possible relationship between the target variable (GPP) and relevant climate drivers (Temperature, precipitation, short wave radiation, and long wave radiation) using linear regression model(Figure 3). The correlation directly informed the strength of linear relationship. We found that GPP was significantly correlated with temperature and shortwave radiation at three out of the four selected sites. FLUX_US-Var site is a exceptional grass land site that GPP was not correlated with any climate variables. Surprisingly, GPP was not correlated with precipitation at any site, given that water availability is one of the most important regulators of vegetation dynamics.

IOP Conf. Series: Earth and Environmental Science **291** (2019) 012016	doi:10.1088/1755-1315/291/1/012016

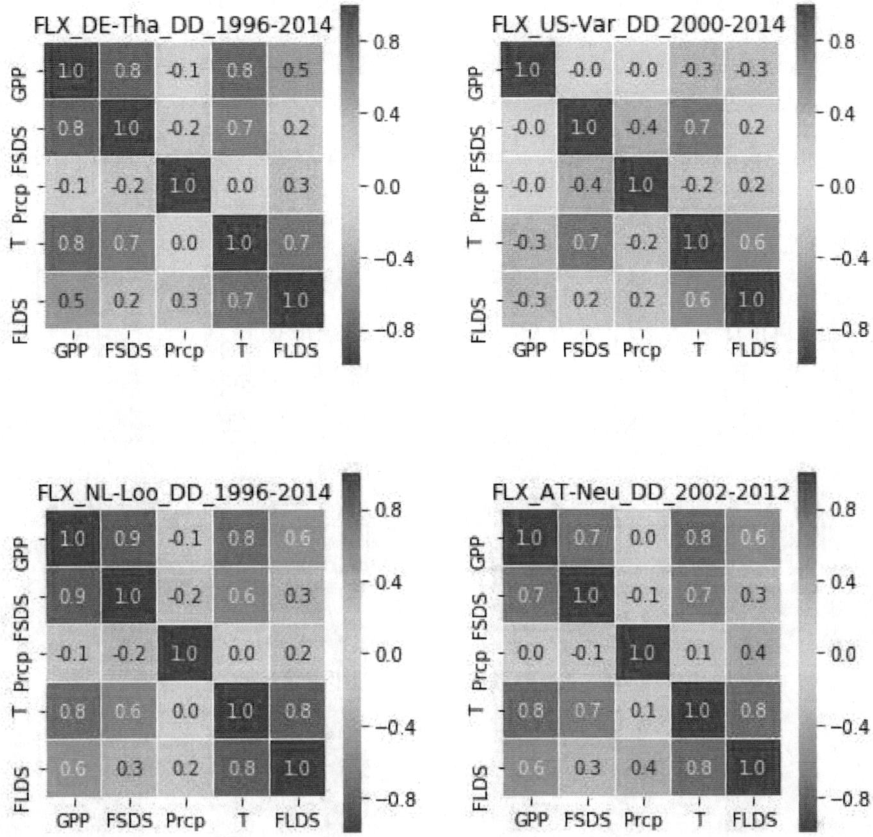

Figure 3. Linear correlation between target variable GPP and climate drivers (T, Prcp, FSDS, FLDS represent air temperature, precipitation, short wave radiation and long wave radiation)

3.2 Non-linear relationship

Given that the vegetation response to climate conditions are highly non-linear, we further analysed the relationships between GPP and climate drivers using the non-linear mutual information metrics, which were depicted in Figure 4. Different color bars represent two probability estimation techniques (see more discussion in section 3.3). In general, high mutual information indicated a strong relationship. We found that precipitation was the major controller of US-Var site GPP, which was not revealed by traditional linear regression model (section 3.1). Overall, short wave radiation and air temperature tended to be important for forest GPP (DE-Tha and NL-Loo), while precipitation tended to be more important at grass site (US-Var and AT-Neu), which is consistent with our theatrical understanding that deep rooting trees are less sensitive to the variation of water availability.

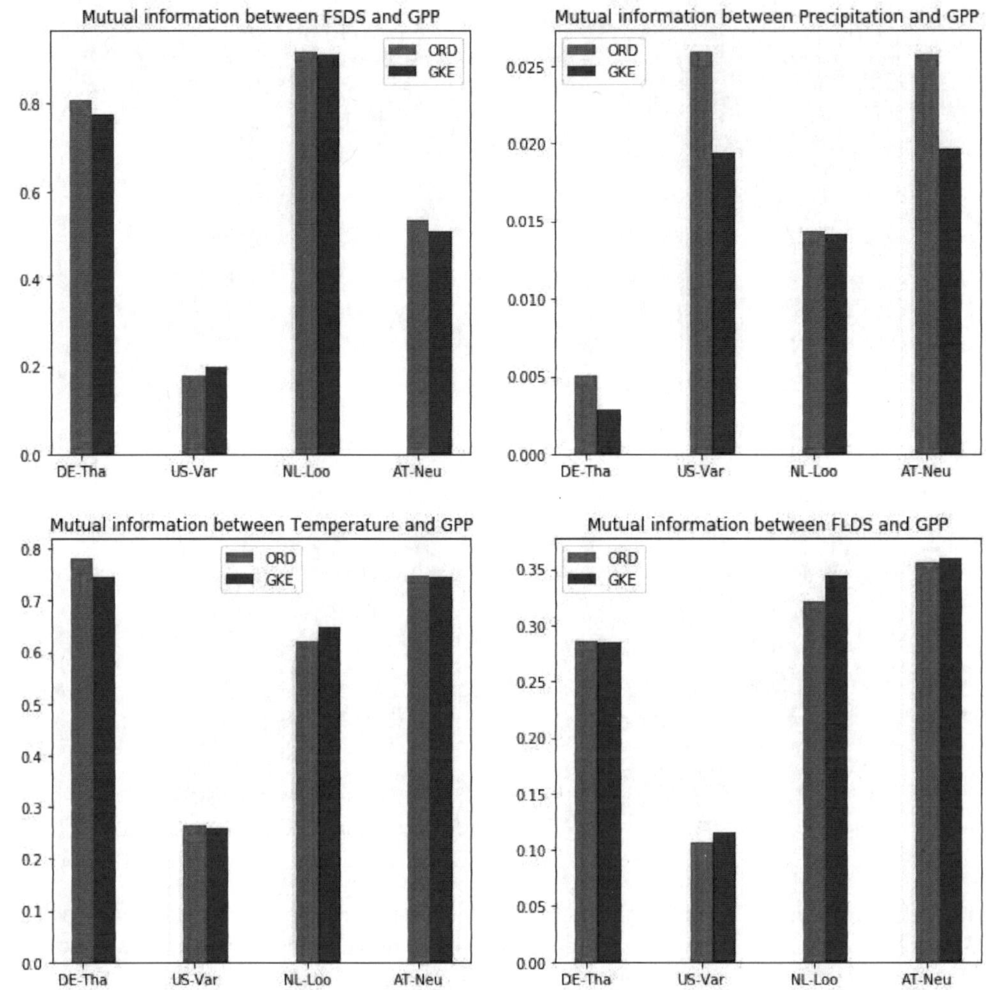

Figure 4. Non-linear relationship between GPP and climate conditions at four sites. Green and blue bars represent different realization of probability estimation.

3.3 Uncertainty analysis

One of the critical uncertainties of mutual information technique is the estimate of Probability Densities Function (PDF). Eqn 1-3 require robust estimation for marginal PDF of X, Y and the joint PDF of X and Y. We first tested classic method of PDF estimation called ordinary ranking (ORD in Figure 4). It divides the variable into equal-distant bins and estimates the density distribution within each bin. Alternatively, we also tested the Gaussian Kernel Estimate (GKE in Figure 4) method that estimate probability density for each bin based on not only the density for the target bin but also considers impacts from adjacent bins given a certain width of influence. Our explorative simulations divided 1D random variable into 8, 9, 10, and 11 bins (Figure 5) and 2D co-varied random variables into 8, 9, 10, and 11 bins (Figure 6 and 7). They consistently showed that GKE method gave a much reliable and consistent estimate of both marginal PDF and joint PDF (Figure 5 red line and Figure 7), while ORD was highly sensitive to the number of bins (Figure 5 blue line and Figure 6). Therefore, the estimate of mutual information between GPP and climate drivers (Figure 4) using GKE (blue bar in Figure 4) should be more reliable.

IOP Conf. Series: Earth and Environmental Science **291** (2019) 012016 doi:10.1088/1755-1315/291/1/012016

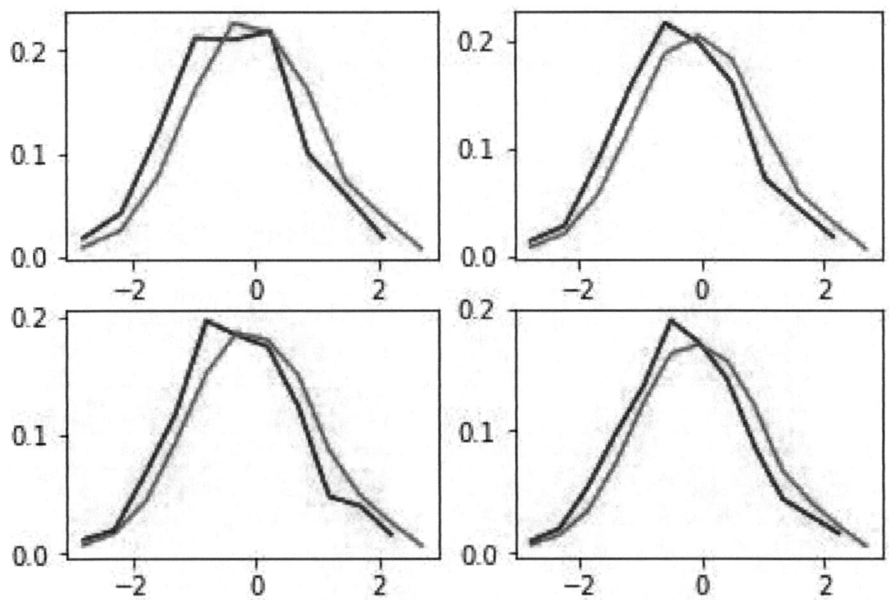

Figure 5. Explorative modelling of PDF with 8 (upper left), 9 (upper right), 10 (bottom left), and 11 (bottom right) bins using ORD method (blue line) and GKE method (red line).

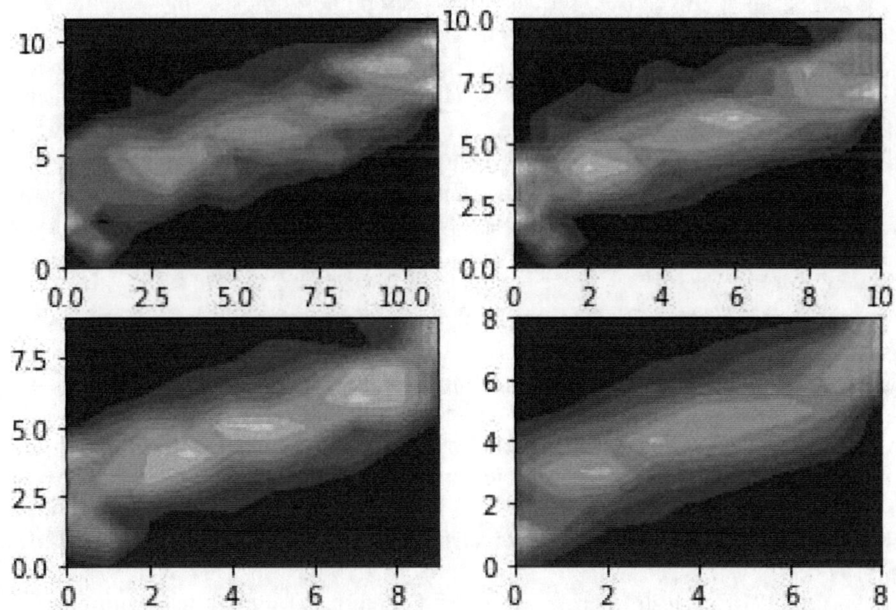

Figure 6. Explorative modelling of joint PDF with 8 (upper left), 9 (upper right), 10 (bottom left), and 11 (bottom right) bins using ORD method

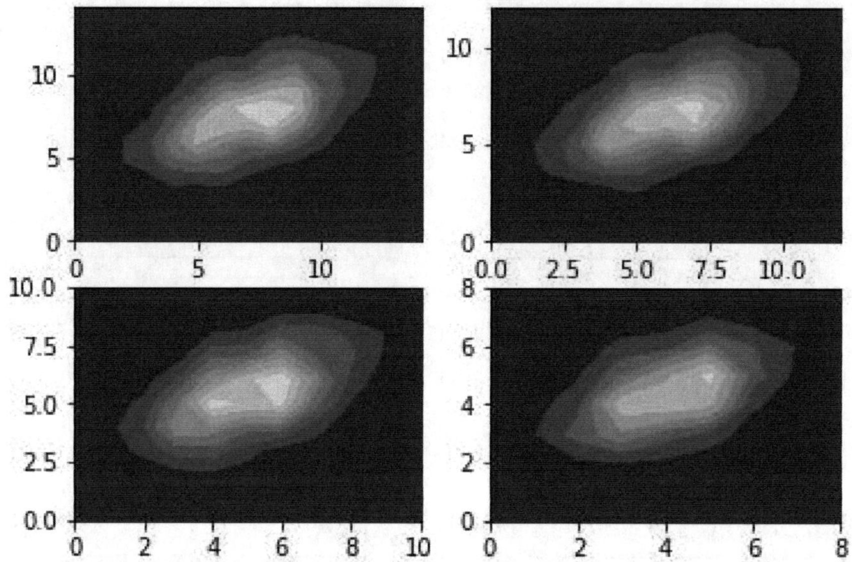

Figure 7. Explorative modelling of joint PDF with 8 (upper left), 9 (upper right), 10 (bottom left), and 11 (bottom right) bins using GKE method

4. Conclusion

Human induced land surface CO_2 emissions and consequently the warming of climate could be partly mitigated by terrestrial plant carbon sequestration via photosynthesis. In this study, we aim to establish possible relationship between plant photosynthesis and climate conditions, thus could help predictive modelling of photosynthesis in the figure. Using mutual information as a proxy of magnitude of relevance between paired variables, we found that temperature and short wave radiation affect photosynthesis significantly at forest sites, while precipitation tended to affected grass land site. We also conducted uncertainty analysis tests using two different density estimate approaches and concluded that GKE is more reliable and consistent than ORD in estimate both marginal and joint PDFs.

5.References

[1] Le Quéré, Corinne, *et al.* 2009 Trends in the sources and sinks of carbon dioxide. *Nature geoscience* **2** 831-36

[2] Feldman, Daniel R, *et al.* 2015 Observational determination of surface radiative forcing by CO2 from 2000 to 2010. *Nature* **519** 339-55

[3] Hansen, J, *et al.*, 2010 Global surface temperature change. *Reviews of Geophysics* **48**

[4] Tkemaladze, *et al.*, 2016 Climate changes and photosynthesis. *Annals of agrarian science* **14** 119-26

[5] Williams, M, *et al.* 2009 Improving land surface models with FLUXNET data. *Biogeosciences* **6** 1341-59

[6] Beer C, *et al.* 2010 Terrestrial gross carbon dioxide uptake: global distribution and covariation with climate. *Science* **329** 834-38

[7] Zhao M, and Running S. W 2010 Drought-induced reduction in global terrestrial net primary production from 2000 through 2009. *Science* **329** 940-43

[8] Shannon, C E1948 A Mathematical Theory of Communication. *The Bell System Technical Journal* **27** 379–423, 623–56

IOP Conf. Series: Earth and Environmental Science **291** (2019) 012017 doi:10.1088/1755-1315/291/1/012017

Taiwan Brownfield Redevelopment & Ecological Restoration Indicator Analysis

S J Feng[1,2], N C Chao[2] and Y J Chen[2]

[1] Department of Architecture, Chaoyang University of Technology, 168, Jifeng E. Rd., Wufeng District, Taichung, 41349 Taiwan
[2] Department and Graduate Institute of Architecture, Chaoyang University of Technology, 168, Jifeng E. Rd., Wufeng District, Taichung, 41349 Taiwan

fengshyhren@yahoo.com.tw

Abstract. Adjustments in global industrial structure accompanied with urbanization, lead in result of derelict sites and lands initiated as manufacturing and production industrial. In which underlies not only creation waste in resource and energy, contaminated soil, damaged ecosystem, urban landscape changed, but also inhibits urban persistent development. Under the extensive aegis of environmental protection, people are highly attentive on brownfield impacts from surrounded life sphere; In Taiwan spatial reutilizing planning approaches are the mainly impetus for brownfield redevelopment programs, having very little or no incorporation on ecological methodology rehabilitation into the program. In this paper, we will elaborate the possibilities of sustainable development for brownfield redevelopment utilizing ecological restoration methodology in 1. preserving the industrial asset, in the meantime restored the ecological urban environment, 2. Initiate a structure framework suitable for brownfield's (industrial sites) ecosystem redevelopment evaluation indicator to maximize the value of brownfield in spatial & urban planning.

1. Introduction

Taiwan, till today, experienced decades of high economic grown and urban development, yet the structural had changed[1]. Back in 1950 ~ 1980, under the fast-industrial growth in Taiwan, lots of industry buildings and sites were implemented lead to continuous land development in the urban with ignoring the importance of environmental protection had cause displacement pressure moving peripheral areas during the urban sprawl process resulting turn downs in the urban metropolitan. Along the ages and industrial structural transformations, with lots of industries had migrate offshore, derelict industrial plants and sites with contaminated soil are derelict or underused. Overviewing on recent researches and literatures on landscape ecological theory, industrial heritage, ecosystem restoration, further discussion and analysis to explore brownfield's ecological restoration (henceforth ER) evaluation hierarchical structural framework inducing important indicators ER under contextualization process.

2. Literature review

2.1. Brownfield's Formation and Definition

Initially the first era of industrialization, In Europe at early 18th century, USA at late 18th century[2], under general urban development trend, industrial building and sites were set urban locations; taking into account the ongoing consumption of lands increase urban sprawl to meet up the fast growth of in-

IOP Conf. Series: Earth and Environmental Science **291** (2019) 012017 doi:10.1088/1755-1315/291/1/012017

dustrialization. Under urban economic and culture change, derelict or underused buildings, lands, fields are the irrevocable consequences from the extensive industrial development of the early stage. Early brownfield problems didn't bring attentions and concerns to the society, with the increase of environmental protection awareness, more and more attentions on restoration and redevelopment brownfield are being discussed. Brownfield management and redevelopment not only provide community financial revenue increase, employment opportunities but also benefits socio-economic and environmental health development creating a win-win outcome for both economic and environmental[3].

Due to offshore migrate industries causing derelict and underused sites exacerbating urban development accompanied further urban structural changes (community unemployment raises) , industrial restructuring(decline in traditional industries) , contaminated environment(toxic chemicals remained in soil). For the past decade, Taiwan were very aggressive in promoting secondary & tertiary industrial sectors which led to major polluted incidents and sites. According to Taiwan's industrial development characteristics and features of brownfields, there is a significant difference in between Taiwan and foreign brownfield's dimension and locations. In foreign, extensive scope of metropolis area likely are regarded as potential brownfield sites[4]; As per Taiwan, brownfield are sole derelict factories and sites that are located in peripheral locations of city development planning zones.

2.2. Regeneration of Brownfield

The existence of brownfield implies problems in shattering urban spatial and function disappearance, waste on urban development resources, increase in crime rates, economic depression, contaminated environment. The aim of brownfield sustainable redevelopment approach is to incorporate urban spatial development restoration and functions to benefit city's public facility services and disaster prevention function, meanwhile fulfill today & future economic and social needs. "European Urban Environmental Brownfield Sustainable Development Plan" describes brownfield sustainable redevelopment approach as a managing and restoring usage of land in order to ensure humanity's current and future demand under acceptable environmental sensitive, economic and social system within its location scope. The benefit of brownfield redevelopment releases pressure of overconsumption of lands and reutilizing lands, stimulates economic revitalization & growth, and improves social environment quality, aid positive effect on ecological environment and humanity health[5]. Consolidation of diverse researches' aspects of brownfield redevelopment features, referred in Figure 1:

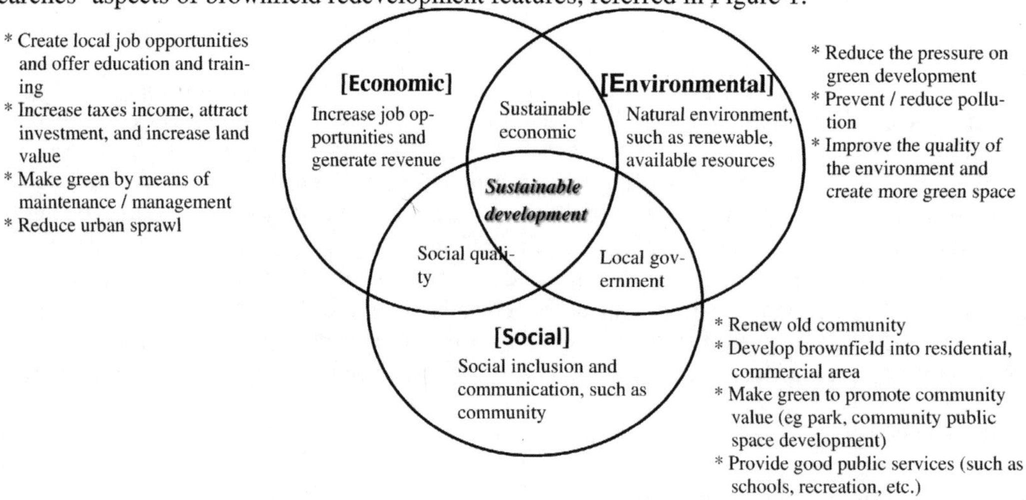

Figure 1. Regeneration of brownfield

2.3. Ecological Rehabilitation

The term "ecological rehabilitation" includes forest, ocean, wetlands and natural environment. Extant ecological rehabilitation structure framework is built up with full-sided, completed, and combined aspects indicators. Creating or restoring a historic environment yet excluding its initiative ecosystem function, the functions/approaches are just artificial structures which does not convey the 'rehabilitation' feature. Thus, the term 'rehabilitation' ought to include human historic heritage and natural environment restoration.

3. Methodology

This study, through literature review, interviews, case evaluation and analysis, constructs the reviewed regeneration evaluation frameworks of brownfield, and applies Fuzzy Delphi Method for induction on reviewed indicators, and afterward analyzes and probes into the relative weights between the indicators, for the reference of the follow-up study on construction of evaluation indicators for brownfield regeneration. Furthermore, the study, through the in-depth interviews with experts in various relevant fields, explores their opinions and advice on brownfields regeneration policy, and collects the first-hand information from their abundance of knowledge to enhance the depth and breadth of this research.

4. Industrial derelict ecological rehabilitation case analysis

4.1. Ruhr Industrial Area, Germany

Table 1. Ruhr Industrial Area, Germany rehabilitation methodology analysis:

Methodology	Description
Securing sewage and drainage of floodwater	Establishing bio-sewage water filtration station and utilizing micro-biochemical to purify raw & dirt waters.
Wind energy utilization	Utilizing wind energy to form a water recycling circulation program such as pumping and discharge purified water through water channel to the higher landscape for dry season irrigation; restore additional waters into overflow retention tanks.
Industrial disposable recycling	Utilizing industrial waste residue to pave roads, city squares, and rebuild clean water.
Plant protection	Organic plants are special selected for a indulging biological succession to remediate the environment brought by the industrialization damage.
Soil remediation	Removing the 2.5 meter depth of contained soil, burn it then refill with new soil.

4.2. Gas Works Park, Seattle

Seattle's Gas work park is an example of "cleaning and greening" approach, introduced by Richard, Haag, implementing the idea of bio-phytoremediation method includes detoxify soil, utilizing natural laws optimizing ecosystem, selected plants, incorporate and reuse existing industrial features. Transform existing industrial remnants from old and obsoleted plant and machinery into aesthetic value. Having "minimized intervene, self-recovery" as basic concept and valuing the historical significance of the site, the designed concept preserved the industrial structure reaching man-nature balance status and restoring its 'historic, esthetic and utilitarian value'. The recovery methodology as per table 2:

Table 2. USA Seattle, Gas Works Park recovery methodology analysis:

Methodology	Description
Soil remediation	Analyzing the contaminant substance and introduce enzyme that will able to breakdown contaminant and increase the natural bio-organism nutrient back to the soil. Utilizing biological and chemical to clean the pollutant area.
Utilizing nature bio-ecosystem remedy system	Introducing a nature way, ecology system, instead of expensive technological techniques to remedy the contaminate soil.
Plant Selection	Installing special plants, nitrogen-fixing plants are fast growing, with strong adap-

	tion and excellent survivability, that help to reduce, recover or remove contamination from the soil.
Industrial remnants reutilization	Having "minimized intervene, self-recovery" as basic concept and valuing the historical significance of the site.
Resource and substance utilization	Respect of historic heritage and landscape and ecosystem development effect by utilizing existing resources and substances when redeveloping the derelict remnants

5. Ecology Restoration methodology and indicator framework establishment

5.1. Ecology Restoration Evaluation

Going through literatures and research analysis on ecological integration, according to "The Society for Ecological Restoration (SER)" definition, including bio-diversity, ecological process and structure, territorial and historic transformation, persisting environmental practice on brownfield's ecosystem rehabilitation can be categorize in to 3 aspect, 'environment greening', 'water vitalization', and 'soil purification'. See table 3

Table 3. Ecosystem Rehabilitation Indicator Assessment

Goal Layer	Layer 2	Description
Ecosystem Rehabilitation Indicator	environment greening	Reburned plants are result from nature plant species transformed, competed and adapted environment. These plants will attract wild animals to habitat. Respecting the natural ecosystem mechanism balance in the field.
	water vitalization	Utilizing plants, animals or chemicals to filtrate and purify sewage. Reuse industrial water channels to form a natural circulation river channel can improve flooding control ability and restore underground water source, providing a habitant environment for the wild animals.
	soil purification	Waste reutilizing with landscape design. Disposable material reuse to reduce impact to the environment. Planting plants that help to get rid or reduce contaminant. Through bio and chemical actions to purify the contamination and revitalize the soil.

5.2. Industrial Derelict Ecosystem Restoration Indicator Establishment

Through above Ecosystem Rehabilitation indicator assessment, generalized from literature and cases analysis on indicator factors and subjected to expert open interview under restrictions, we convergence 12 assessment indicators, describe as follow:

5.2.1. Environmental Greening Indicators

- Green master plan: Utilizing green plants to fully cover the building walls and rooftop.
- Nitrogen-fixing plant: Installing special plants, nitrogen-fixing plants are fast growing, with strong adaption and excellent survivability, that help to reduce, recover or remove contamination from the soil.
- Greening landscape: utilizing nature plants to cover and bound with the soil providing plants and animal growth.
- Biodiversity Land: planting plants that will attract butterflies, and establishing a top-mid-bottom structural (top green, bottom woods, ground planting on layers), considering a complete environmental 'ecological pyramid'. For Example, making sure that the survival spatial on bottom living being organism environment is suitable for toads or centipedes which depends on fungi or moss grown on dead trees. Ensuring the flow of ecological pyramid circulation, provides abundant food basis and assess biodiversity environment.

5.2.2. Water Vitalization Indicators

- Restoring natural river channels: reuse industrial water channels to form a natural circulation river channel can improve flooding control ability, restore underground water source, formation of ecosystem habitation environment and environment landscape, and recreating natural leisure parks.

- Water preservation station: The ability to preserve water from natural or artificial method, the better the preservation station establishment is, helps the activation of microorganism on the soil and nourish plants improve organic qualities, and maintain base natural ecosystem environment balance.
- Sewage recycling circulation plan: establishing a complete sewage and water recycling system to filtrate industrial waste and urban sewage water reaching a recycling and reuse of water.
- Waters biodiversity: cross-interaction of river corridor and trough, beach and river edges forms an ecosystem; river corridor ecology quality reflects riverbed's biodiversity and water quality forming an rich water biodiversity habitant.

5.2.3. Soil Purification Indicators

- Disposable handling: categorizing and handling should be taking extremely care to prevent any disposable leaking sewage that pollute ground soil and underground water and enhance living environment quality.
- Recycling on waste: waste that are not handling properly and buried underground, contaminate soil/land. In order to reduce the waste impact to the environment, waste materials should be reutilized.
- Plant rehabilitation: Utilizing the plants characteristic to absorb contaminant and heavy metal aiming to reduce, remove and restore the contaminations on land/soil.
- Improve soil contamination: using biochemical regenerating and improve polluted soil by adding humus accumulation, increasing microorganisms, soil mixing methodology, chemical stabilization techniques, aciding soil method and soil dressing method.

5.3. Weighting Indicator Analysis

Through expert censuses, participant from industrial, government, and scholars with abundant understanding and knowledge on urban brownfield redevelopment, ecosystem rehabilitation, and industrial derelict & environmental landscape planning, introducing AHP into analysis procedure result data from the weighting each indicator are shown below, table 4. (total of 13 censuses, 11 received completed, 2 invalid)

Table 4. Indicator Level weighting and ranking

Aspect	Indicators	Relative Weight	Relative Rank	Absolute Weight	Absolute Rank
environment greening 0.3980	Green master plan	0.2043	12	0.0813	8
	Nitrogen-fixing plant	0.2535	5	0.1009	4
	Greening landscape	0.2575	4	0.1025	**3**
	Biodiversity Land	0.2847	3	0.1133	**1**
Water Vitalization 0.2386	Restoring natural river channels	0.2364	8	0.0564	10
	Water preservation station	0.2245	10	0.0536	11
	Sewage recycling circulation plan	0.3227	1	0.0770	9
	Waters biodiversity	0.2163	11	0.0516	12
soil purification 0.3634	Disposable handling	0.2394	7	0.0870	6
	Recycling on waste	0.2433	6	0.0884	5
	Plant rehabilitation	0.2311	9	0.0840	7
	Improve soil contamination	0.2862	2	0.1040	**2**

6. Conclusion

6.1. Brownfield Recovering Impacts

For the past decades, industrialization had contribute country growth significantly, set up the first mile-stone on industry economic. Yet, through industrial structure changes, many industries are facing the decision of shutdown and offshore migrate, resulting contaminated brownfield. Considerations should be taken on redevelopment and transformation of brownfields, meanwhile, soil recovery and purification should be following foreign standards under environmental protection; letting brownfield redevelopment program bring good influences to its community development. Considering Taiwan land limitation and throughout literature studies and experts' view, understanding the formation of Taiwan brownfield and its redevelopment impact, not only concentrating on influence environmental quality and increase polluted problem-solving skills, but also reconsidering rigorous plans and developments on restoring the land, enhance economic activities and social impact.

6.2. Ecosystem Indicator Restoration Methodology

- Biodiversity Land: creating soil porosity environment, not influence or interference by humanity, with layers of eco-green environment creating habitant for biological organisms, plants, animals to protect and rich green infrastructure.
- Improve soil contamination: regenerating land and soil with 2 methods: 1. Change exsiting-heavy metal type to stabilize and reduce its mobility. 2. Remove heavy metal by any method. The best method is to dig the contaminate soil then utilize the micro-biochemical to sink edible bacteria for biochemical treatment. Or remove and burned the polluted soil.
- Green master plan: Leaving big space for greenfield, then plant palm trees and shrubs. Density of planting can be increased to fulfill multilayer of greening function.
- Selection of nitrogen-fixing plant: nitrogen-fixing is a nutritious circulation method. Lots of plants, specially leguminous plants, contains nitrogen that nourish and fertilize other organism. Nitrogen also help to adjust critical ecosystem environment. Large nitrogen trees such as: acacia, albizzia julibrissin, senna, golden shower, pigeon pea, king wook, cockspur coral, muntingia calabura. Small nitrogen plants: cassia minosoides, horse bean, peanuts, soy bean, mimosa, lucern.
- Recycling on waste: utilizing industrial waste residue to pave roads, city squares, and rebuild clean water.
- Disposable handling: providing spatial environment landscape to handle disposal and establish recycling categorizing system program.
- Plant rehabilitation: Utilizing the plants characteristic to absorb contaminant and heavy metal aiming to reduce, remove and restore the contaminations on land/soil.
- Green master plan: Utilizing green plants, such as longiniaceae, to fully cover the building walls and set gardening on rooftop to increase greening.
- Sewage recycling circulation plan: establishing a complete sewage and water recycling system separating the sewage and river stream, utilizing micro-bioorganisms to purify sewage will help to reduce contaminant and prevent harmful and adverse effects to ecosystem quality.
- Restoring natural river channels: remove concrete from all the river channel, utilizing natural material for buildings and cooperate with ecosystem construction legislation to recover the initiative natural ecosystem mold.
- Water preservation methodology: adding soil area, greenfield helps to preserve rainwaters and also nourish plants will be the most natural way of environmental protection; increasing water penetration pave, excellent penetration pave functions as soil/land; restoring penetration design, a pool/dam/station that facilities preserving waters from rainfalls, outflow back to soil will help increasing soil density in prevention of flood and mudslide.
- Waters biodiversity: cross-interaction of river corridor and trough, beach and river edges forms an ecosystem; river corridor ecology quality reflects riverbed's biodiversity and water quality forming an rich water biodiversity habitant.

7. References

[1] Shiu J Y, Research of Policies of Industrial Lands and Production Efficiency in Taiwan, 2004, National Chengchi University.
[2] Hsu Y C, A study on the development of brownfield - the new direction of urban industrial zones renewal in urban planning, 2002, National Taiwan University.
[3] Guo P, Analysis of American Brownfields Cleanup and It's Revelation of China, *Environmental Protection Science*, volume 36, no3, Jun 2010, Pages 73-76.
[4] Cheng Y H, A Study of the Industrial Heritage Regeneration as Local Revitalization: An Application on Taiwan Cement Factory in Kaohsiung, 2008, National University of Kaohsiung.
[5] Teng Y C, The Evaluation and Potential Analysis of Turning Brownfield into Urban Habitat—A Case Study of Yungkang City, 2006, National Cheng Kung University.

Key strategies of sustainable and energy-saving development for green universities

Yi-Xin Cai[1,a], Sheng-Jung Ou[2], Hsuan-Hung Chen[3], Chien-Chen Chiu[4]

[1]Graduate Student program in Landscape and Urban design, Department of Landscape and Urban Design, Chaoyang University of Technology, Taiwan
[2]Department of Landscape and Urban Design, Chaoyang University of Technology, Taiwan
[3]Graduate Student program in Theme park management and analysis, Department of Architecture, Chaoyang University of Technology, Taiwan
[4] Adjunct Lecturer of General Education Center, Chaoyang University of Technology, Taiwan

Email:[a]caiyixin840731@gmail.com

Abstract. Sustainable environment development has become a global and popular issue. In Taiwan, regardless of public authorities or private organizations, departments often promote green buildings, green campus and other practices according based upon different sustainable plans. In 2007, the Ministry of Education promoted the green universities and formally launched the Green University Development program. Subsequently, in 2009, 13 schools jointly signed "The Tallories Declaration" to promote the sustainable work of the Green University. Up to 2018, eight universities have entered the ranks of the world's largest green universities. For finding better sustainable development strategies to improve the Green university evaluation and environmental benefits, this study analyzes the current situation of the Green University in Taiwan and presents a trend analysis of the development of Green University assessment indicators in Chaoyang University of Technology (CYUT) over the years. Finally, through the analysis, the study proposes the concrete strategies for achieving green university's requirement, and proposes the key strategies of further achieving sustainable energy saving of green campus for Chaoyang University of Technology.

1. Introduction

As climate change intensifies, the world attaches great importance to the theme of "Sustainable development". The sustainable development and energy conservation issues are also widely discussed in university. The Green University originated in the 1972 *The Stockholm Declaration,* which was the first declaration of the Sustainability of higher education. In 1990 *The Talloires Declaration* was the first to express the high school reputation of the management class, declaring on the need for a sustainable development obligation and an indicator to the green university of the global university [1]. In 1992, *Agenda 21* addressed the world's lack of environmental awareness, it was necessary to address the state of consumption of Earth resources through education. In 1993, *the CRE-Copemicus Charter* thought the university should become the leader of sustainable Society. In 2000, *Earth Charter* applied the sustainable concept into life, combined with formal education and lifelong education to promote the arts, humanities and science in the field of sustainable development of

IOP Conf. Series: Earth and Environmental Science **291** (2019) 012018 doi:10.1088/1755-1315/291/1/012018

education practice. In 2002, *The Ubuntu Declaration* input the key sustainable development into higher education, and asked education resources to lead sustainable development of relevant knowledge, education, and promote the development of a sustainable environment in primary, secondary, tertiary institutions and institutions [2]. In 2002, The Global Green University developed the indicator system, including environmental policy management, architecture equipment energy saving system, environmental education, and daily energy-saving behavior [3]. This shows that the preparation of the Green University must start from campus infrastructure to school environment education, the behavior of campus users, from top to bottom, from inside to outside to implement higher education responsibilities and obligations.

2. Method & data

2.1 Current Situation of Green University promotion in Taiwan

In 2007, Taiwan formally promoted the Green University program. In 2009, 13 universities signed *The Tallories Declaration,* which originated from French in 1990 and signed by presidents and leaders of the ten participating universities during the International Symposium held by the Tuft University in the United States. Since then, *The Talloires Declaration* has become the most valuable document for the university to promote sustainable development [4]. Up to date, Taiwan has 48 universities signed, the number of universities signed ranked after the United States and Brazil. To fulfill the social responsibility and task of higher education and put *The Talloires Declaration* in practice, the Ministry of Education held "the first Green University demonstration school levy activity" in 2009 and selected 13 universities as Taiwan's first green University Demonstration School. At the same year, the Ministry of Education commissioned the National Taiwan Normal University (NTNU) to establish the Green University indicator evaluation system that included environmental system, environmental management, environmental education, and so on. The evaluation of Green campus not only focused on school's energy resources, hard equipment, but also included the measures of the soft system and the common resources of the surrounding community. By doing so, the concept of green can become a regional type of sustainable development extending from campus to the adjacent community.

At present, the green university evaluation system of the world includes six categories: "Setting and Infrastructure", "Energy and Climate Change", "Waste", "Water", "Transportation", and "Education". In comparison with the current evaluative system of Green University in Taiwan, the World Green

University focused more on the basic equipment, resources and other factors that have a lasting impact on the environment. In 2012, NTNU was actively preparing for the work related to the alliance of the Taiwan Green University. The Alliance established in 2013. Up to date, 51 universities have applied to join the Alliance to jointly strive for and promote the sustainable environment of Green University every year [5]. According to the 2017 *UI Green Metric World Universities Ranking*, in the overall rankings, eight universities in Taiwan have been listed within the ranks of hundred [6]. In 2010, only three Taiwan universities were in the list, which showed there were preliminary results in the promotion of green universities in Taiwan. At present, there are 163 education schools in Taiwan, only 51 of them involve the work of green university. It is expected that through the strategic analysis of Green University development, we will have more universities participating in this great mission and achieve the goal of becoming an effective energy-saving and reducing carbon campus.

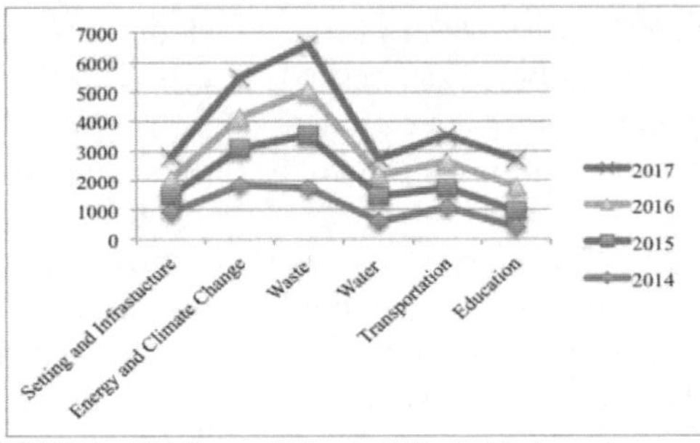

Figure1. The line chart of CYUT in UI Green Metric World University quantity assessment in the past.

2.2 Development status of Green University in Chaoyang University of Technology (CYUT)

CYUT founded in 1994. In 2009, after the signing *The Talloires Declaration*, CYUT started the development of sustainable campus environment in order to move towards a low-carbon campus. From promoting the strategies of "Management System Introducing", "Facilities Improvement", "Sustainable Activities Promotion" and "Sustainable Curriculum Planning", CYUT had paid more efforts on the dimensions of "Environmental Management", "Energy Resources Management", "Safety and Health Management" and "Disaster Prevention Management" to achieve a sustainable campus environment [7]. As shown in Figure 1, from 2014 to 2017, the rank of participating in *UI Green Metric World Universities*, the scores in the worldwide university ratings showed that there is a significant upward trend in the number of projects per year.

CYUT is located in the hillside area. In order to practice ecological campus and Green University, CYUT adopted the basic blueprint of campus to construct hardware facilities, plan campus Landscape Greening and plant a large number of trees so that CYUT achieve the goal of lowering carbon emission. At present, the campus greening area is about 94.43%, in addition to greening and lowering carbon emission, CYUT put a number of sustainable environmental facilities in the campus including:reclaimed water facilities, related energy management systems, rainwater facilities, roof insulation facilities, and solar photovoltaic systems. In addition to hardware facilities, in recent years, CYUT are also committed to environmental education, such as environmental volunteer training, environmental camp, environmental education facilities, and environmental education advocacy. CYUT also integrates curriculum features such as energy-saving and water-saving buildings, water resources, natural environment and disaster prevention. The related actions not only enable students to understand the importance of sustainable campus through environmental advocacy and curriculum communication, but also encourage teachers and students participate in sustainable energy-saving green University together. In order to improve the evaluation result of CYUT in green universities, this study compares the performance of the 2017 *UI Green Metric World Universities ranking* among National Pingtung University of Science & Technology (NPUST), National Chi Nan University (NCNU), and CYUT. From the comparison, CYUT can find out its own shortcomings and think of the future strategic directions towards a better green University and sustainable campus.

Table1. Comparison of 2017 *UI Green Metric World University* quantity assessment among NPUST, NCNU, and CYUT[6]

Dimension	NPUST	NCNU	CYUT
Setting and Infrastructure	**933**	**977**	**757**
Energy and Climate Change	**976**	**1049**	**1349**
Waste	**1101**	**1452**	**1527**
Water	**825**	**515**	**557**
Transportation	**1162**	**1063**	**863**

Education	**1030**	**966**	**954**

Table 1 shows that CYUT has lower scores in the dimension of Setting and Infrastructure, Water, Transportation, and Education in comparison with other two universities. However, CYUT has better performances on the project of energy and climate change, and waste. The differences are resulted from campus' geographical environment, basic facilities, and professional areas. Although the assessment only shows the final score, the practices of Green University are obviously different. The following content will examine the differences of dimension among universities:

- **Setting and Infrastructure**: In this dimension, CYUT obviously has a score lower than other schools. The assessment contains the basic campus environment and infrastructure. The campus size of other two schools is bigger than CYUT, so that the operation of the Green University field is wider in other schools. In addition, the other schools' manpower participating in the green university is also higher than the CYUT. Since it is limited by the status quo, this dimension is less able to get better performances in the future.
- **Energy and Climate Change**: Compared to other schools, CYUT has the highest score in this dimension. Reviewing the literature, we found that NPUST built a large number of solar panels into teaching of agricultural facilities, so that facilities become the key to the development of renewable energy. In addition, because of the advantages of agricultural specialty and the development of biomass, renewable energy is diverse and the power supply of the original resources in NPUST has been rationally utilized. CYUT is expected to make more use of the campus space and professional resources to develop a more diversified energy system in the future.
- **Waste**: Compared to other schools, CYUT also has the highest score in this dimension. CYUT has an effective way to handle waste disposal. In addition to having new facilities, CYUT advocates campus users to maintain environmental policy is also an important key to minimize the risk of pollution. Every CYUT member has a strong intention to get together and protect the green campus environment. In this dimension, CYUT should continue its original actions.
- **Water**: NCNU used the reclaimed water recovery system to create an ecological pool and create a biological habitat. It not only doubles the benefits, but also enriches the campus landscape. NPUST also introduced the conservation of the natural environment into the concept of water recycling, and set up a variety of habitat management systems for groundwater pools, lakes and ponds. In addition, the organic wastewater for agriculture is treated naturally and is used in the campus air-cooling water tower. CYUT also has a system for the recovery of reclaimed water, which will eventually use in toilets. For having a better performance in this dimension, CYUT needs to create more ecological habitats to enhance overall sustainability.
- **Transportation**: Because the geographical environment of the campus is different, the transportation evaluation is obviously different. CYUT is located in the hillside, so that promoting the use of bicycles is highly impossible. Other than that, in the part of public transportation and other related transportation policies are all similar in three universities. CYUT can communicate with city government to increase the bus frequency and encourage teachers and students to take buses by the rewarding strategy.
- **Education**: Participating in the Green University's higher education schools all pay more attention to and promote the development of school software system. In 2016, NPUST obtained a full score in this dimension. Its specialization in the field of the relevant environment and sustainable education courses also received the budgetary support from government. NPUST's good performance revealed that using education as a core concept to carry on sustainable development is worthy of reference. CYUT needs to learn from that

and take systematic actions as soon as possible.

3. Results & discussion

To achieve the goal of sustainable campus development at Green University, the following content will compare the development strategies of the three universities in each dimension and take CYUT as an example to explore the future operating systems to improve the performance in each dimension.

3.1 Analysis on the development strategy of Green University

NPUST: In accordance with the goal of green and sustainable university, the following strategies can be considered including the environmental maintenance vision of the world, the effective use of waste resources, reduce of the non-environmental transport in school access, and education as sustainable development as the core, the implementation of the concept of Green University [8].

NCNU: To achieve the goal of sustainable environment of ecology, energy saving, safety and health, the following strategies such as promotion of *Green University White Paper,* comprehensive promotion of environmental education, maintaining the campus ecology and humane environment, maintaining environmental quality, energy saving and carbon reduction, implementing campus safety and health management, and building green university management system as the framework should be considered [9].

CYUT: To achieve the goal of low carbon campus environment, the strategies such as using environmental management, energy management, safety and sanitary management, and disaster control and prevention management, positively promote the introduction of management system, improvement of facilities, promotion of sustainable activities, and planning the sustainable courses should be considered [7].

The driving factors for making better development of green university for CYUT can be divided into three dimensions, building sustainable environment, introducing management system, and promoting sustainable activities and planning sustainable curriculum. This study also proposed six strategies and six operational actions to correspond with each driving factor (see Table 2).

Table2. Strategic development of Green University for CYUT

Project	Driving the situation	Strategy	Operational Actions
Setting and Infrastructure	Building Sustainable Environment	Building Sustainable Environment	Improving the impermeable pavement and making full use of the topography to create the ecological environment.
Energy and Climate Change	Introducing Management System	Effective use of school internal energy resources	Taking advantage of all resources as equipment for sustainable development and improving energy-saving system rationally and comprehensively.
Waste		Maximizing the pollution reduction	Implementing environmental protection work and enhancing environmental safety.
Water		Diversifying the Use of Water Resources	Strengthening water recycling systems and creating biodiversity habitats.
Transportation		Building Green Transportation System	Increasing intelligent transportation system, reducing non-green transportation access to campus.
Education	Promoting Sustainable Activities and Planning Sustainable Curriculum	Implementing the environmental education system	Integrating sustainable development issues into interdisciplinary curricula and developing more effective environmental education venues.

4. Conclusion

After literature comparison and discussion, this study proposes the development strategies of operating Green University for CYUT in the future. The study also pointed out the weakness of CYUT in the performances of different dimensions. The study result revealed that sustainable development could not only rely on school facilities as the main development tools. In fact, education is the core of sustainable development. Through education, the concept of energy-saving development for green university has the possibility to spread. It is expected that 163 universities in Taiwan can work together in implementing the green university actions in the future. If we continue to use university as the core and expand to the surrounding neighborhoods, the overall green environment area will be foreseen.

5. References

[1] Tsung-Han Tsou, 2016, *Survey on Environmental Literacy of Students in Green Universities: Taking Green Universities in Cental Taiwan fir Example*

[2] Shin-Cheng Yeh, 2003, *Development of a Green University Evaluation Indicator SytemNational Science Council of the Executive YuanMember support project research project*No.NSC 94-2511-s-017-002- not publish

[3] Zih-Hao Shen, 2017, *Research on Taiwan Sustainable Campus: Construction of esources and Energy Recycle Index*

[4] *Asscoiation of University Leaders for a Sustainable Future*, 2015a, *The Talloires Declaration Institutional Signatory List.* Retrieved April 12, 2015

[5] Green University Union of Taiwan, 2018, OUR ACHIEVEMENTS, http://www.guut.org.tw/ach_list.php

[6] UI Green Metric World Univeristy Ranking, 2017, Overall Ranking 2017, http:// greenmetric.ui.ac.id/overall-ranking-2017/

[7] Chaoyang University of Tenchology, 2018, Green University Information Network, http:// 163.17.8.215/cygu/

[8] National Pintung University of Science & Technology, 2018, Explore Green NPUST, http:// green.npust.edu.tw/bin/home.php?Lang=zh-tw

[9] National Chi Nan University, 2018, Green University, https://www.docs.ncnu.edu.tw/green/

An Environmental Improvement Strategy for an Urban Green Farm Garden: A Case Study of Dongguang Green Garden Road, Taichung City

Meng-Li Lin[1] ,Rui-Xing Yang[2,a]

[1] Assistant Professor, Department of Landscape and Urban Design, Chaoyang University of Technology, Taiwan
[2] Graduate Student, Department of Landscape and Urban Design, Chaoyang University of Technology, Taiwan

Email: [a] 610472191@qq.com

Abstract: Amid rapid urban growth, the inhabitants of Taiwan's 'concrete jungles' have tended to neglect the protection of natural resources, and to engage in activities that emit large amounts of carbon dioxide. The urban heat island effect also remains a serious problem, only partially addressed by urban green spaces created or protected via intergovernmental urban planning. Nevertheless, Taiwan's cities still contain many small areas of wasteland – both the sites of demolished buildings, and former green spaces that have been allowed to die off. If transformed into urban gardens and/or farms, this waste ground could meet the public's need for additional green spaces and experiences, and help to gradually improve their quality of life. Plans and mechanisms to that effect have existed since 1990, and Taipei City was the first place in Taiwan where farmers provided land for urban dwellers to cultivate. However, due to insufficient governmental support, the operation and management of such initiatives has been largely ineffectual, and thus their development has not been sustainable. Therefore, this paper uses Dongguang Green Garden Road in Taichung as a case study of this sustainable-development bottleneck, with the wider aim of increasing the quantity and quality of urban agricultural gardens in Taiwan and beyond.

1. Introduction

Considerable empirical and anecdotal evidence indicates that the quality of Taiwan's urban living environments is deteriorating as their population densities increase, and that residents feel a profound need to escape their day-to-day environments and experience natural beauty elsewhere [1]. Amid unbalanced economic development, most human capital in agricultural production has become concentrated in the outer suburbs of big cities, rather than in traditional agricultural villages. Beginning in the late 1970s, Taiwan's orchards began to mature, and those that were opened to the public emerged as tourist attractions where visitors could pick and taste their own fruit as well as simply enjoy the green space . Then, at the turn of the 1980s/1990s, initially through the Taipei City Farmers Association, citizen farm gardens appeared, organically combining the functions of agriculture, leisure, and education on land provided by farmers for city-dwellers to cultivate themselves.As early as 1983, to guide what they already perceived as the healthy development of urban agriculture, Taiwan's agricultural and political authorities formulated a Demonstration Plan for the Development of Tourism Agriculture, which formalized such projects' . In 1994, the agriculture committee of the Executive Yuan followed this up with the Urban Agriculture Pioneer Project, which supported the development of citizen farm gardens as exemplary living experiences and educational centers; specific initiatives that flowed from this project included the Farm Garden Flower Park and Yin Fa National Farm Park.

At present, the main forms of urban agriculture in Taiwan include sightseeing agricultural gardens, leisure farms, citizen farms, holiday flower markets, sightseeing and fishing grounds, agricultural parks, educational agricultural gardens, forest amusement parks, and rooftop agriculture. Such spaces' levels of development, maturity, and government support vary considerably. One of the goals of the low-carbon eco-cities currently being promoted by the governments of Taiwan and other countries is to divert cars away from residential streets and neighborhoods, leaving them clear for pedestrians and cyclists. However, citizen farm gardens are often located in distant suburbs and accessed via motor vehicles, complicating analysis of their overall environmental impact. In tightly configured cities, residents have easier access to employment opportunities and public services, and generally support the development of mass transit systems [2]. This paper is a case study of the Taichung Dongguang Green Garden Road as an embodiment of urban green land and agricultural gardens, with the aim of revising and improving the prevailing general model of such sites and their associated food culture.

2. The present state of urban agricultural garden development in Taiwan

Taiwan's early economic strategy relied on farming as the base for the development of other sectors, especially manufacturing. By the early 1970s, however, the country's agriculture had begun to stagnate. To raise farmers' incomes, Taiwan accelerated the upgrading and transformation of its agricultural sector to include sightseeing and other leisure components; and "experience" is increasingly seen as the main characteristic of the country's urban agriculture [3]. Whether in the suburbs or the urban core, farm gardens plan 3-5 or 20-50 pints of land, and then rent it out for the purpose of planting flowers, grass, vegetables and fruits, not merely to make a profit but so that the tenants can enjoy the experience of cultivation [4].

3. Research scope and instruments

3.1. Description of Dongguang Green Garden Road, Taichung

3.1.1 Location characteristics

The research site is located to the north of Fuxing East Road, south of Asia Street, east of East-West Kwong Yuen Road, and west of West Dongguang Garden Road. As such, it includes Dongguang Green Garden Road, Tai Chi Park, Zaoxi Ma Zu Park, Jianguo Market and other vacant spaces. The annual average temperature in Taichung is 23.3 degrees centigrade, and remains fairly stable across the four seasons; however, lower temperatures from December to February make those months the most suitable for planting various kinds of crops. Taichung's rainfall is lower than the Taiwanese average, at just 1773 mm in total, and it has more hours of annual sunshine, both of which factors are beneficial to the growth of crops.

3.2. Baseline status

3.2.1 Dongguang Green Garden Road

Dongguang Green Garden Road, built of red brick and 3.4 km long, is the coolest of the cycleways that ring Taichung City, being heavily shaded by blackboard trees along its entire length. In addition to its cycling facilities, it features pavilions and benches. Near this road's junction with Dongfeng Park there is a me-BIKE bicycle rental point; and an additional such point can be found in the Dazhi Primary School. These have a high rental capacity, and their usage rates are highest in the evening. Rented-bike utilization rates are lowest at the junction of Yueye. In addition to Dongguang Green Garden Road itself, the immediate area contains five green spaces: Lacheng Park, Zusheng Park, Leye

Pet Park, Tai Chi Park, and Dongfeng Park. However, the somewhat wider area represented by the local blue-belt system also includes East Hanxi and the West Taiwan Sugar Ecological Park.

3.3. Theoretical context and research problems
Traditional urban planning has been vehicle-oriented, and geared especially toward the expansion of cities via residential suburbanization and the avoidance of serious traffic problems[5]. Inevitably, recent national, international and grassroots calls for the development of low-carbon cities are not fully compatible with this planning tradition as it stands. Therefore, mass-transit-oriented development (MTOD) and people-oriented transportation-development planning have emerged as basic building blocks of low-carbon urban transportation planning [6]. The combination of MTOD and land-use planning is closely related to Urban planning[7] .As such, in low-carbon urban planning, the emphasis is on green, people-oriented transportation planning, coupled with mass-transit systems and mixed land use, enabling residents to operate as pedestrians, cyclists, and mass-transit users, and thus reducing the use of private motor vehicles [8].

The close proximity of Dongguang Green Garden Road to a number of parks, as noted above, means that the research site's green-space resources are very rich. It is also close to primary and secondary schools. Its shortcomings are threefold. First, it can only be used by bikes, and its housing is mostly already occupied; second, its resources are too concentrated in the west; and third, the low utilization rate of its green space means that it cannot reach the standard of urban green space.

4. Improvement Strategy for the Dongguang Green Garden Road Urban Agricultural Garden

4.1. Spatial planning (see Fig. 1)

- Use rich green-space resources to expand the proportion of edible landscape.
- Develop abandoned land into agricultural gardens, and maintain the base inside and outside the series.
- To build a low-carbon community, cultivate the interest of the city's people in learning about agriculture.
- Reduce the scale of development and fill; maintain the waterway system and intertidal zone inside and outside the area; and construct a community that accords with the green environment index.

4.2. Architectural design (see Fig. 2)

- Reduce urban sprawl through mixed land-use zoning, land-use concentration and efficient management strategy, and thus reduce unnecessary traffic and energy consumption.
- Full use of rainwater recovery, along with wastewater storage and utilization devices.
- Closure of the entire road to all motor vehicles to achieve environmental benefits including carbon reduction.

4.3. Community organization (see Fig. 3)

- The agricultural gardens around the study area should be jointly maintained by the community's residents, with the wider aim of achieving community self-sufficiency.
- The wetland – which has the functions of storing and purifying water, protecting species, and creating a diverse biological habitat – will also be managed and utilized sustainably at the community level.

IOP Conf. Series: Earth and Environmental Science **291** (2019) 012019 doi:10.1088/1755-1315/291/1/012019

Figure 1. Artist's impression of citizen farm garden

Figure 2. Artist's impression of I-BIKE ecological cycleway

Figure 3. Artist's impression of semi-open active rest space

Table 1. Sizes of green areas before and after the creation of urban farms

Total area (m^2)	72,967	
Project	Before improvement	After improvement
Trees (m^2)	14,780	29,514
Shrubs (m^2)	9,750	13,580
Ground cover (m^2)	10,755	20,124
Total greenery (percentage)	47.5	86.6

4.4. Sizes of green areas before and after the creation of urban farms (see Table 1)
Before the improvement, trees accounted for 20.2% of the research site's total area; shrubs, an additional 13.3%; and ground plants, another 14%. After the improvement, the tree cover doubled, to 40.4% of the total area; the shrubs made up 18.6%; and the ground cover (including low-lying plants such as grasses and vegetables planted in agricultural gardens), 27.6%. Such an increase will enable residents to directly obtain the vegetables, fruits and other crops needed in their day-to-day lives, reducing the carbon footprint of vegetable transport while adding to carbon sequestration, achieving carbon reduction, and going some way toward reversing the heat-island effect.

5. Conclusion and Recommendations

5.1. Conclusion
Planting, in combination with the creation of new recreation areas, can potentially re-connect urban Taiwanese people with the fun and community spirit of the countryside, while improving both their individual health and the sustainable development of the country and its cities. Amid increasing popular concern about environmental issues, the creation of an urban green garden along the Dongguang Green Garden Road in Taichung will improve the environment, increase the proportion of green cover, and effectively utilize sewage, waste water, and other resources that would otherwise be lost. In addition to making more efficient use of the area's environment, and protecting and increasing urban biodiversity, this initiative will enhance popular participation and, ideally, lead to autonomous management of the community. Prior research has found that the establishment of a sustainable urban agricultural garden can not only enhance the environment in the immediate vicinity, but also improve the quality of life in surrounding areas, and drive up the value of nearby land [9]. Through urban planning, original green space can be improved and connected, rendering the whole city more environmentally sustainable and less prone to the urban heat island effect and other urban climate issues, thus improving not only the material quality of life, but also the residents' emotional states.

5.2 Recommendations

Innovative thinking and plural urban-planning strategies should be used to explore the space of green systems and how sustainable design techniques could improve them further: for instance, through low-impact green architecture and construction.

6. References

[1] Gu, Dawei (2004). *Discussion on the Policy Framework for the Development of Ecological Cities in Taiwan.*
[2] Yan, Hui-qi, 2014, *The Development and Demonstration of Urban Agriculture in Taiwan.*
[3] Peng, Wu-xiang & Peng, Wu-male (year). *Study of the exhibition of the Guilin City of the People's Republic of China in 2001 and Jiang.*
[4] Lin, Yann-Jou & Shen, Li (2001). *A study on the Demand of Citizens' Farm Gardens.*
[5] Gene, W. G. & Frederick, J. D. (1978). *Urban Forestry.* New York: ASME.
[6] Department of Construction, Ministry of the Interior (2010). *Operational Manual for Eco-urban Urban Design.*
[7] Tuma, R. (2001). *Money: The New Face of Ecology.*
[8] Grahn, P. & Stigsdotter, U. (2003). *Landscape Planning and Stress.*
[9] Haaren, C. & VonReich, M. (2006). The German way to greenways and habitat networks. *Landscape and Urban Planning*, 76, pp. 7-22.

Application of Ecovillage for the Redevelopment Strategy of Rural Communities in Taiwan. A Case Study of Jianguo Military Dependents' Villages

Jhong-Ping Xie and Kuo-Wei Hsu

Chaoyang University of Technology, Taiwan

xiejhongping@gmail.com

Abstract. Agriculture in Taiwan is facing challenges of issues related to industrial pollution, ground subsidence, agricultural safety as well as rural regeneration. Yunlin County has among the highest output of agricultural products in Taiwan. To achieve "Taiwan agricultural technology corridor" policy, Yunlin County promoted sustainable development model for creative water and energy saving agriculture and low-carbon environment by hosted the 2013 "Agricultural Expo." Jianguo military dependents' village, located nearby the Agricultural Expo Ecological Park, is the representative agricultural military dependents' village in Taiwan which is evaluated to be redeveloped. Previous studies approved sustainable development was benefited with residents. Relative literatures indicated that imported brought positive effects for local redevelopment. This study, therefore, reviews the relevant literature on eco-village, and proposes a redevelopment strategies framework for eco-village based on that review and a two-stage questionnaire administered to experts. The first stage utilized the Fuzzy Delphi Method, which focuses on impact factors, and the second, the Analytical Network Process Method, deals with performance factors. The results indicate that the key impacts on the ecovillage redevelopment strategies for Jianguo military dependents' village were (2.1) water and energy efficiency, (2.2) community symbiosis, and (2.3) cultural preservation. Water recycle, energy saving, and agricultural technology are the most important criteria for eco-village redevelopment strategies of Jianguo military dependents' village.

1. Introduction

Agriculture in Taiwan is facing challenges of issues related to industrial pollution, ground subsidence, agricultural safety as well as rural regeneration. Yunlin County has among the highest output of agricultural products in Taiwan. On the other hand, Yunlin County suffers from trend of youth population outflow and aging in rural area seriously.

The government in Taiwan announced to build a "golden agricultural corridor" in areas suffering from land subsidence in central Taiwan to solve the problem of land subsidence in the region. The policy aims to make farming in this area more water-conservative and energy-saving to deal with the over-pumping of underground water in the region. Yunlin County promoted sustainable development model for creative water and energy saving agriculture and low-carbon environment by hosted the 2013 "Agricultural Expo" which remoted the environmental quality and agricultural development in rural areas. The major goals of these actions are to satisfy residents' living demand simultaneously on economic and environmental improvement. However, challenges remain, including the question of

how to promote systematic endeavours, both in policies or research on aging, and how to encourage greater involvement of nongovernment organizations in the rural revitalization issue [1].
According to relative researches, the concept of ecovillage provides rural community redevelopment strategies to integrate ecology, social, culture and economy aspects. In order to investigate success factors behind rural revitalization, this study explores a project in Jianguo military dependents' village [2]. located nearby the Agricultural Expo Ecological Park, is the representative agricultural military dependents' village in Taiwan which is evaluated to be redeveloped as a demonstration of ecovillage.

2. Literature Review

Robert and Dianne Gilman (1991) defined ecovillage: an intentional community, which is human-scaled, full-featured, harmlessly integrated with nature, supports healthy human development and is sustainable [7]. Previous researches indicated the rural ecological environment protection, economic sustainable development and eco-village construction need to be explored. Top -down approaches of rural community redevelopment typically focus on government policy and legislation for public investments. Bottom-up approaches generally focus on social movements and civil society directing their actions toward sustainable development. Ecovillages are consciously designed through locally owned, participatory processes in ecologic, social, cultural, and economic dimensions of sustainability to regenerate their social and natural environments [3][4].

2.1. Ecology

Ecovillages allow people to experience their spiritual connection to the living earth. They provide for their daily needs-food, shelter-while respecting the cycles of nature, which include water recycle, energy saving, and new agricultural technology.

2.2. Society

Ecovillages are communities in which people feel supported by and responsible to those around them. Ecovillages provide a deep sense of belonging to a group which include participants' self-identity, social value, community participation, and local organization operations.

2.3. Culture

Ecovillages have a particular significance for the prospect of cultural transformation on a larger scale offering concrete demonstrations of both a different way of life and the lived embodiment of a holistic worldview which include preservation of cultural heritage, cultural values, and promotion of cultural activities.

2.4. Economy

Ecovillages demonstrate a unique economic dimension to generate income not only through products and services but also improvement of every participant in the communities which include local brand marketing, satisfaction of tourists' requirements, local industrial development, and environmental improvement.

3. Research Methods

This study aims to contribute to local-redevelopment policies and strategies by arriving at a clearer picture of ecovillage redevelopment strategies, with particular reference to local government and non-governmental organizations in the Jianguo Military Dependents' Villages. Based on the foregoing review of the relevant literature on ecovillage and rural redevelopment strategies, its research approach was divided into two main phases. First, using the Fuzzy Delphi Method (FDM), experts were invited to answer questionnaires on ecovillage redevelopment strategies indicators; and second, having summarized the indicators arrived at in the first phase, it adopted the Analytical Network Process Method (ANPM) to base a local-redevelopment strategy around Jianguo Military Dependents' Villages [5]. The ANPM was created to provide interdependency and feedback in response to a broad range of decision-making problems in the real world. In contrast to the analytic hierarchy process (AHP), it has a decentralized pattern similar to that of the Internet, and its goal is to foster optimal decisions via internal interdependence among all goals, criteria, and plans through the use of ratio scales.

This study, through a review of rural revitalization literature related to ecovillage development, explores the framework of the interview issues, which is mainly conducted via in-depth interview, in order to understand the current situation of the planning strategies. This study used in-depth interviews to collect data that cannot be obtained from secondary sources. While purposive qualitative research samples are not representative of a whole group, they are nevertheless the best source of abundant, in-depth, first-hand information about the real-world experiences of the members of a pluralistic society. Semi-structured interviews can also be applied to obtain detailed answers to similar questions. In this method, the interviewees respond using standardized words and in a standard order, facilitating rapid and straightforward organization, comparison, and analysis of all the respondents' answers. Based on a review of the relevant literature, a preliminary interview framework and context was developed, and interviews were conducted. Based on the results of our pilot interviews, the structure and content of the interview protocol were revised slightly.

4. Research Object: From Abandoned Military Dependents' Village to Ecovillage Demonstration

Military Dependents' Village is a unique cultural landscape in Taiwan. The original purpose of these buildings was to serve as provisional housing for soldiers and their dependents after the Government from Mainland China retreated to Taiwan in 1949. Some of the military buildings suitable for living in the Japanese troops' war readiness ruins in Taiwan were converted into military houses. These settlements were designed to integrate the family experience with the military experience for the soldiers. A unique subculture was created with the cultures from various Chinese homeland [6].

Jianguo Military Dependents' Villages in Huwei, Yunlin County were built in 1943 under Japanese colonization. After World War II, military family-built extensions into the original buildings and organically transformed to liveable space. Jianguo Military Dependents' Villages were related to the Japanese troops' war readiness relics in Taiwan were preserved in the name of "military housing community culture preservation". The Japanese colonial war-remnant facilities intended for residential use by the military housing communities had the dual identity of "relics" and "military housing community". Jianguo Military Dependents' Villages were scattered villages and village I and village II were preserved to be cultural heritages, since village III transformed as an elementary school and village IV are now a prison (see the figure 1.2.3.4.). The condition of unique culture and rural agricultural environment provide the opportunity for Jianguo Military Dependents' Villages to redevelopment as a ecovillage demonstration in Taiwan.

Figure 1. status of Jianguo Military Dependents' Villages I

Figure 2. status of Jianguo Military Dependents' Villages II

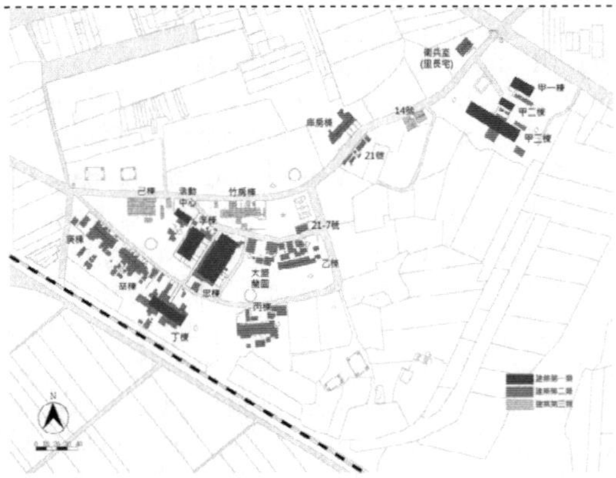

Figure 3. Configuration diagram of building level classification of Jianguo Military Dependents' Villages I

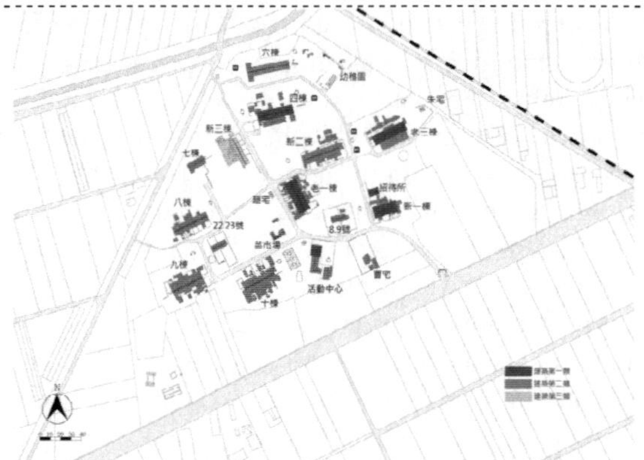

Figure 4. Configuration diagram of building level classification of Jianguo Military Dependents' Villages II

The research objective is to conduct a research on changes in the space and use pattern of "ruins of military housing villages" in order to explore the evolution of how residents in Taiwan used different types of Japanese colonial war-remained facilities and converted them into living spaces during the military housing community period. For redeveloping with ecovillage strategies, the research scope was extended to the correlation of the evolution of the space of the ruins of military housing communities, the 39 retained buildings, and agricultural environment [8].

5. Conclusion and Discussion

5.1. Analysis of the FDM Questionnaire Results

The FDM is a stepwise problem-solving procedure. This study constructed a hierarchical structure of 3 dimensions and 9 impact factors of ecovillage redevelopment. 12 questionnaires of experts were analysed to verify the impact factors of ecovillage redevelopment. All of this study's FDM items were scored from 0 to 10; the higher its score, the more important an influence factor is, according to the expert's individual professional experience. A threshold value of 6 was set as the pass score, and the remaining factors below the threshold of 6 were deleted. The results are shown in Table 1.

Table 1. The impact factors of ecovillage redevelopment

Objective	Dimension	Impact Factor	Expert Consensus Value (Gi)
Ecovillage Redevelopment	Water and Energy Efficiency	Water recycle	8.9167
		Energy saving	9.0833
		Agricultural technology	8.5000
	Community Symbiosis	Resident identification	8.1667
		Community participation	8.1667
		Local organization operations	7.3333
	Cultural Preservation	Preservation of cultural heritage	8.0833
		Presentation of cultural values	7.3333
		Promotion of cultural activities	7.0833

5.2. Analysis of the ANPM Questionnaire Results

This study proposes a taxonomy model that applies ANPM to a specially designed questionnaire. In this phase of the study, a total of 9 expert questionnaires were collected. After passing the consistency check, all were deemed valid. Finally, we present the weights for ecovillage redevelopment. The multi-criterion decision software. The geometric averaging method was used to arrive at the experts' relative importance of each indicator. The ranked weights of Water and Energy Efficiency, community symbiosis, cultural inheritance, and regional revitalization are shown in Table 2. Water and energy efficiency (weight 0.4107) was again found to be the most important, followed by community symbiosis (weight 0.3436), and cultural preservation (weight 0.2457).

Table 2. The dimensions ranked by weights

Dimension	Weight	Ranking
Water and Energy Efficiency	0.4107	1
Community Symbiosis	0.3436	2
Cultural Preservation	0.2457	3

5.3 Discussion

This study sought to provide insights of redevelopment of Jianguo military dependents' villages via the Fuzzy Delphi Method and indicated economy dimension are relatively unimportant in the planning stage since public investment fund provided reconstruction costs. Thus, the major issue is how to integrate ecology, community, and culture and establish a platform for the interactions between dimensions of ecovillage redevelopment for achieving sustainable community development of military dependents' villages in Taiwan.

6.References

[1] Bossel, H. 1999. Indicators for Sustainable Development: Theory, Method, Applications. Canada: International Institute for Sustainable Development.

[2] Huang, Yu-Hsiu 2012. Establishing the evaluative system of developing eco-village in rural communities in Taiwan. National Chung Hsing University.

[3] Hamstead, M. P. & Quinn, M. S. 2005. Sustainable Community Development and Ecological Economics: Theoretical Convergence and Practical Implications. Local Environment, 10(2), 141-158.

[4] Karen Litfin 2009. Reinventing the future: The global ecovillage movement as a holistic knowledge community, Environmental Governance, 124-142.

[5] Peat, K. 2007. The Ideal of a Sustainable Community, In. Clay, C. J., Madden, M. & Potts, L. (eds). Towards Understanding Community: People and Place, 97-110, London: Palgrave Macmillan.

[6] Peng-Cheng Ho 2014. A Research of the Transformation of Living Space on Japanese Colonial War-remained Facilities. Chung Yuan Christian University, Taiwan.

[7] Robert and Diane Gilman 1991. Ecovillages and Sustainable Communities: A Report for Gaia Trust.

[8] Tim Strasser 2013. Cultures of (Un)Sustainability: Ecovillages as Seedbeds for a Cultural Transition. Maastricht University Journal of Sustainability Studies, Vol.1.

IOP Conf. Series: Earth and Environmental Science **291** (2019) 012021 doi:10.1088/1755-1315/291/1/012021

A Study on Application of Ecological Engineering Methods to a River Pollution Remediation– Case study of Liuchuan River

Che-Yu Hsu [1], Sheng-Jung Ou [2]

[1.] Assistant professor, Department of Landscape and Urban Design, Chaoyang University of Technology, 168, Jifeng E. Rd., Wufeng District, Taichung, 41349 Taiwan, R.O.C.,
[2.] Professor, Department of Landscape and Urban Design, Chaoyang University of Technology, 168, Jifeng E. Rd., Wufeng District, Taichung, 41349 Taiwan, R.O.C.

Email:cyhsu428@cyut.edu.tw ;sjou@nchu.edu.tw

Abstract. The Liuchuan River is an urban river in Taichung City of Taiwan. It was been affected by the pollution of domestic sewage and commercial wastewater for many years, resulting in the degeneration of water quality in rivers. Because of the hard bank revetment along the coast, the urban blue ribbon are unable to interact with residents. In order to improve water quality pollution and waterway environment, the Taichung Water Resources Bureau prioritizes water quality improvement. Under the goal of ensuring flood control and drainage, it actively diversifies the waterway space to enhance the city's local culture and life value. The purpose of this study is to analyze the improvement of water quality through the discussion of low-impact development engineering remediation projects. The water permeability of the riverbank before repair is almost 0%. After the improvement, the water permeability is increased by 50%, the green area is increased by 3,322 square meters, and the water quality RPI index is reduced from 7.25 "serious pollution" to 5. "moderately pollution". This result shows that the Liuchuan River remediation project has successfully improved the water quality and river channel. Re-injecting green elements to create a hydrophilic environment for urban blue-green belts, improving the quality of residents' living environment, and promoting the benefits of surrounding businesses and tourism.

1. Introduction

The problems of urban river pollution and ecological damage are becoming more and more critical. According to the statistics, by the early 20th century, there is not almost a complete natural river in the world [1]. Therefore, it is urgent to develop a cost effective technique to manage the river water quality. Low impact development (LID) is a land development strategy for managing storm water at the source with decentralized micro-scale control measures. Since the emergence of LID practices, they were been used successfully to manage storm water runoff, improve water quality, and protect the environment. [2]. Liuchuan River, is an urban river with functions such as flood discharge and sewage discharge channels. In the past, due to the impact of the surrounding population and economic growth, the channel water quality has deteriorated. It has only gradually improved in recent years. In 2012, the Environmental Protection Agency approved the Liuchuan River pollution remediation and environmental improvement plan, and built a pebbly contact oxidation facility and a low-impact development method in Liuchuan River. The four-year construction period has officially launched in

2016 and received great repercussions. After the renovation project in Taichung, Liuchuan River became a well-known attraction. The Goal of the research is to explore the pollution remediation and environmental improvement in Liuchuan River and the methods of its use. The results show that water quality improvement is the core, and the urban blue belt space creates a hydrophilic environment to improve the living environment quality of residents. This historical water mill has a different look. Liuchuan River passes through the city and has been polluted by domestic wastewater in the past, becoming an urban drainage ditch. In 2014, the Taichung City Government carried out the "Liuchuan river Pollution Remediation and Environmental Improvement Project" to build a river that can be hydrophilic. The improvement of water quality has become the primary task of remediation. The city uses parking space, set up a sewage treatment plant, and filters through the cobble layer to provide a clean water source. With many polluted urban river affecting the environment of the cities, improvement work have become necessary. Therefore, Literature reviews, method development, water quality and specification analysis were conducted to further understand the core of the project.

2. Study area

Liuchuan River is an urban river with functions such as flood discharge and sewage discharge channels. In the past, due to the impact of the surrounding population and economic growth, the channel water quality has deteriorated. It has only gradually improved in recent years.

3. Strategy analysis of river renovation engineering

Analysis of river ecological engineering rectification project. Ecological engineering was defined as the practice of joining the economy of society to the environment symbiotically by fitting technological design with ecological self-design. [3]. This ecological engineering introduce the low-impact development method. The concept of low-impact development (LID) integrates environmental concerns with land development, focusing on water and pollutant balances. LID represents a fundamental change in the way residential, commercial, and institutional properties are developed and minimizes negative impacts on the environment and local ecology.[4]. This method is mainly by widening the river to 8 meters on both sides, changing to the gentle slope type, using the near-natural method to increase the permeable area, reducing the ecological impact, and reducing the disturbance of the land. It hopes to transform the waterfront environment of Liuchan River by strengthening sewage interception, water quality treatment, and waterfront environment construction. The project has turned the canal into a popular recreational area.

3.1 Permeable pavement

Permeable pavements have commonly used around commercial and industrial buildings to reduce the environmental impact. Permeable pavement is a specific type of pavement with a high porosity that allows rainwater to pass through it into the ground below.

3.2 Infiltration trench

A percolation trench, also called an infiltration trench, is a type of best management practice (BMP) that have used to manage storm water runoff, prevent flooding and downstream erosion, and improve water quality in an adjacent river, stream, lake or bay. [5].

3.3 Rain garden

A rain garden is a shallow planted depression designed to hold water until it soaks into the soil. A key feature of eco-friendly landscape design, rain gardens—also known as bio-infiltration basins—are gaining credibility and converts as an important solution to storm water runoff and pollution. Rain gardens were been recommended as a best management practice to treat storm water runoff. Rainwater garden, which is an ecological facility used for storm and flood control and utilization, is widely used for retention and purification of storm and flood runoff from buildings, parking lots and roads in public spaces, urban residential areas, commercial areas and industrial areas.

3.4 Dry well / seepage pit

A dry well, or seepage pit, is a small underground pit filled with stone that collects storm water from roof gutters and allows it to absorb into the surrounding soil. Underground piping can connect the dry well to the roof downspout. A filtration system have provided for a dry well in the form of a buffer tank made of concrete to remove sediment and pollutants from water supplied from a header to an inlet pipe in the tank, so that the water exiting from an outlet pipe is free of particulate matter and pollutants [6].

3.5 Vegetative filter strip

Vegetative filter strips have only intended to provide runoff treatment and have therefore situated between the pavement surface and a water collection system, pond, wetland, or river. Vegetated filter strips (VFS) are used recently for removal, at or near the source, of sediment and sediment-bound chemicals from cropland runoff. Vegetation within the flowpath increases water infiltration and decreases water turbulence, thus enhancing pollutant removal by sedimentation within filter media and infiltration through the filter surface.

3.6 Vegetated swale

A vegetated swale is a gently sloping landscaped depression that collects and conveys storm water runoff. Vegetated swales are an accepted and commonly implemented sustainable urban drainage system in the built urban environment. Event mean concentrations of suspended and bed load sediment have been calculated using current best analytical practice, providing single runoff event specific sediment conveyance volumes through the swale.

4. Data resource

We *collect the data on Water Quality to discover the trend of the river pollution over these years*

This research collects relevant monitoring data information on water quality monthly from the Yuan EPA National Environmental Quality Monitoring Information Network providing information which it has collected in a total five year from 2012 to 2016, in which the information was obtained from Taichung County Water Pollution Control Inspection and basin management plan. Every monthly water quality monitoring data builds relevant data, and builds a water quality database [7]

5. Result & discussion

The research is to adopt Taichung City's metropolitan river (Liuchuan) pollution remediation and Environmental Improvement Demonstration Project, and to evaluate the Project.

5.1. The low-impact development method

The low-impact development method includes Permeable Paving, Infiltration Trench, Rain Garden, Dry Well / Seepage Pit, Vegetated Filter Strip, and Vegetative Swale, etc. Through the filtration, adsorption and microbial decomposition of the soil to achieve the effect of water purification, according to the application area. The pollution reduction amount have shown in Table 1. It was expected that the daily pollution of suspended solids, total phosphorus and nitrates will be 20.6, 0.085 and 0.04 kg respectively. It was expected that the suspended solids, total phosphorus and nitrates will reduce the pollution rate. Each is 69%, 67% and 24% (see table 1).

Table 1. Low-impact development method to reduce pollution benefit analysis.

Structural project	Area (m^2)	Facility infiltration volume(m^3)	Collection volume (m^3)	sum of SS decreased (kg)	Phosphorus Decreased (kg)	Nitrate decreased (kg)
Permeable Paving	626	11.268	41.316	3.020	0.013	0.006

Infiltration Trench	224	0.027	53.787	3.932	0.016	0.008
Rain Garden	250	3.6	165.600	12.105	0.051	0.023
Dry Well	0.32	0.00576	0.326	0.024	0.000	0.000
Vegetative Swale	201	3.96	3.960	0.001	0.289	0.001
Vegetated Filter Strip	2,647	47.646	47.646	1.229	0.003	0.002
total	3,948.32	66.507	312.635	20.600	0.085	0.040
Pollution Reduction rate	-	-	90.23%	69.13%	67.75%	24.32%

5.2. Water quality improvement benefits
To improve the water quality of the Liuchan River. Clean-up efforts will be devoted to three major problems: nutrient over enrichment, toxic substances, and the decline of submerged aquatic vegetation. In order to improve the efficiency of water quality, we compared the four pollution indicators such as water quality DO, BOD, SS and (NH3-N).

5.3. River pollution before remediation
Liuchuan River pollution mainly comes from domestic sewage. According to Taichung City Environmental Protection Bureau's water quality monitoring stations in Liuchuan River, although the sewage sewer construction has actively promoted in the basin, the sewage take-over rate is still not popular, and most of the sewage has not been properly disposed and directly discharged. Liuchuan River, resulting in poor water quality for many years, except for the water quality of the downstream stations is between moderate to severe pollution, the other 2 stations are serious pollution.

5.4. After river remediation
From 2016 to March 2018, the river pollution (RPI) index of the rectification section of Liuchuan River was reduced from 7.25 to 3.5, from "severe pollution" to "light and moderate pollution" .From 2014 to March 2018, the downstream river pollution (RPI) index of Liuchuan River was reduced from 8 to 3.75, from "severe pollution" to "moderate pollution" (see Table 2). It is obvious that the Liuchuan River remediation project has successfully improved the water quality.

Table 2. The outcomes of sampling and analysis in the LiuChuan River.

Monitoring stations	Sampling date	SS (mg/L)	DO (mg/L)	BOD (mg/L)	NH3-N (mg/L)	RPI	Degree of pollution
Upstream Midstream	2014	16.95	3.14	7.82	5.38	8	Serious pollution
Downstream	March 2018	12.2	7	4.9	4.5	3.75	Moderately polluted

6. Conclusion
Urban river renovation is a way to solution floodwaters and river pollution. In the short term, building a riverbank which getting instant results to protect our safety is a very common way. However, this

way does not involve in the concept of ecological environment and all things on earth environmental symbiosis.
The completion of the rectification of Liuchuan River in Taichung City is the first scenic riverbank tobe developed with low impact. The fishery in the river were be seen in the river. Even the "damselfly fish", which has extremely high water quality requirements, has also recirculated. The value of the Liuchuan River pollution index has dropped significantly and the water quality will be safe. After the transformation, Liuchuan River, the concrete revetment has greatly widened. The original willows had removed and replaced with large green plants and bright flowers. The river and the bridge were set with hanging LED lights. The installation art, however, is this urban highlight really in line with residents' expectations? Most people hold a positive attitude and think that the rectification of Liuchuan River has indeed become beautiful, and the stench of the scatter also a lot weaker. However, some people think that the artificial landscaping of Liuchuan's LED lights not only destroys the ecology, but also drives away the animals that inhabit them. It also makes the quaint elegance of Liuchuan River in the past disappear. When the city and rivers begin to be valued and rehabilitated, under the beautiful riverbank, how much ecology can be restored, so that the river is not only a landscape park, but also a biological habitat.

References

[1] Ahiablame LM, Engel BA, Chaubey I, 2012. Effectiveness of Low Impact Development Practices: Literature Review and Suggestions for Future Research, Water, Air, & Soil Pollution, Volume 223, Issue 7, 4253–4273
[2] Perrow MR, Davy AJ. Handbook of Ecological Restoration. The United Kingdom: Cambridge University Press; 2002
[3] Odum HT, Odum B, 2003,-Concepts and methods of ecological engineering, Ecological Engineering, 20(5), 339-361
[4] Allen P Davis 2005, Green Engineering Promote Low- Principles Impact Development, 39 (16), 338A–344A
[5] Infiltration Trench- Wikipedia https://en.wikipedia.org/wiki/Percolation_trench
[6] Lester R. Mathews, John A. Work - US Patent 4,689,145, 1987 - Google Patents, Dry well filtration system
[7] Environmental Protection Administration Executive Yuan, R.O.C. (Taiwan) 2012-2016 River Water Quality History Data, Environmental Water Quality Information.

IOP Conf. Series: Earth and Environmental Science **291** (2019) 012022 doi:10.1088/1755-1315/291/1/012022

Hybrid-approach for sinkhole occurrence risk mitigation in urban areas

A A Malinowska[1] A Guzy[1] R Hejmanowski[1] P. Ulmaniec[2]

[1] AGH University of Science and Technology, Faculty of Mining Surveying and Environmental Engineering, Al. Mickiewicza 30, 30-059 Cracow, Poland
[2] Wieliczka Salt Mine, Park Kingi 1, 32-020 Wieliczka, Poland

e-mail: amalin@agh.edu.pl; pawel.ulmaniec@kopalnia.pl

Abstract. Presented research focused on the development of a novel methodology for sinkhole risk assessment above shallow caverns in a salt mine. The research was carried out for the Wieliczka Salt Mine, which is registered on the UNESCO list and visited by near around 2 million tourists every year. The main assumption of the investigation was to estimate root cause of sinkhole occurrence on the surface. Based on the arch pressure theory the vertical stress in the roof of salt caverns was established. Constructed three-dimensional model of underground mine allowed to determine the stresses between the caverns. Furthermore, the caverns which were hazarded by roof collapse were indicated. Hybrid solution was supported by multicriteria risk analysis based on Analytic Hierarchy Process carried out in Geographical Information System. Spatial analysis led to the identification of caverns potentially influenced by other risk factors. Developed final risk maps were based on four the most significant risk factors leading to sinkhole occurrence. Vulnerability maps developed with support of Analytic Hierarchy Process indicated areas where habitants and infrastructure are exposed to sinkhole occurrence. Merging risk maps with vulnerability maps led to final hazard map, where urban areas susceptible to sinkhole occurrence were shown. In conclusion, the research carried out proved that combined spatial analysis with theoretical solution may pave the way for reliable sinkhole risk assessment above shallow caverns.

1. Introduction

In 1960 a sinkhole over 100 m in diameter occurred as a consequence of collapse of the salt cavern in the Wieliczka Salt Mine. As a result two houses were completely destroyed. The sinkhole was formed within 15 minutes. Similar events take place all over the World, frequently causing danger to urban public space. The problem of sinkhole-prone caverns is extremely complex and despite many years' investigations has not been satisfactorily solved yet. It is known that old and shallow caves have to be backfilled, though it is not known in what sequence. Moreover, such salt mines as Wieliczka are World's heritage, therefore their preservation and maintenance for future generations is one of the priorities. On the other hand, security measures must be taken to provide safety for the citizens of Wieliczka with the oldest part of the mine under the town centre.

The course of deformation process in the salt rock mass is influenced by geomechanical properties of various types of salt present in the Wieliczka Salt Mine and also a number of mining and geological factors being potential causes of sinkhole-formation. Studies presented in this paper are focused on determining the most important risk factors responsible for the formation of weaker zones in the rock mass and, consequently, opening ways to the formation of caving processes. In the past, most of evaluations of the risk of sinkhole formation concentrated on empirical solutions or numerical geomechanical calculations [1-4]. However, those methods turned out not to be fully reliable for a number of reasons:

- strength tests cannot reflect so strongly variable properties of the rock mass composed of several types of salt deposited in strata of various geological origin;
- numerical models can neither account for changing tectonic conditions in the rock mass, nor interrelations between caverns and inter-cavern beams in the mine's rock mass;
- empirical models simplify the geometrical model of the caverns, without accounting for interrelation of caverns in the rock mass.

Critical approaches to previously applied methods of estimating the sinkhole risk inspired the concepts presented in this paper. It was stated that the sinkhole risk evaluation methods should be principally based on the spatial analysis encompassing the present technical state of caverns, mutual bedding, applied exploitation systems and interrelations between the caverns. Application of Geographical Information System (GIS), as well as Analytic Hierarchy Process (AHP) in order to gather all information regarding geological, mining, hydrological conditions in mine, and factors related to urban development, supported analysis of zones prone to sinkhole occurrence. Presented solution significantly improved sinkhole risk mitigation in Wieliczka town. Geographical Information System also supported safety management of historical UNESCO object.

2. Geological and mining conditions of urban area

Over twenty sinkholes took place during the mine's life. Sinkholes formed in the study area varied in their range. One of the biggest ones destroyed the Małdrzyk cavern in 1582, and 25 buildings on the surface.

The formation of sinkholes in Wieliczka town can be mainly attributed to:

- bad technical state of roof and pillars;
- poor condition of cavern roof support;
- fires;
- migration of water into the mine;
- loading, caused by development of new buildings or vibration caused by heavy traffic.

All caverns, which caused sinkholes in the Wieliczka Salt Mine were done in typical green salt blocks. Collapse of the roof parts of the salt crust and the lack of overburden above the caverns resulted in the migration of the empty void towards the surface and formation of a sinkhole. On the basis on many year world's investigations and observations, it was proved that other factors which can be also significantly attributed to the formation of sinkholes are:

- Height of caverns:

World's studies and Polish experience [1-6] revealed that the height of caverns is one of the most important factors responsible for sinkholes. The height of caverns in the Wieliczka Salt Mine ranged between 4 and 50 metres.

- Thickness of roof beam:

The thickness of the roof beam is understood as a difference between terrain altitude and depth of cavern roof. The thicknesses of the roof beam in the area of investigation varied depending on the cavern location depth (the roof beam for caverns on the first level ranges between 23 and 300 meters).

- Quaternary sediments:

Quaternary strata are considered as a risk factor increasing the probability of occurrence of a sinkhole. The filtration of water to salt cavern increases when fractures occur in the rock mass strata. Quaternary strata in the Wieliczka area are thick with inserts and water-bearing lens. Water infiltrates through fractures, which are formed as a result of rock mass deterioration [7].

- Type of caverns:

Salt mining method is related to the type of salt and the way it was deposited. Mining in the Wieliczka area was carried out with various methods. Basically, caverns can be divided into: made in a block of salt, in salt bed, leached, and hole ones. Most sinkhole-prone caverns are the shallowest ones of irregular shape.

- Spatial distribution of salt caverns:

Spatial distribution of caverns is the most complex issue, requiring advanced three-dimensional spatial analysis. The most hazardous are complexes of caverns located close to each other, particularly when the size of the caverns is considerable (figure 1).

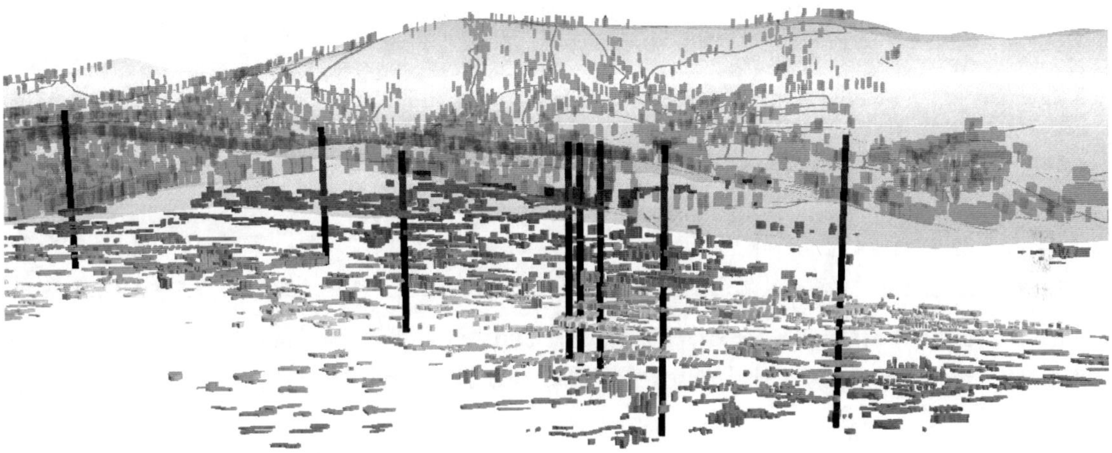

Figure 1. Underground salt caverns and shafts located below Wieliczka town - 3D perspective. Spatial distribution of salt cavern located close to each other.

Afore-mentioned factors will be the criteria of final selection of sinkhole-prone caverns and development of final risk map.

When risk map is strictly related to mining and geological risk factors, vulnerability map is related to factors like:

- urban development (housing, public buildings);
- land use (build-up residential areas, areas ready for new development, green areas, industrial zones);
- geotechnical conditions (characteristics of shallow geotechnical conditions, where ground is stable or unstable. That factor should be taken into consideration when new buildings or infrastructure is planned to be built).

Based on these three factors vulnerability map was developed.

3. Sinkhole risk mitigation in urban areas

The first stage of the investigation was to design the architecture of functional database. Currently, Geographical Information Systems (GIS) are commonly used for spatial and attribute data integration and risks analysis [8-12]. Moreover, Analytic Hierarchy Process (AHP) is a multicriteria method commonly applied in order to resolve complex problems, which are burden with uncertainty [13-15]. Combination of GIS and AHP is a powerful tool supporting facing the problems related to geohazards. In the data base information was gathered regarding geological, mining, hydrological condition, and urban development in the study area. The information about following risk factors was gathered in the database: shape of caverns and their technical condition, characteristic of geological strata and their geomechanical properties, hydrological conditions like watered layers and observed water leakages, digital terrain model, land development, landuse, and geotechnical conditions (figure 2).

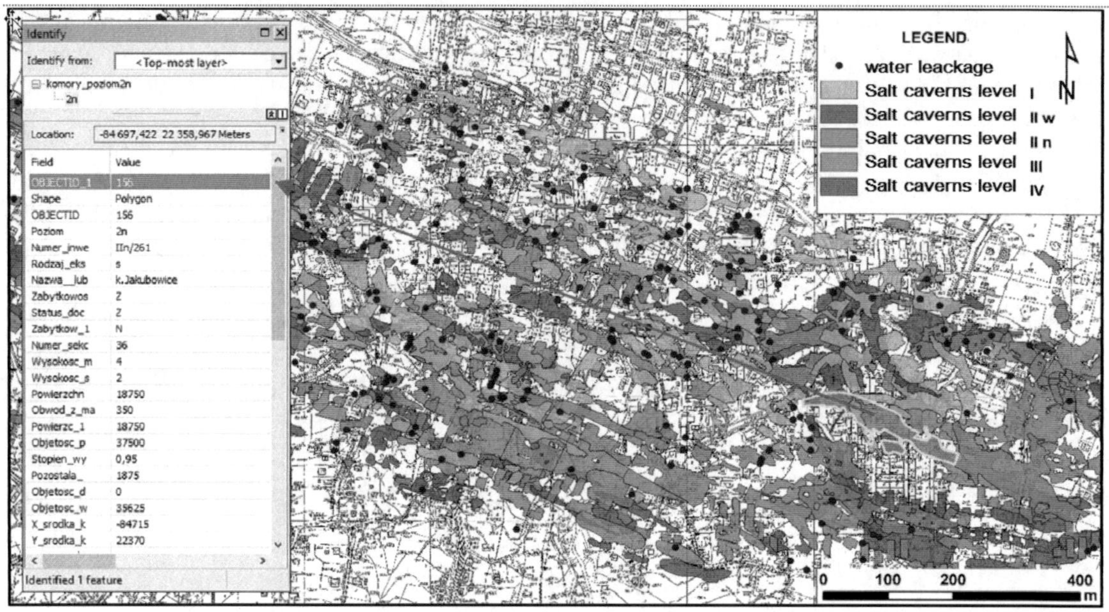

Figure 2. Database content in Geographical Information System (GIS). Salt caverns and observed water leakages located at different depths. depths and observed water leakages.

Based on collected data, further spatial analysis were carried out. Research was divided into a few steps (figure 3a). First step was to select risk and vulnerability risk factors. Furthermore, a raster maps were created for every factors. Then, based on AHP, the risk and vulnerability maps were developed. Final step of the research was to merge both the vulnerability and the risk map. A final sinkhole susceptibility map was created based on combination of information about vulnerable zones and hazarded areas'.

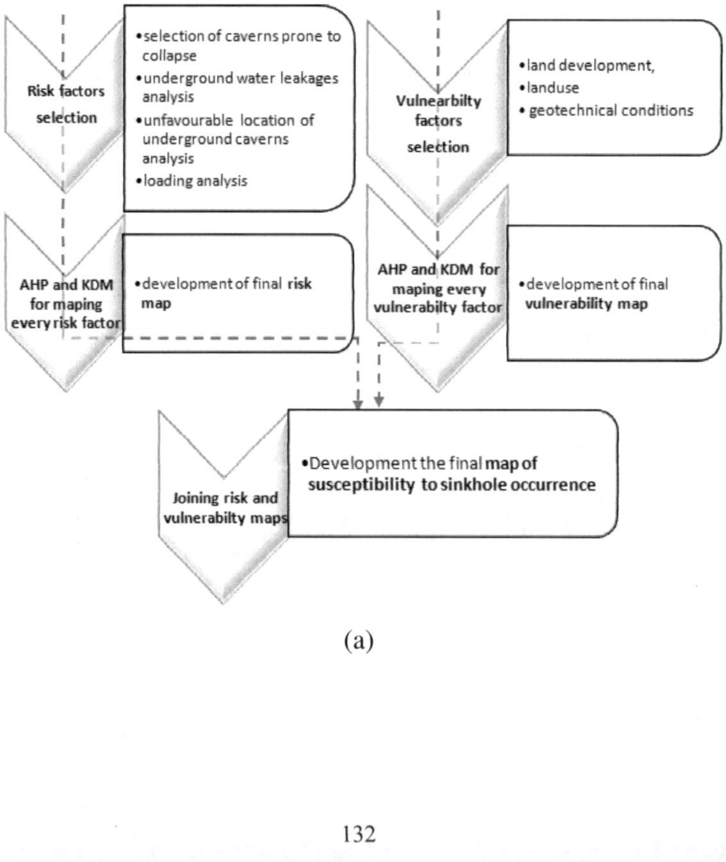

(a)

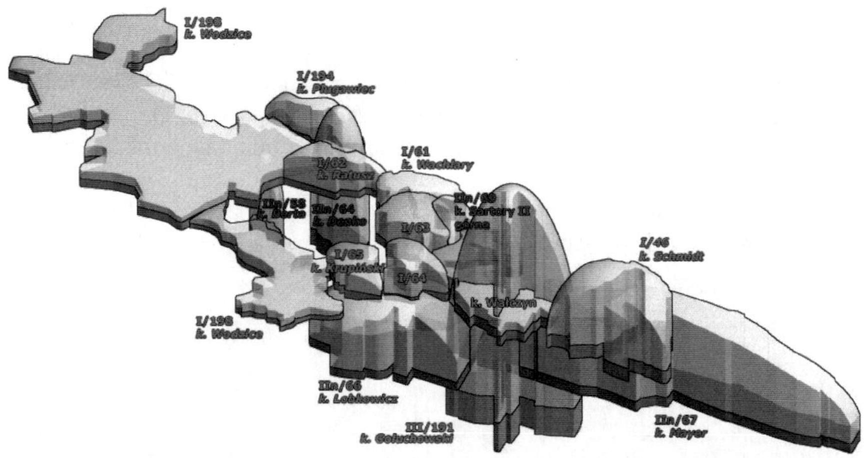

(b)

Figure 3. (a)Flow chart of the research ; (b)Salt cavern in unfavourable spatial distribution located at different depths. Applying Ryncarz method for every salt cavern probability of roof collapse was estimated.

In the next chapters steps of the research will be explained in the details.

3.1. Risk map development

For salt rock deposits a novel approach has been developed in order to assess probability of sinkhole occurrence on the surface. Presented approach was based on combining four risk factors with the use of the AHP method. Every risk factor was unify to the values from 0 to 100, were 0 meant low risk and 100 meant high risk.

First factor related to the selection of cavern prone to collapse was based on the Ryncarz theory. That theory was applied in order to simulate the vertical stresses in the roof of salt caverns [3]. That solution enabled us to separate caverns, for which vertical stresses were regarded as critical ones. Depending on the risk of exceeding critical values, cavern prone to collapse were chosen. Moreover, depending on the zones in the rock masses, which are influenced by cracking roof of salt caverns areas prone to sinkhole occurrence were found (figure 4 a).

The second factor was related to unfavourable spatial location of mining cavern. Based on spatial analysis areas where underground fissure may occur, influencing cavern stability, were selected (figures 3b, figure 4 b)

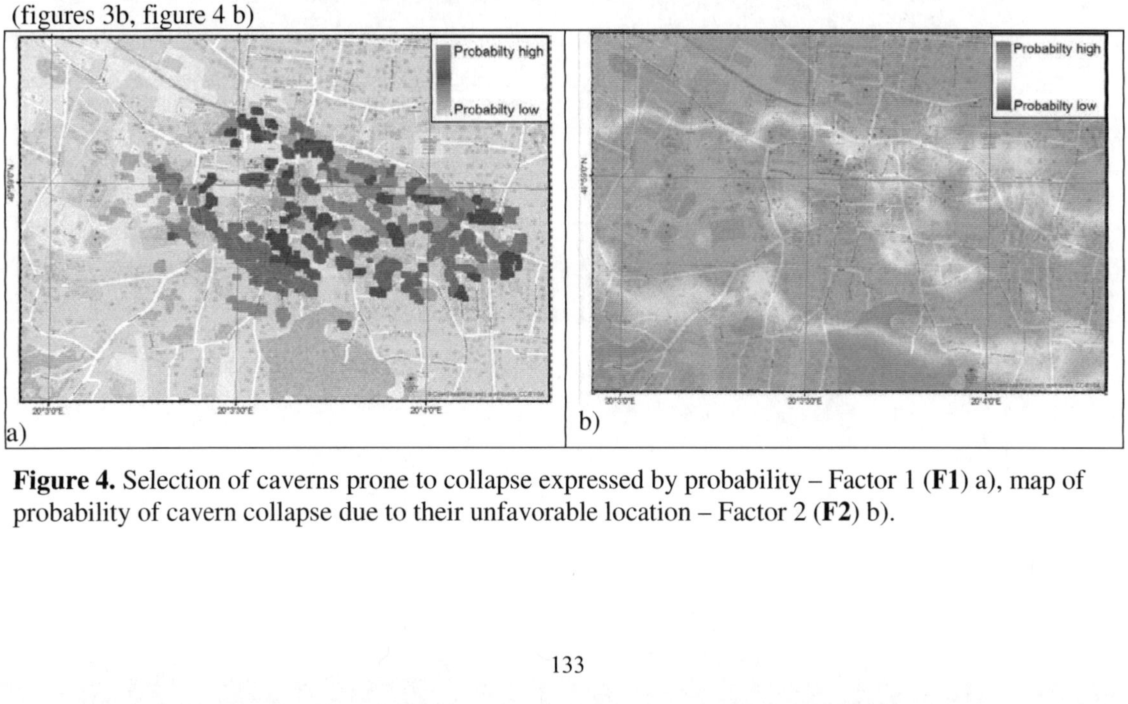

Figure 4. Selection of caverns prone to collapse expressed by probability – Factor 1 (**F1**) a), map of probability of cavern collapse due to their unfavorable location – Factor 2 (**F2**) b).

The next step was to evaluate hydrological risk factors, which could contribute to sinkhole development. For every area, where a water leakage was observed, a buffer zone was created. The radius of the buffer zone was 20 meters. In that area water leakage could initiate salt leaching and contribute to reducing the stability of the salt excavation (figure 5 a). Moreover, geological and hydrological conditions were analysed taking into consideration a rock mass strength and their permeability.

The last analysed risk factor was ground vibration caused by heavy traffic. An estimation of possible rock masses stability disturbances due to heavy traffic was investigated. For significant city roads a map of intensity of possible vibration was created (figure 5 b).

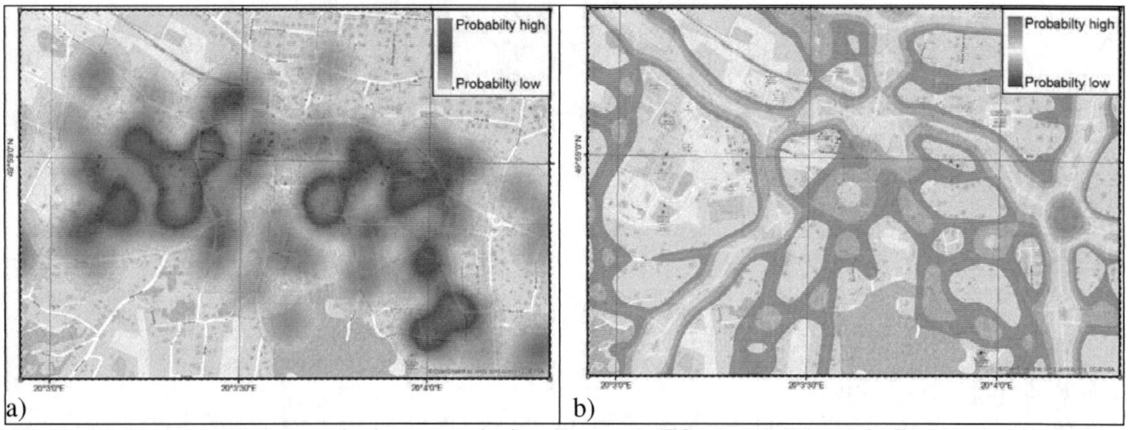

Figure 5. Underground water leakages analysis – Factor 3 (**F3**) a), loading and vibration analysis due to heavy traffic Factor 4 (**F4**) b).

Finally, multicriteria analysis were provided for all above discussed factors (figure 6).

The Pairwise Matrix

	F1	F2	F3	F4
F1	1	3.0	5.0	9.0
F2	0.333	1	3.0	7.0
F3	0.2	0.333	1	3.0
F4	0.111	0.143	0.333	1

AHP Indicators

λ = 4.088

CI = 0.029

CR = 0.032

Weighted Linear Combination (WLC)

Layer Weights

	Layer Name	Weight
1	F1	0.566
2	F2	0.274
3	F3	0.113
4	F4	0.046

Figure 6. Weights for risk factors estimated based on AHP pairwise matrix comparison.

Based on weighted risk factors final risk map was generated (figure 7).

IOP Conf. Series: Earth and Environmental Science **291** (2019) 012022 doi:10.1088/1755-1315/291/1/012022

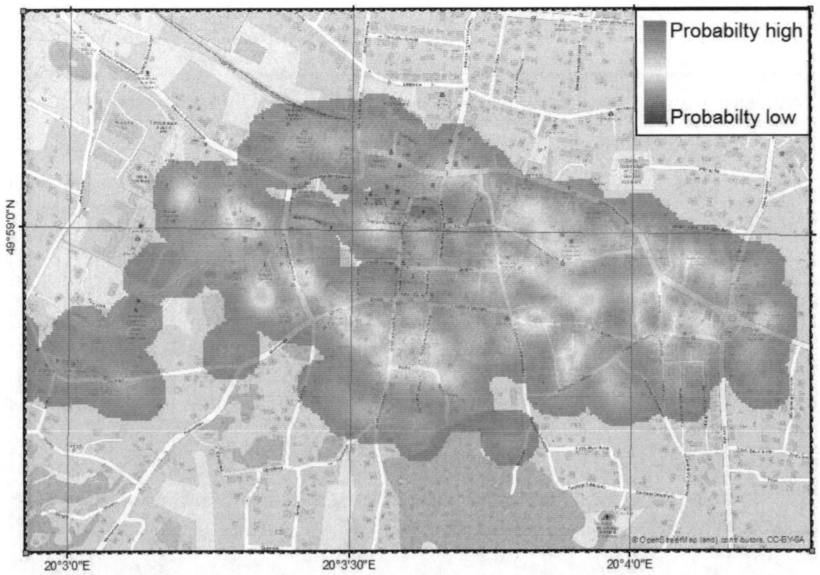

Figure 7. Risk map of probability of sinkhole occurrence.

3.2. Vulnerability map development

Similar approach was established for the vulnerability map development. Three factor were taken into consideration:

- land development;
- landuse;
- ground geotechnical conditions;

First vulnerability factor was related to land development. Location of buildings and their type was taken into consideration (factor 5, **F5**). Next factor was related to landuse. Area where new buildings may be developed were highlighted (factor 6, **F6**). Last factor was associated with technical condition of ground according to its stability (factor 7, **F7**). If the ground conditions are unstable, then damages to buildings may occur with high probability. Moreover, when a new buildings are constructed, additional buildings reinforcements are needed. For that three vulnerability factors based on AHP method weights describing their importance were assigned (figure 8).

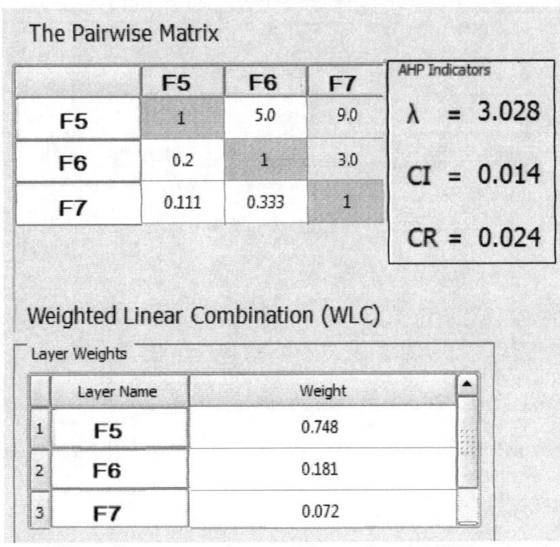

Figure 8. Weights for vulnerability factors estimated based on AHP pairwise matrix comparison.

Based on weighted risk factors final vulnerability map was generated (figure 9).

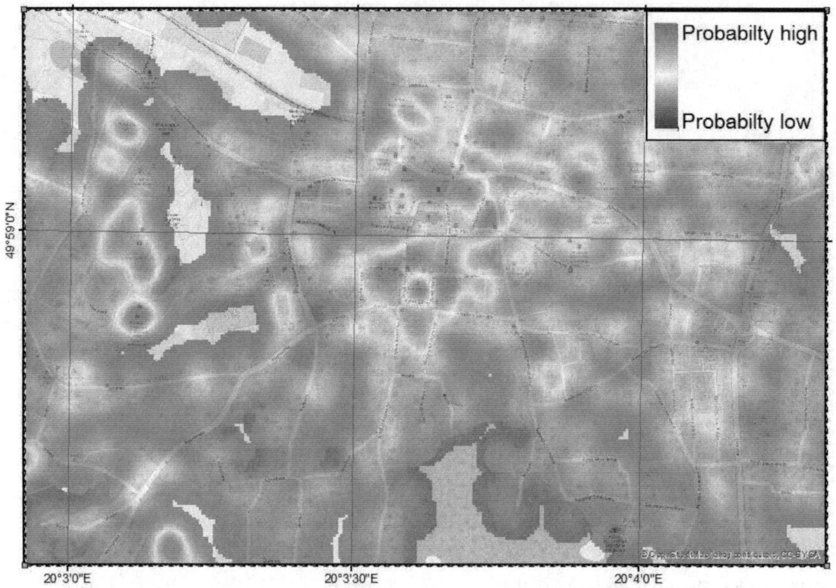

Figure 9. Vulnerability map to sinkhole occurrence.

The final map indicating urban areas susceptible to sinkhole occurrence was established as merging both the risk and the vulnerability map. Only a few areas in the city were significantly hazarded (figure 10). Probability of sever damages to building covered around 4% of all urbanised area. Medium risk was estimated for around 15% of all the city. The rest of urbanised area should not undergo sinkhole treads (figure 11).

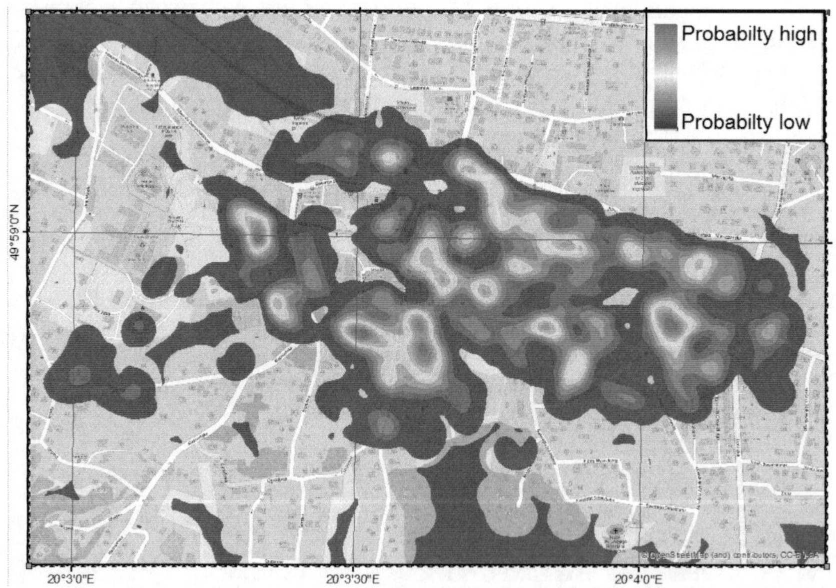

Figure 10. Final map indicating urban areas susceptible to sinkhole occurrence.

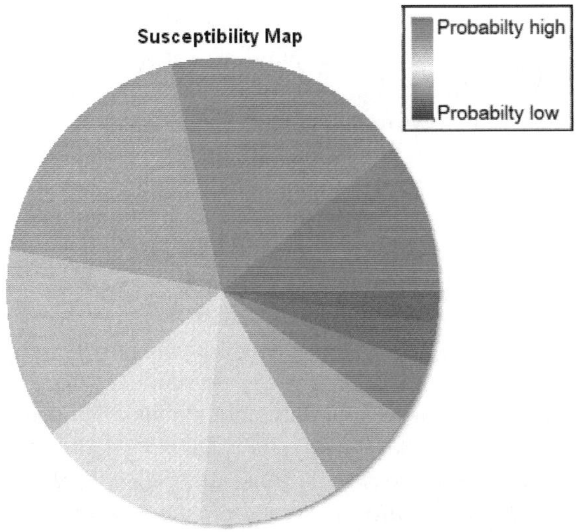

Figure 11. Graph representing percentage of the areas with high, medium and low hazard for the urban areas caused by sinkhole occurrence.

As a result of performed analysis the mining company and the local authorities may incorporated additional monitoring systems in the areas with high hazard of sinkhole occurrence.

4. Results discussion

Based on the obtained computation results, it can be stated only small part of Wieliczka town is substantially susceptible to sinkhole occurrence on the surface. Probability of sever damages to buildings covers around 4% of all urbanised area. Medium risk was estimated for around 15% of all the town. The rest of urbanised area should not undergo sinkhole threats (figure 10, figure 11).

Although the highest probability of sinkhole occurrence is predicted in a relatively small area, it covers, however, Wieliczka downtown, which is characterised by dense building, as well as highest density of population. In addition, these areas are rich in important objects of surface infrastructure and monuments. Nonetheless, it should be noted, the areas of the town center are located directly above salt caverns. These caverns belongs to the oldest and shallowest mining operations. For this reason, they are already a priority in the case of carrying out the monitoring and protection works by mine's measuring department.

In the second area with highest probability of sinkhole occurrence, which is located east to the center of Wieliczka town, only one risk factor prevails. This factor is coupled with transportation. It can imply vibrations contributing to the weakening of the strength of the surface rock layers overlying salt caverns. However, it should be recognized the sinkhole threat in this region in the context of the existing infrastructure and inhabitants is relatively small. It is due to the negligible influence of other factors that were taken into account when establishing the final map of sinkhole risk occurrence.

In other regions of the study area, the main risk factor regarding sinkhole occurrence is the unfavourable mutual position of salt caverns. The obtained computation results indicate that a relatively large number of caverns may be prone to unfavourable deformation forces caused by deeper localized salt caverns. In addition, in areas where water leakages are observed, the incidental threat may be greater. The reason for that is migration of water in rock layers. As a results of water migration, deformation processes may accelerate, especially in the case of small thickness of rock layers separating adjacent salt caverns.

5. Conclusion

The assessment of sinkhole related hazard is a crucial issue encountered in a number of active and closed mines. Sinkholes occurring on the surface generate a special risk in urbanized areas. The

Wieliczka Salt Mine is world heritage object. Moreover, area above that sallow salt mine is inhabited by 23 thousand of people. Especially, in areas similar to mentioned above aware of risk of sinkhole plays a pivotal role. That is why, an author's made and attempt to develop reliable method for mapping susceptibility of sinkhole occurrence in urban areas.

The proposed research methodology is based on multistage algorithm. This algorithm enables the selection of threatened areas due to geological and mining factors, as well as related to the development and land use of the surface. Based on AHP method and advanced GIS analysis significant risk and vulnerability factors were chosen. The analyses carried out revealed that the most significant vulnerability factors were: buildings location and a type of building. The most crucial risk factor was shallow salt caverns in unfavourable mutual location. AHP method supported weighted linear combination of all selected risk factors. As a final result of the investigation areas prone to sinkhole occurrence and urban vulnerable areas were mapped. Achieved results proved that less than 20% of the urbanised area is highly or partly vulnerable to sinkhole occurrence. Those areas should be under monitoring of ground movements. What is more, technical conditions of existing buildings and infrastructure should be also established. In addition, landuse on hazarded areas should be developed under restrictive control of ground stability.

However, the presented methodic has its limitations. In the case of underground exploitation of mineral resources, full recognition of geological and mining conditions is often not able to carry out. Hence, only limited parametrisation of these conditions is possible. For this reason, values of parameters such as: the height of salt cavern, the thickness of rock layers, the location and volume of water leakages are characterized by considerable uncertainty. In addition, the use of AHP method allows for solving complex decision problems, however, due to the significant subjectivity of individual scales of assessing given decision parameters, the obtained results of calculations may be randomness and uncertain. Therefore, when carrying out the research using this method, knowledge and empirical experience in the field of solving particular research issue are highly recommended.
Nevertheless, the obtained computation results indicate the ease and universality of the application of presented research methodology. Thus, it can be used as an effective tool to support activities aimed at assessing, managing and prediction of sinkhole occurrence on the surface induced by underground exploitation of mineral resources.

6. References

[1] Malinowska A A, Misa R and Tajduś K 2018 Geomechanical modeling of subsidence related strains causing earth fissures *Acta Geod Geomater* **15** pp 197–204

[2] Delle Rose M, Federico A and Parise M 2004 Sinkhole genesis and evolution in Apulia, and their interrelations with the anthropogenic environment *Nat Hazard Earth Sys* **4** pp 747–755

[3] Ryncarz T 1978 *Analysis of the impact of filling ventricle excavations at levels I-III of the Wieliczka Salt Mine on surface deformation and stability of lower-lying cavern subject to protection* (Cracow: AGH University of Science and Technology) (in Polish)

[4] Ryncarz T 1993 *Outline of rock mass physics* (Katowice: Silesian Technical Publisher) (in Polish)

[5] Brudnik K 2010 The complex hydrogeology of the unique Wieliczka Salt Mine *Polish Geological Review* **58** pp 787–796 (in Polish)

[6] Wieworka W and o'Byrn K 2012 Selection of backfilling technology works in the Ksawer caverns complex of the Wieliczka Salt Mine *AGH Journal of Mining and Geoengineering* **36** pp 107–115 (in Polish)

[7] Kleczkowski A 1993 Groundwater in the vicinity of Cracow – potential and threats *Archives of the Faculty of Hydrogeology and Geological Engineering of AGH University of Science and Technology* pp 35–38 (in Polish)

[8] Malinowska A A 2010 A fuzzy inference-based approach for building damage risk assessment on mining terraing *Eng Struct* **33** pp 163–170

[9] Mancini F, Stecchi F and Gabbianelli G 2009 GIS-based assessment of risk due to salt mining activities at Tuzla (Bosnia and Herzegovina) *Eng Geol* **109** pp 170–182
[10] Hejmanowski R and Malinowska A A 2010 Building damage risk assessment on mining terrains in Poland with GIS application *Int J Rock Mech Min* **47** pp 238–345
[11] Thierry P, Prunier-Leparmentier A M, Lembezat C and Vanoudheusden E 2009 3D geological modelling at urban scale and mapping of ground movement susceptibility from gypsum dissolution: The Paris example (France) *Eng Geol* **105** pp 51–64
[12] Yilmaz I 2007 GIS based susceptibility mapping of karst depression in gypsum: a case study from Sivas basin (Turkey) *Eng Geol* **90** pp 89–103
[13] Saaty T L 1980 *The analytic hierarchy process* (New York: McGraw-Hill)
[14] Malinowska A A and Dziarek K 2014 Modelling of cave-in occurrence using AHP and GIS *Nat Hazard Earth Sys* **14** pp 1945–1951
[15] Malinowska A A and Matonóg A 2016 Spatial analysis of mining-geological factors connected with discontinuous deformation occurrence *Acta Geodyn Geomater* **14** pp 159–172

Acknowledgments

This research has been financed by the National Science Centre, Poland, grant No.UMO2014/15/B/ST10/04892.

IOP Conf. Series: Earth and Environmental Science **291** (2019) 012023 doi:10.1088/1755-1315/291/1/012023

Land Use Evolution and Simulation Prediction of Pyongyang Municipality in North Korea from 2000 to 2020

Qinchen Zhang[1,2], Pingyu Zhang[2] and Dongyan Wang[1]

[1] College of Earth Sciences, Jilin University, Changchun 130061, China
[2] Northeast Institute of Geography and Agroecology, Chinese Academy of Sciences, Changchun 130102, China

Abstract: Using Global Land30 standard product as data source, combined with CA-Markov model to quantitatively study the land use evolution of Pyongyang municipality in North Korea from 2000 to 2010 and simulate the land use situation of the region in 2020, the research found: (1) From 2000 to 2010, the area of forest land and bare land in Pyongyang was greatly reduced, and the cultivated land, grassland, water body and artificial surface increased to varying degrees, and the expansion of cultivated land area was the most significant. (2) In the first decade of 21st Century, the comprehensive utilization of land use in Pyongyang increased, the information entropy declined weakly, and the land use system remained in a stable and orderly state. (3) In 2020, cultivated land in Pyongyang is still the main type of land use, but the area will be significantly reduced, and the artificial surface will continue to expand, becoming the main land area for area transfer. (4) The land use change in Pyongyang is mainly driven by the three aspects of consumer demand, economic benefits and government decision-making. With the development of society, the influence of economic interests will become more and more prominent.

1. Introduction

Land use change is an important manifestation of material and energy interaction between human beings and the global environment [1]. It occurs at any time and space scale, affecting not only the geographical distribution pattern of terrestrial ecosystems and its productivity, but also objectively reflecting the changes in the earth's biochemical cycle, the structure and function of ecosystems, and the supply of products and services, and also reproducing the temporal and spatial surface of the land [2]. Therefore, land use change has always been the core field of regional system research on human-land relationship [3]. With the rapid development of social economy, the speed and scale of land use change has changed greatly. It is a hot issue to study the temporal and spatial changes of land use and predict its future speed and direction. By analyzing the characteristics of land use change, reconstructing past and predicting future land use status, we can assess the impact of various factors on land use development, reveal the interaction mechanism of the complex system of human-land relationship, and provide decision-making for regional sustainable development [4-6].

Pyongyang is the capital of the Democratic People's Republic of Korea. It is located at the junction of the Pyongyang plains and hills in the lower reaches of the Datong River. The east, west and north are undulating hills. Pyongyang has a long history and has been the political, economic and cultural center of the northern part of the Korean peninsula since ancient times. After the Korean War, the industry developed rapidly and has become the industrial center of the country. Pyongyang is also a modern city with a beautiful scenery. The urban construction area accounts for 20%, and the remaining 80% is green land for parks and so on. It is one of the cities with the largest proportion of green areas in the

world. In the 21st century, North Korean leader Kim Jong Il proposed 'building a strong and powerful country in accordance with the requirements of the new era'. The face of Pyongyang has also changed, and the pace of construction has accelerated markedly. It has become one of the regions where the human system in the Korean Peninsula and even the entire Northeast Asian region is more intense. This paper takes the entire regional perspective of geography research and the entire Pyongyang administrative scope (Pyongyang municipality*) as the research object, through the analysis of land use structure, degree and system change from 2000 to 2010 and the simulation prediction of land use status in 2020. It is intended to reveal the characteristics of land use evolution and the future development trends in typical urban areas under the special social system, and to explore the driving mechanism behind land use change. This study is helpful to deepen the understanding of land use change in Pyongyang. It is of great significance to understand the status quo of North Korea's national land control and its impact on the resources and environment of Northeast Asia.
* The scope of the Pyongyang municipality is based on the newly adjusted administrative boundaries around 2010, with specific reference to the North Korean map of the North Korean Map Publishing House in 2015.

2. Data and methodology

2.1 Data source
This study is based on the world's first 30m resolution global surface coverage product, the GlobeLand30-based data source, including products for the two base years of 2000 and 2010. The data adopts WGS-84 coordinate system, UTM projection, 6° banding, and the surface coverage types are divided into ten categories such as cultivated land and forest land [7]. The product has high spatial resolution and precision. Compared with the 1000m and 300m resolution products in Europe and the United States, the spatial resolution of the surface coverage product has been improved by one to two orders of magnitude. According to third-party verification, the classification accuracy is 83%. At present, the research application based on GlobeLand30 has been widely carried out. It has been proved that the data can not only meet the application needs of scientific research in land resources management, environment, agriculture, urbanization, etc., but also provide basic geographic data support for global change and Earth system model research.

2.2 Land use dynamic simulation
Land use process modeling and simulation is an important tool to explore land use drive mechanisms, support urban planning and policy development, and assess the impact of land use on the ecological environment. At present, the dynamic simulation model of land use change can be divided into two categories: retrospective macro model and forward-looking micro model [8]. The retrospective macro model focuses on the macroscopic model of land use temporal and spatial change process, and quantifies its driving factors based on land use temporal and spatial data. The mechanism of action explores the characteristics and laws of changes in its temporal and spatial scales, including geographic weighted regression, logistic regression, Markov prediction, and System Dynamics (SD). The forward-looking micro-model focuses on the interaction mechanism between micro-factors such as land mass, population and environment, and formulates the type transition rules between the plots and the influence function of the driving factors, by analyzing the relationship between the plot unit and the driving factor [8]. The microscopic effect is obtained as a simulation result of land use change on a macroscopic whole, represented by Multi-Agent System (MAS) and Cellular Automata (CA). This paper uses the CA-Markov model integrated in IDRISI software to simulate the land use status of Pyongyang municipality in 2020 with reference to the data of 2000 and 2010. The model has both Markov and cellular automata (CA). The advantage of the macro-top-down and micro-bottom-up coupling modeling method can not only express the overall evolution mode of land-use space-time process, but also quantify the micro-interaction between the block unit and the driving factor. It has a

good predictive effect on the spatial and temporal changes of land use, especially in the mesoscale basins and cities, and is therefore widely used [9-11].

2.3 Analysis of land use change

2.3.1 Land use change rate

Through the area tabulation in ARCGIS 10.4, the area transfer of various regions from 2000 to 2010 was counted, and the extent of land use change was analyzed. From the change of the area of a single land type and the change of the total land use quantity, the rate of land use change is measured [12], and the speed of structural change is analyzed.

$$V_{t,i} = \frac{u_{bi}-u_{ai}}{u_{ai}} \times 100\% \qquad (1)$$

$$V_t = \frac{\sum_{i=1}^{n}|u_{bi}-u_{ai}|}{2\sum_{i=1}^{n} u_{ai}} \times 100\% \qquad (2)$$

Where: $V_{t,i}$ is the rate of change of land use type i in the study area; V_t is the overall rate of change of all land use types; u_{ai} is the area of land type i in the starting year a; u_{bi} is the area of its ending year b; t is the number of years from start to end; $n$is the number of all land types in the area.

2.3.2 Comprehensive Index of Land Use Level

By classifying the land use types of the four ideal states and assigning values to their own categories, four classification indexes of land use levels are obtained (Table 1). Since the four types in reality are mixed in the same area, each occupying a different proportion of the area, and making local contributions to the extent of local land use, according to their own weight [13-15]. Therefore, the comprehensive quantitative index of land use degree is based on mathematical synthesis (Equation 3), forming a continuous distribution of comprehensive index, the value of which comprehensively reflects the degree of land use in a certain area.

Table 1. Land use degree classification value table

category	unused land level	natural land level	agricultural land level	town level
land use type	unutilized or hard to use	forest, grass, water	cultivated land, garden	towns, residential areas, industrial and mining land, transportation land
graded index	1	2	3	4

$$L_a = 100 \times \sum_{i=1}^{n} A_i \times C_i \qquad (3)$$

Where: L_a represents the comprehensive index of land use degree, A_i represents the grade index of land use degree of the i-th grade, and C_i is the percentage of the graded area of the land use degree of the i-th grade.

2.3.3 Land use information entropy

Information entropy is based on the concept of information theory used to measure the amount of information and uncertainty of the system [16]. The information entropy value of the land use structure can be used to reflect the order degree of the land use system in a region [17, 18]. Generally speaking, the larger the information entropy value, the lower the degree of order of the land use system, and vice versa.

$$H = -\sum_{i=1}^{n} P_i \ln P_i \qquad (4)$$

Where: H is the information entropy, P_i is the ratio of the area of the i-th land-use type to the total area of the area, and n is the number of all land types in the area.

3. Results and analysis

3.1 Temporal and spatial variation characteristics

During the period from 2000 to 2010, the area of forest land and bare land in the Pyongyang municipality was greatly reduced (Table 2). Among them, the forest land decreased by nearly 30% to 119.14km^2, and the bare land decreased by nearly 70%, but the absolute amount was only 1.21km^2. Correspondingly, the area of cultivated land, grassland, water body and artificial surface has been significantly expanded. In terms of magnitude, the grassland area increased by up to 37.91%, the artificial surface was 10.79%, and the cultivated land and water body were again 5.85% and 3.97%. However, from the perspective of changing areas, the arable land expansion reached 69.8km^2, the grassland and artificial surface were 29.8 and 18.39km^2, respectively, while the water body area growth was only 2.37km^2. From the perspective of structural changes, the conversion of forest land-cultivated land, forest land-grassland, cultivated land-artificial surface, bare land-water body is most obvious (Table 3). In space, woodland-cultivated land, woodland-grass mainly occurs in the northern and eastern hilly areas near Pyongyang. In order to ensure a stable food supply in the capital, large-scale deforestation and land reclamation have occurred in these areas, and the regional ecology has gradually deteriorated. Cultivated land-artificial surface occurred mostly in the south of the urban area and along the Datong River, mainly due to population growth and urbanization development. Artificial land expansion has encroached on a large number of urban suburbs. The change of bare land-water body is mainly affected by climatic factors. Relevant research shows that the average annual precipitation of North Korea has increased since 2000, which turns most of the bare beaches into rivers and lakes. a part of. In addition, the conversion of forest land-artificial surface in Pyongyang city is also very obvious. It can be seen that the urban construction of Pyongyang is accelerating and the building density of urban areas is increasing.

Table 2. Changes in the area of different land use types from 2000 to 2010

Index	cultivated land	forest land	grassland	water body	artificial surface	bare land
area in 2000（km2)	1193.48	403.52	78.60	59.51	170.47	1.73
area in 2010(km2)	1263.28	284.38	108.40	61.87	188.86	0.52
single rate of change(%)	5.85	-29.53	37.91	3.97	10.79	-69.94
comprehensive rate of change(%)	6.31					

Table 3. Land use type area transfer matrix from 2000 to 2010

Category	cultivated land	forest land	grassland	water body	artificial surface	bare land	transfer out
cultivated land	116532.72	113.85	160.11	499.41	2041.29	0.36	2815.02
forest land	8302.05	28094.13	3179.61	95.49	681.03	0.00	12258.18
grassland	151.02	208.44	7459.11	30.15	10.98	0.00	400.59
water body	511.83	18.99	18.00	5370.48	31.05	0.36	580.23
artificial surface	828.90	2.61	23.04	71.01	16121.34	0.00	925.56
bare land	1.17	0.00	0.00	120.87	0.00	51.30	122.04
transfer in	9794.97	343.89	3380.76	816.93	2764.35	0.72	17101.62
net transfer	6979.95	-11914.29	2980.17	236.70	1838.79	-121.32	

IOP Conf. Series: Earth and Environmental Science **291** (2019) 012023 doi:10.1088/1755-1315/291/1/012023

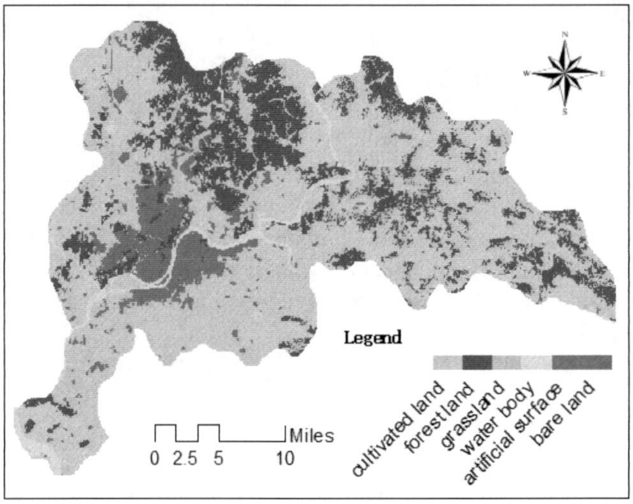

Figure 1. Distribution of land use in Pyongyang municipality in 2000

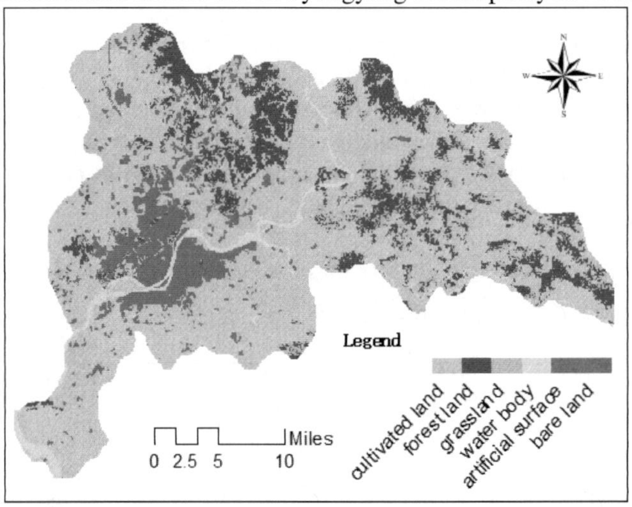

Figure 2. Distribution of land use in Pyongyang municipality in 2010

The land use degree index formula is used to calculate the comprehensive land use degree of Pyongyang municipality from 2000 to 2010. The comprehensive index for 2000 and 2010 was 280.36 and 286.01, respectively, and increased by 5.65 in 10 years (Table 4). The expansion of land use indicates that Pyongyang is in the development stage, the urbanization process is gradually accelerating, and the role between human and land systems will become more intense. According to the calculation of land use information entropy formula, the information entropy of the land system in 2000 was 1.08, and in 2010 it was 1.06, a decrease of 0.02, and the change was very weak. It can be seen that the land use situation in Pyongyang is still in an orderly use range. For the system, there is no obvious imbalance in the conversion between land types.

Table 4. Land use comprehensive degree and information entropy

Index	2000	2010	difference
land use degree	280.36	286.01	5.65
information entropy	1.08	1.06	-0.02

3.2 Simulation and prediction analysis

The land use change probability transfer matrix of Pyongyang municipality from 2000 to 2010 was obtained by means of the CA-Markov module in IDRISI software, and the distribution of land use in Pyongyang municipality in 2020 was predicted by 10a (Fig. 3).

According to the forecast results, the area of cultivated land, forest land and grassland will decrease to varying degrees in 2010-2020 (Table 5). The reduction of cultivated land is most obvious to reach 121.95hm^2, but it is still the main type of regional land use, accounting for 66.17% of the total area. The reduction of woodland was 48.78hm^2, and the area ratio fell to 14.89%. The artificial surface is the main type of area growth, which has expanded by 183.96hm^2 in 10 years, and the proportion of the total area has increased to 10%. In addition, there is a weak transition between the water body and the bare land, and the area remains basically stable.

Comparing the land use change in Pyongyang municipality from 2000 to 2010 and 2010-2020, it can be found that the cultivated land and grassland gradually change from the early expansion to the attenuation, and the cultivated land will replace the forest land as the main type of net transfer, and the artificial surface will become the region. The main type of net transfer into the area. Changes in water and bare land area will gradually become weaker and remain at a dynamic level of stability. From the perspective of structural transformation, the cultivated land-artificial surface will surpass the forest land-cultivated land, and the forest land-grass becomes the most important feature of regional land use change. To a certain extent, this indicates that in the future, in the Pyongyang region, urbanization expansion will replace the traditional agricultural cultivation as the main way of interaction between human and land systems.

Table 5. Statistics of land use type area in 2020

Land use category	Cultivated land	Forest land	Grassland	Water body	Artificial surface	Bare land	Total
Area (hm^2)	126232.47	28396.89	10828.89	6188.22	19069.83	51.93	190768.23
Proportion (%)	66.17	14.89	5.68	3.24	10.00	0.03	100.00
Change* (hm^2)	-121.95	-48.78	-13.23	0.27	183.96	-0.27	0

*The amount of change here refers to the status comparison with 2010

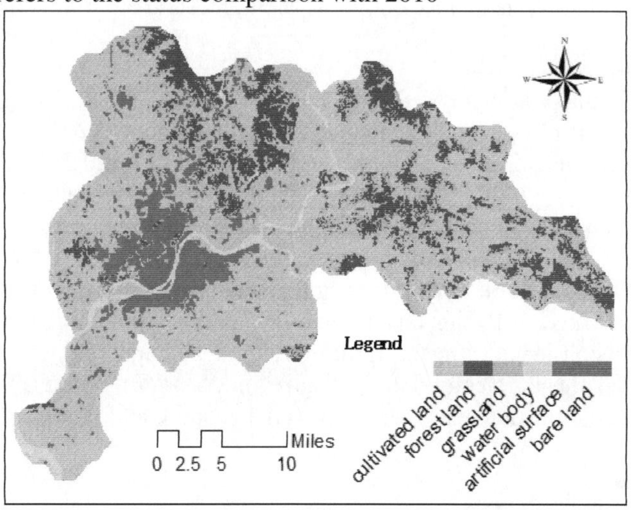

Figure 3. Simulation of land use distribution in Pyongyang municipality by 2020

4. Discussion

From the historical analysis of human society development, consumer demand, economic interests and rational decision-making are three humanistic driving forces that affect land use change at the present stage and in the future [3]. Since the 21st century, the transformation of land use types in Pyongyang has also been affected by the synergy of the three driving forces.

One of the core of consumer demand is food, and land is always the most important carrier for humans to access food. When food supplies are in short supply, it is the only option to survive without breaking new land to meet basic food needs. Since the early 1990s, North Korea has been facing a food shortage and has experienced a 'great famine'. Opening up new arable land and expanding planting area have become the most direct way to alleviate the contradiction between the supply and demand of the grain in the capital. The increase in arable land area at the expense of forest land has become a distinct feature of land use change in Pyongyang from 2000 to 2010. On the other hand, as the capital of North Korea, the country's population, industry and capital gathering places, economic interests also have an impact on land use in Pyongyang. The industrial structure succession and the urban-rural settlement change centered on the maximization of the economic benefits of land use are the two main driving forces, resulting in the adjustment of land use in various industries within the city and the expansion of marginal urban space. The proportion of construction land in Pyongyang's jurisdiction has been rising, and as the future food supply improvement and consumer demand tend to weaken, it will become the main form of regional land use transformation.

In addition, government decisions are also important factors affecting land use change in Pyongyang. On the one hand, Pyongyang has always been a model for North Korea to 'build a strong country in accordance with the requirements of the new era'. The three generations of North Korean leaders pay special attention to the development and construction of the capital region, with the intention of building Pyongyang into a prosperous socialist modern city. On the other hand, North Korea pays special attention to the protection of the ecological environment in the management and development of land planning, and puts forward five principles for the construction of cities, industrial and mining, etc.: 1.It is not allowed to invade farmland; 2.The size of the city should not be too large, mainly in small and medium-sized cities;3.According to natural conditions and local characteristics; 4.Construction should be combined with national vision planning; 5.Prevent environmental pollution and protect the ecological environment. Therefore, Pyongyang's urbanization development is generally in an orderly state. Although the construction land will inevitably encroach on cultivated land, forest land and grassland, it has not caused obvious ecological and environmental problems until now.

5. Conclusion

The essay quantitatively analyzes the land use change from 2000 to 2010 in the Pyongyang municipality of North Korea and simulates the land use status in 2020. The results show that the forest land in Pyongyang has decreased significantly from 2000 to 2010, and the cultivated land area has increased the most. Obviously, the degree of land use has increased, and the role between human and land systems has become more intense, but it still belongs to the scope of orderly utilization. The area of cultivated land will gradually decline from 2010 to 2020. Instead, the artificial surface will expand. Urbanization will become the main driving force for regional land use transformation. On the macro level, land use change in Pyongyang is mainly driven by food consumption, economic benefits and government decision-making. In the future, the gradual improvement of the food supply situation in North Korea and the acceleration of the pace of economic construction will inevitably make the driving role of economic interests on land use change more and more significant.

6. References

[1] Wang C, et al 2006 *Progress in Geography*.**25** 76
[2] Zhuang D F and Liu J Y 1997 *J. Natural Resources*.**12** 105

[3] Fan J, Xu Y D and Shao Y 2003 *Progress in Geography*.**22** 1

[4] Liu J, Li H and Ma Y G 2014 *Soil and water conservation research.* **21** 51
[5] Yang Y, Ma X Y and He C Y 2016 *China Land Science.***30** 72
[6] Xu Y L and Le G P 2018 *Economic Geography.***38** 69
[7] Cheng J, et al 2014 *Acta Geodaeticaet Cartographica Sinica.***43** 551
[8] Li S Y, Liu X P, Li X and Cheng Y M 2017 *J. Remote Sensing*.**3** 1007
[9] Hu B S and Zhang H Y 2018 *Resources and Environment in the Yangtze Basin.***6** 1207
[10] Zhou X, et al 2018 *J. Northwest University (natural science edition)*.**2** 291
[11] Li Z M, et al 2017 *J. Chinese Agricultural Resources and Regional Planning*.**12** 41
[12] Jing C, Wang Y J, Jiang T, Cheng J and Zhai G Q 2016 *J. Arid Land Resources and Environment*.**12** 1003
[13] Huang Y B and Liao S B 2016 *Geographical Research*.**35** 1433
[14] Tan Y Z and Wu C F 2003 *J. Natural Resources*.**18** 112
[15] Jing Y Q, Zhang F and Zhang Y 2016 *Chinese J. Applied Ecology*.**27** 3649
[16] Song G and Wang Y 2016 *Transactions of the Chinese Society of Agricultural Engineering*.**32** 255
[17] Xu J W 2002 *Northeast Asia Forum*.**3** 57
[18] Pan Y, et al 2018 *Soil*.**2** 391

A customization-oriented carbon footprint service for mechanical products

Yajun Zhang[1], Jianjun Yi[1], Zhuoran Wang[1] and Liang He[2,3]

[1] Department of Mechanical Engineering, East China University of Science and Technology, Shanghai 200237, China
[2] Shanghai Aerospace Control Technology Institute, Shanghai 201109, P. R. China
[3] Shanghai Key Laboratory of Aerospace Intelligent Control Technology, Shanghai 201109, P. R. China

Email: yajunzhang1993@163.com; jjyi@ecust.edu.cn; 18616814218@163.com; aerospace@vip.sina.com

Abstract. As the growing concerns of global warming, carbon footprint of a product becomes an important designing criterion. In recent years, the issue of product carbon emission has been extensively studied and most research focus on the computing method, life cycle assessment and optimized design of single category product. However, a lack of versatility in these existing research programs make them difficult to be applied in other product. Furthermore, in many cases, target customers are not concerned about carbon emissions throughout the life cycle of the product and full-cycle calculations of carbon emissions make no sense. To fill this gap, a service mode to provide quantification of carbon footprint for mechanical products is presented in this work. The multi-scale and full-direction system, which is based on process chain employing multiple scale task model is proposed to meet the different carbon footprint requirements, and then encapsulate it as a service to the user.

1. Introduction

With the issue of atmospheric warming becomes increasingly intensified, concerns about climate change rose sharply[1]. The climatic change brings threaten to the human survival and development. Sustainability, especially environmental sustainability, has emerged as a crucial issue amongst states, policymakers, researchers, and the public[2]. Manufacturing industry plays an important role all over the world and it is the key industry in every modern market economy[3]. However, converting materials into industrial parts causes a large amount of energy consumption and carbon emissions[4]. Several reports reveals that, products are responsible for about 84 % of energy-related greenhouse gases (GHGs) emissions and 90 % of the energy consumption in the industrial sector[5]. Carbon emission has become an important topic in the industry, especially in the manufacturing industry. It is critical to quantify product carbon footprints so that appropriate reduction actions can be taken[6].

Over the past two decades, many scholars and researchers have devoted to the studies on the calculation and optimization of carbon footprint, the carbon footprint of products has become an important design criterion[7]. As a result, many companies are focusing on a servitization pattern towards calculating carbon emission[8]. Those companies calculate the carbon footprint of the product during the life cycle, and the results are encapsulated into services available to the users. The service can be recognized as a new kind of environment friendly service, which can effectively reduce greenhouse gas and provide strong support for the realization of green manufacturing[9]. However, the service of calculating carbon emission provided by many companies are still in the primary level and unable to support the diverse requirements of users[10].

In modern industry, the production process is regarded as a chain of processes. Normally, a process chain is a set of independent production processes made up of various processing equipments, raw material and operators[11]. However, the existing services of carbon emission rarely take the process chain as the unit of calculation, which leads to the problem that the accuracy of the calculation results is often hard to control, and cannot provide users with diversified services according to their requirements. Sometimes, the target customers are not focusing on the carbon emission throughout the life cycle of the product. Another remarkable problem is that the optimization of manufacturing process on low-carbon is often considered inadequate[12]. Currently, carbon services are limited to the calculation, but the ultimate goal is reducing carbon emission not calculating. It is worth to investigate how to combine optimization with services.

To solve the problem mentioned above, this paper proposes a closed-loop scheme of carbon footprint service based on the process chain. According to the user requirements, it would provide multidimensional dynamic and customized carbon footprint service.

2. Methodology

The general framework of the proposed method is depicted in figure 1. There are three keys related to the carbon service, namely, user requirements analysis, the establishment of task set, and carbon footprint service results. User submit information related with mechanical product to the system, and the information includes designing information, manufacturing information and the requirements of carbon service. The related production process chain is added to the system according to the requirement analysis, and the execution task set is automatically generated by semantic matching. The results of calculation and optimization are combined as decision schemes which are finally delivered to the user of the carbon service.

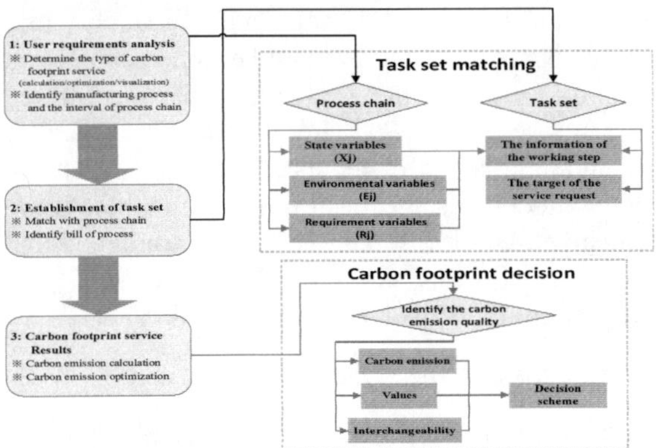

Figure 1. General methodological framework.

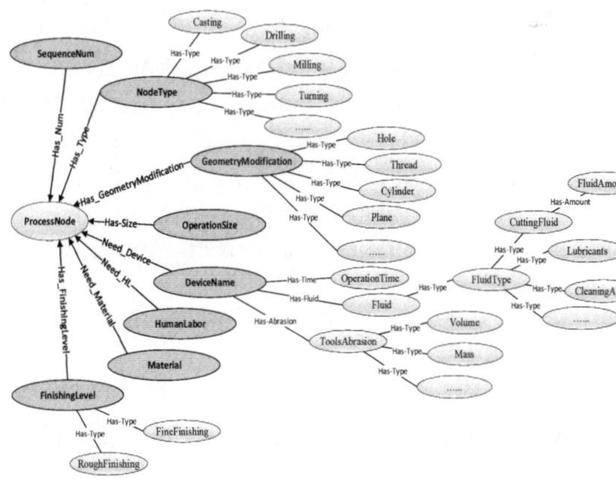

Figure 2. The ontology model of carbon footprint variables.

2.1. Requirements analysis

First, users submit the product related information to the system to request the corresponding carbon footprint service, and the information submitted includes the designing information, the manufacturing information and the requirements for carbon service.

The processing of the product is the process that makes rough pieces with certain geometric shape to meet certain mechanical properties[13]. It is made up of a series of independent machining steps. Each step has its unique properties, so considering resource consumption and carbon emission from the viewpoint of process are more reliable than traditional methods[14]. The process-cycle is a dispersed manufacturing system with multiple-input multi-output. The status of the workpiece, processing requirements, equipment and personnel will all influence the output of the carbon emissions. Therefore, the carbon emission of the product is a dynamic process based on the process chain.

In this paper, the process chain is used as the reference template to integrate the related information of the product. It also divides the factors that influence carbon footprint into three kinds of variables: state variables (Xj), environmental variables (Ej) and requirement variables (Rj). Then, the related information are reorganized by ontology model as shown in figure 2. Ontology is an explicit formal specification of shared conceptual model[15]. It greatly improve the efficiency of the search, accumulation, and sharing of information. Finally, user requirement is transformed into a set of ontology model based on process chain, which would provide support for generation of task sets.

In this paper, the variables that affect carbon footprint are classified into three categories. The state variable describes the related variables in the process of changing the current state of the workpiece into the target state. It includes the serial number of the step (SequenceNum), the type of the step (NodeType), the size of the operation (OperationSize), etc. The environment variable describes the external variables in the process, including device (Device), humanlabor (HumanLabor), material (Material), etc. Requirement variables vary dynamically according to user requirements, which consisted of accuracy, quality and even alternative working steps can be selected. The ontology class diagram shown in Figure 2 is incomplete, but based on the characteristics of ontology, the model does not need too much modification can support the evolution of ontology. It can be quickly extended according to real-time requirements.

2.2. The task model of user requirement

The demands of carbon services are diverse and the target customer sometimes is not focusing on carbon emissions throughout the life cycle of the product. For example, designers are concerned about the carbon emissions of product features, manufacturers pay more attention to the carbon emissions generated during the manufacturing process, while different technologists care more about different machining steps. The demand for carbon footprint service is not the same even in the same company at different department. Plainly, the life cycle method could no longer satisfy the diversified requirements of users.

A task set model is proposed in this paper, which contains a series of independent tasks which are corresponding to the machining steps in the ontology model. As the basic unit of manufacturing processing, it can reflect the carbon emission at each stage. And it also can be further combined and subdivided.

The description of the task node of mechanical products or components is given by equation (1).

$$et^i = \{PF^i, A\}, \quad A = \{Query, Calculation, Visible, Decision_{making} \cdots\} \tag{1}$$

Where et^i is the ith task in the whole task set. PF^i is corresponding to geometric structure, material, and other information of the working step, which is obtained by matching with the process chain model introduced in the section 2.1. A is the target of the service request, which includes query, calculation, data visualization, service decision, etc.

All carbon service tasks are arranged in order, the set of service tasks can be formed as equation (2).

$$\mathrm{ET} = \{et^1, et^2, et^3, \ldots, et^n\} \tag{2}$$

Where n represents the total number of tasks containing user requirements.

When users input their carbon service requirements, the system automatically assigns an empty set of service tasks. By matching the corresponding process chain resources in the database, the task set obtains the carbon knowledge which is used for the carbon footprint calculation of the process. The length of service task set is not fixed, it can be adjusted automatically according to the user requirements, and the corresponding link is automatically locked according to the matching algorithm.

2.3. The calculation model of carbon footprint

Carbon emissions reflect the impact the manufacturing process of products on the natural environment. With the increasing problem of global warming, low carbon manufacturing has become the common goal of global manufacturing. It is one of the main purposes of this paper to study the carbon footprint of the product through the process chain and refine the carbon footprint of every node in the manufacturing process. In this paper, the process chain is matched with the task model, and the carbon emissions of single node is calculated by task unit. In the calculation process, three main aspects are considered, there are device, materials, transportation, etc. In terms of device, this paper mainly considers the power consumption, machine wear, the use of cooling fluid and lubricating fluid, etc. in the manufacturing process. The material mainly refers to the carbon emissions from the part of the waste that is not converted into the product during the process. The transportation of raw materials is an essential link between working steps. Many literatures have pointed out that the carbon footprint of transportation process is an important component of product carbon footprint, so this is a problem that cannot be ignored. Other factors include manpower, substandard products, and other uncertainties. In general, these parts account for a relatively small proportion. So, these parts are considered by making a reasonable small scale amplification of the total carbon footprint in the foregoing aspects.

The carbon emission of single task C_{et^i} is constructed as equation (3):

$$C_{et^i} = \left\{\sum_{r=1}^{n} D_{ir} \cdot \left[\left(\int_0^{t_{ir}^{load}} P_{ir}^{load}\, dt + \int_0^{t_{ir}^{no-load}} P_{ir}^{no-load} dt\right) \cdot CE_f^{ele}\right] + \sum_{r=1}^{n} D_{ir} \cdot \frac{t_{ir}^{load}}{T_r^{tool}} \cdot CE_f^{tool} + \sum_{r=1}^{n} F_{ir} \cdot m_{ir}^{fluid} \cdot CE_f^{fluid} + \sum_{r=1}^{n} M_{ir} \cdot m_{ir}^{material} \cdot CE_f^{material} + \sum_{r=1}^{n} Tr_{ir} \cdot t_{ir}^{transport} \cdot CE_f^{transport}\right\} \cdot \mu_i \quad (3)$$

Where D_{ir}, F_{ir}, M_{ir} and Tr_{ir} are the types of device, fluid, material and transport involved in ith task, respectively. t_{ir}^{load}, $t_{ir}^{no-load}$ and P_{ir}^{load}, $P_{ir}^{no-load}$ are the time and power when device is under load or no load running. t_{ir}^{tool} is the total time of the device working in this task, T_r^{tool} is defined as the total service time of this device. m_{ir}^{fluid} is the mass of fluid consumed in ith task, $m_{ir}^{material}$ is the mass of the material that fails to be converted into the product in the process of processing. $t_{ir}^{transport}$ is the time that spent in material transportation. If it is the transportation between two nodes, half of the process is calculated. CE_f^x is the carbon emission factor of class x. Finally, μ_i is the factor that adjusts the overall carbon emissions according to other factors.

2.4. The optimization model of carbon footprint

Carbon footprint optimization is the ultimate goal of this research. At present, the related research is mainly concentrated in three aspects: equipment level, product level, manufacturing system level. The optimization results mainly include simplification of manufacturing process and resource conversion. But these methods are lack of guiding significance in real life, and cannot be effectively realized in daily production. This paper proposes a model similar to the RFM model in customer relationship management, which evaluates the carbon footprint of each manufacturing node, and realizes the classification of carbon footprint. Low carbon manufacturing can be achieved by adopting targeted methods for different types of nodes.

Similar to the RFM model, three key parameters are proposed for each node in this paper: carbon emissions, values, and interchangeability. The carbon emissions of nodes have been discussed in detail in section 2.3. Node value refers to the contribution of the node to the whole product value in the manufacturing process. Due to the limited space, the specific methods are not repeated here. Interested readers are encouraged to read Song et al[8]. In his paper, the concept of node value is described in detail. Node interchangeability refers to whether the process of the node can be replaced by other process, and the feasibility of the implementation is also considered. Then, the scores of three parameters are obtained through expert evaluating method and further divided into high or low parts. Finally, the types of nodes are divided into 8 categories and table 1 shows several typical categories. Different measures can be taken for different types of carbon emissions to achieve green production.

Table 1. Several typical categories.

Node type	Carbon emission	Values	Interchangeability	Scene corresponding to machining process
High value replacement node	↓	↑	↑	The node can be replaced by other low carbon process.
High value maintenance node	↑	↑	↓	The node is low carbon and high value, but the interchangeability is not high, and it is an important maintenance node.
Removable node	↓	↓	↓	The three parameter are low and this node can be removed.

3. Conclusion

There are many factors affecting the carbon emission in the process of production, and the diversity of user needs also makes the traditional life cycle based evaluation method no longer feasible. In order to meet the demands of different users for carbon emissions at different stages of production, a multi scale task model of carbon footprint based on process chain is proposed in this paper and encapsulated as service for the user. This fine-grained carbon footprint assessment method would help users have a deeper cognition of carbon emissions in the production process, and also makes the proposed process optimization plan more practical.

4. References

[1] Chen J X and Chen J 2017 Supply chain carbon footprinting and responsibility allocation under emission regulations J. Environ. Manage. 188 255–267
[2] Čuček L, Klemeš J J and Kravanja Z 2012 A review of footprint analysis tools for monitoring impacts on sustainability J. Clean. Prod. 34 9–20
[3] Wang Q, Liu F and Li C 2013 An integrated method for assessing the energy efficiency of machining workshop J. Clean. Prod. 52 122–133
[4] Jeswiet J and Nava P 2009 Applying CES to assembly and comparing carbon footprints Int. J. Sustain. Eng. 2 232–240
[5] He B, Wang J and Deng Z 2015 Cost-constrained low-carbon product design Int. J. Adv. Manuf. Technol. 79 1821–1828
[6] Dong H, Geng Y, Xi F and Fujita T 2013 Carbon footprint evaluation at industrial park level: A hybrid life cycle assessment approach Energy Policy 57 298–307
[7] Su J C P, Chu C H and Wang Y T 2012 A decision support system to estimate the carbon emission and cost of product designs Int. J. Precis. Eng. Manuf. 13 1037–1045
[8] Song W and Sakao T 2017 A customization-oriented framework for design of sustainable product/service system J. Clean. Prod. 140 1672–1685
[9] El-Diraby T, Krijnen T and Papagelis M 2017 BIM-based collaborative design and socio-technical analytics of green buildings Autom. Constr. 82 59–74
[10] Zhou Z, Liu F, Zou R, Liu J, Xu H and Jin H 2016 Carbon-Aware Online Control of Geo-Distributed Cloud Services IEEE Trans. Parallel Distrib. Syst. 27 2506–2519
[11] Zhu S, Jiang Z, Zhang H, Tian G and Wang Y 2017 A carbon efficiency evaluation method for manufacturing process chain decision-making J. Clean. Prod. 148 665–680
[12] Shi X and Meier H 2012 Carbon emission assessment to support planning and operation of low-carbon production systems Procedia CIRP 3 329–334
[13] Givehchi M, Haghighi A and Wang L 2017 Cloud-DPP for distributed process planning of mill-turn machining operations Robot. Comput. Integr. Manuf. 47 76–84
[14] Denkena B, Park H, Behrens B, Henjes J, Bertys S, Lüken I and Klassen A 2013 Development of a Concept to Optimize the Energy Efficiency in Forging Process Chains 14, 1229–1236
[15] Gailly F, Alkhaldi N, Casteleyn S and Verbeke W 2017 Recommendation-Based Conceptual Modeling and Ontology Evolution Framework (CMOE+) Bus. Inf. Syst. Eng 59 235–250

IOP Conf. Series: Earth and Environmental Science **291** (2019) 012024 doi:10.1088/1755-1315/291/1/012024

Acknowledgments
This research is supported in part by the National Natural Science Foundation of China (No. 51575186, No. 51275173, No. 50975088 and No. 51210105018), the Science and Technology Commission of Shanghai Municipality (No. 10DZ1500200), Shanghai Science and Technology Innovation Action Plan (No. 18DZ1204000, No. 18510745500) and the Shanghai Software and IC Industry Development Special Fund (No. 180121).

Application and thermal economic analysis of plate low pressure economizer in flue gas waste heat recovery

X Xu[1,2] and G D Ju[2]

[1] Xinglin College, Nantong University. Jiangsu, Nantong, Tongjing Road 999, China
[2] School of Mechanical Engineering, Tongji University. Shanghai, Cao'an Road 4800, China

Email: xinshoux2@126.com

Abstract. Low pressure economizer is an important equipment to reduce exhaust gas temperature and recover waste heat of flue gas in power plants. In this paper, the welded plate heat exchanger is applied to the field of low pressure economizer. The waste heat of coal slag and flue gas is utilized comprehensively to raise the boiler feed water temperature and reduce coal consumption. The thermal economy of plate economizer and finned tube economizer is analyzed by the equivalent enthalpy drop method. The results show that the plate economizer has obvious advantages under the same pressure drop and heat load. Compared with the finned tube economizer, the heat transfer coefficient the of the plate economizer increases by more than 50%, and the volume decreases by about 2/3. After installing the slag cooler, the heat consumption rate decreases by 119.7 KJ/kWh, and the coal consumption decreases by 4.68 g/kWh.

1. Introduction

In recent years, with the increase of the proportion of renewable energy in the power grid, the number of generating hours of thermal power units is further reduced. How to improve the thermal efficiency of the units and reduce coal consumption has become the primary goal of thermal power plants. The heat loss caused by excessive exhaust gas temperature accounts for 70%-80% in the boiler heat loss[1]. When the exhaust temperature increases by 10-15°C, the boiler efficiency decreases by 1% and the standard coal consumption increases by 3-4g/kWh[2]. In China, the exhaust gas temperature of thermal power unit boilers is generally between 120-140°C, and it even exceeds 160°C for some units burning high sulfur coal. High flue gas temperature will not only affect the efficiency of generating units, but also affect the efficiency of dust removal and desulphurization efficiency.

The low pressure economizer system heats the condensate water of the steam turbine by recovering the waste heat from the flue gas, thereby reducing the steam extraction regeneration of the steam turbine, and the crowding extraction steam can be returned to the steam turbine to do work, thus effectively reducing the coal consumption. Scholars have done a lot of research on the flue gas waste heat utilization system. Han et al.[3] analyzed the economic performance of 320MW unit equipped with economizer. The results show that the boiler efficiency is obviously improved and the coal consumption of power supply is obviously reduced after adding low pressure economizer. Wang et al.[4] studied the influence of fin structure and distribution on heat transfer characteristics, wear characteristics and ash deposit performance of H-type finned tube economizer. An et al.[5] analyzed the influence of parallel and series connection of low pressure economizer and thermodynamic system on relative internal consumption rate, absolute internal consumption rate and cycle thermal efficiency of steam turbine.

The above researches are mostly concentrated on the finned tube economizer. However, finned tube economizer covers a large area, has low heat transfer efficiency and is easy to wear. Finned tube is easy to accumulate slag during long-term operation, which causes heat transfer deterioration. At the same time, the space of old units flue is limited, finned tube economizer is not conducive to the transformation of old units. In this paper, welded plate economizer is proposed to replace finned tube economizer. The thermal economy and heat transfer characteristics of plate economizer and finned tube economizer under the same working conditions are compared by equivalent enthalpy drop method. The comprehensive utilization of waste heat and acid dew point corrosion of flue gas were considered. The low pressure economizer and slag cooler are connected in series to raise the temperature of condensate water entering the low pressure economizer so as to avoid acid dew point corrosion. The thermal economy after installing slag cooler was also studied.

2. Thermal economy analysis of low pressure economizer system

2.1. Basic principle

The low pressure economizer system is independent of the main feed water system. The exhaust heat is used to heat some condensate water of lower temperature, replacing some extraction steam of the steam turbine, so that the extraction steam is crowded to the steam turbine and continues to expand and work. After adding low pressure economizers, a large amount of flue gas waste heat enters the regenerative system, which is the additional heat obtained without increasing the amount of boiler fuel, and it is converted into electrical power with a certain efficiency. This new increment is far greater than the power loss caused by the crowding steam extraction and turbine vacuum drop, so the unit economy is improved without exception.

2.2. Thermal economic analysis method

The thermal system analysis methods adopted in this paper are based on heat balance and flow balance. The concept of equivalent enthalpy drop is adopted in the measurement of thermal economy[6]. The equivalent enthalpy drop method[7] is based on the selected initial and final steam parameters and regenerative parameters. It is premised on the fixed values of the new steam flow rate and the fuel heating rate of the unit. The economic effect of the change of the whole thermodynamic system can be obtained by quantitative calculation of the flow and heat of a certain stage of extraction steam. It is accurate and convenient to diagnose the thermal system of power plant and is the basic theory of economic diagnosis of thermal system.

2.2.1. Expression of heat release for heater

For any heater j, the heat release can be classified into three types: the enthalpy rise of feed water in the heater τ_j, the heat release of steam in the heater q_j, and the heat release of drainage in the heater γ_j. In addition, There are other steam brought into the heater, the heat release of this part of the external steam is recorded as q_{sgj}.

The expressions of the heat release of drainage heater are

$$\tau_j = h_{wj} - h_{w(j-1)} \tag{1}$$

$$q_j = h_j - h_{j-1} \tag{2}$$

$$\gamma_j = h_{d(j+1)} - h_{dj} \tag{3}$$

The expressions of the heat release of collection heater are

$$\tau_j = h_{wj} - h_{w(j-1)} \tag{4}$$

$$q_j = h_j - h_{w(j-1)} \tag{5}$$

$$\gamma_j = h_{d(j+1)} - h_{w(j-1)} \tag{6}$$

where, h_{wj}, h_j, h_{dj} are the enthalpy of the outlet water, extraction steam and drainage of heater j respectively.

2.2.2. Equivalent enthalpy drop method for thermodynamic system

For any heater j, the equivalent enthalpy drop is expressed as

$$H_j = (h_j - h_c) - \sum_{r=1}^{j-1} A_r \eta_r \text{ (kj/kg)} \tag{7}$$

where, if the heater j is a drainage heater, $A_r = \gamma_r$ below heater j until the collection heater (including this collection heater), and for the subsequent stages of the collection heater, $A_r = \tau_r$. If the heater j is a collection heater, $A_r = \tau_r$.

The gross equivalent enthalpy drop of 1kg new steam

$$H_{gr} = (h_o - h_c) - \sum_{r=1}^{j-1} \tau_r \eta_r \text{ (kj/kg)} \tag{8}$$

The equivalent net enthalpy drop of the new steam can be calculated after deducting shaft seal leakage and other losses.

$$H = H_{gr} - \sum_{r=1}^{z} \Pi \text{ (kj/kg)} \tag{9}$$

2.3. Analysis of thermal economy after installing low pressure economizer

2.3.1. Rated condition of unit

There are two 55MW extraction heat supply turbines in a power plant. The exhaust gas temperature of boilers exceeds 135°C. The potential of flue gas waste heat utilization is great. With the application of low temperature corrosion resistant steel in heat exchangers, the utilization potential can be further improved. The original thermodynamic system diagram of the power plant is shown in Figure 1.

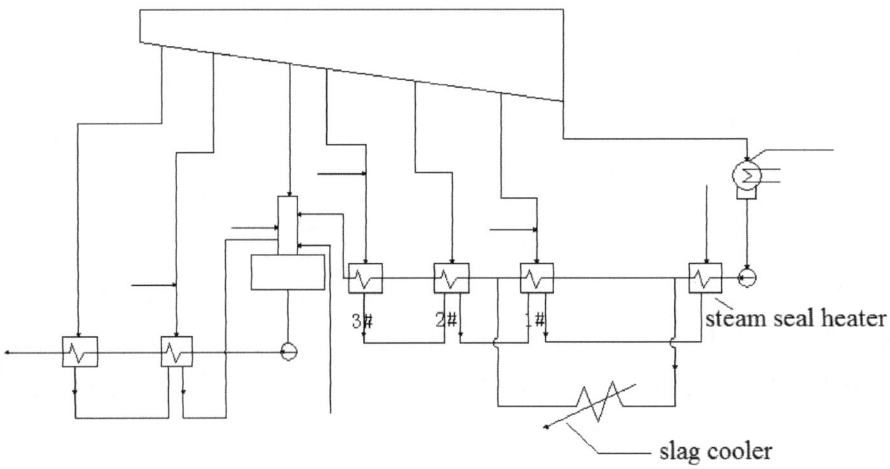

Figure1. The thermodynamic system diagram of original rated condition

The condensed water after the steam seal heater with an enthalpy of 130.88kJ/h and a flow rate of 100t/h is derived and heated to 85°C by the slag cooler and brought to the condensate system before the #2 low pressure heater.

2.3.2. Transformation scheme of unit waste heat utilization

The thermodynamic system diagram after transformation is shown in Figure 2.

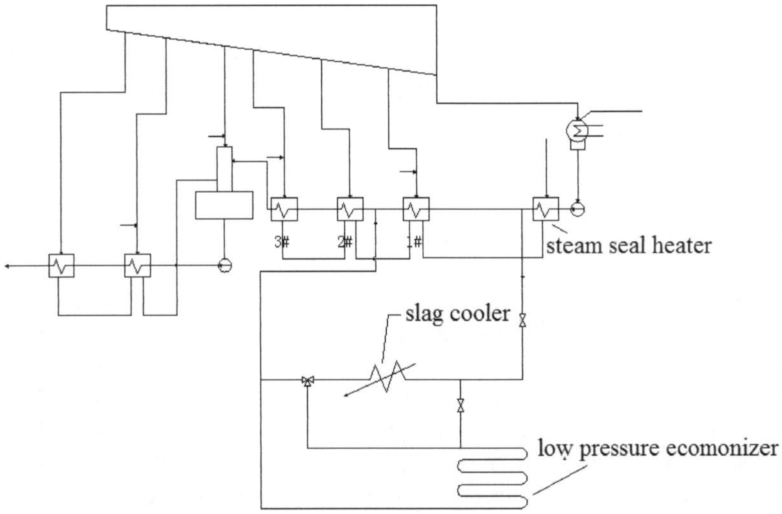

Figure 2. The installation of low pressure economizer and slag cooler

Considering that if the inlet water temperature of low pressure economizer is too low, it is easy to cause low pressure corrosion of the tube, and the high water temperature will make the heat transfer temperature difference too small and increase the heat transfer area, so the low pressure economizer and the slag cooler are connected in series. The rated outlet water temperature of the slag cooler (85°C) is set as the inlet water temperature of the low pressure economizer. Due to the adoption of new sulfuric acid corrosion resistant special steel, the flue gas outlet temperature of low pressure economizer system can be designed to be lower. Under the modified design conditions, the inlet temperature of flue gas side is 145°C, the flow rate is $2.5\times10^5 m^3/h$. The exhaust temperature is finally reduced to about 100°C, which is close to or below the dew point temperature of flue gas.

3. Characteristics of plate low pressure economizer

The plate bundles of the new plate low pressure economizer are stacked together by two identical plates. A circular channel with elliptical cross section is formed by laser penetration welding and charging. The circular channel is a condensate water channel, as shown in Figure 3. The plates are staggered and stacking. The space between the plates is the flue gas ripple channel, as shown in Figure 4.

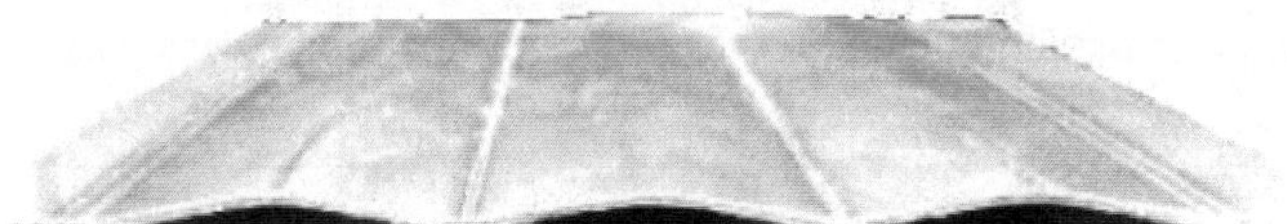

Figure 3. The core body for heat exchange of plate and shell economizer

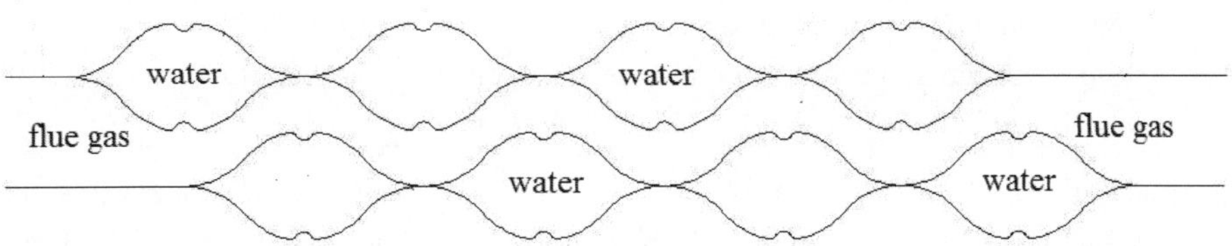

Figure 4. The schematic diagram of channles for flowing medium

Compared with the traditional finned tube low pressure economizer, the advantage of the new type plate low pressure economizer is that its heat transfer component is plate. Its characteristics are: firstly, all the heat transfer surfaces of the medium in the core are primary surfaces, which improves the heat transfer efficiency; secondly, the structure is compact, high temperature resistance and not easy to scale; thirdly, it can be made into multiple modules, easy to remove and clean.

4. Calculation and analysis of thermal economy

4.1. Calculation results

Under the design condition, the condensate water with a flow rate of 90 t/h and a temperature of 85°C at the outlet of the slag cooler is evenly distributed to the tube row of the heat exchanger through the inlet header. After absorbing the heat of the flue gas, it is sent back through the outlet header and collected with the unheated condensate water and then enters the inlet of the #2 low pressure economizer. When the slag cooler needs to be shut down, the low pressure economizer can also be operated separately, that is, the slag cooler is bypassed. The water inlet temperature of the low pressure economizer is still 31.3°C and the flow rate is still 90t/h before the #1 low pressure heater in the rated working condition of the thermal system. The condition of two types of low pressure economizer operating separately and in series connection with slag cooler are simulated by using Fluent numerical simulation software. The results are shown in Table 1.

Table 1. Comparison of two kinds of low pressure economizer operating conditions.

heat exchanger		flue gas temperature/°C		condensate water temperature/°C		heat transfer coefficient/ w(m^2·k)	heat exchange area/m^2	total heat exchange/ kw
		inlet	outlet	inlet	outlet			
run separately	plate	145	78	31.3	73.1	69	1043	4320
	finned tube	145	79.7	31.3	72.1	41.6	1908.4	4267
run with slag cooler	plate	145	103	85	109.6	75.5	1043	2510
	finned tube	145	110.3	85	106.6	42.6	1908.4	2275.8

The total heat transfer of the two kinds of low pressure economizer is basically the same under the condition that the slag cooler is connected in series. The equivalent enthalpy drop method is used to calculate the energy saving state of the system connected with low pressure economizer. The calculation results of the total energy saving effect of the series system are shown in Table 2.

Table 2. The thermal economy when the low pressure economizer and the slag cooler are installed in series

Item	value	unit
Net equivalent enthalpy drop H_{gr}	987.5	kJ/kg
Equivalent heat drop increment d_H	13.22	kJ/kg
Unit thermal economic increase ratio	1.36%	-
Unit heat consumption reduction	119.7	kJ/kWh
Saving of coal consumption rate	4.68	g/kWh
Turbine efficiency	0.3873	-

Because of absorbing of external heat sources, the steam extraction of the steam turbine is saved. When the extracted steam is returned to the steam turbine's regenerative system and worked, the thermal economy of the unit can be increased by 1.36%, the heat consumption of the steam turbine can be reduced by 119.7 kJ/kWh, and the standard coal consumption can be reduced by 4.69 g/kWh.

4.2. Comparative analysis of energy saving effect

Results show that compared with the traditional finned tube economizer, the new plate-shell heat exchanger has the advantages of high compactness, good heat transfer effect and small volume.

Taking the design condition of the slag cooler in series as an example, it can be seen from Table 1 that the heat transfer coefficient of the new type of plate economizer is much higher than that of the finned tube economizer. Under the same transmission pressure drop, the heat transfer coefficient is increased by more than 50%. Therefore, when the heat transfer area is reduced by nearly half, the heat exchange quantity is increased on the contrary. As the heat transfer area is greatly reduced, the volume of plate economizer is also reduced by about 2/3 than that of finned tube economizer. The comparison visual histogram of two kinds of economizer is shown in Figure 5.

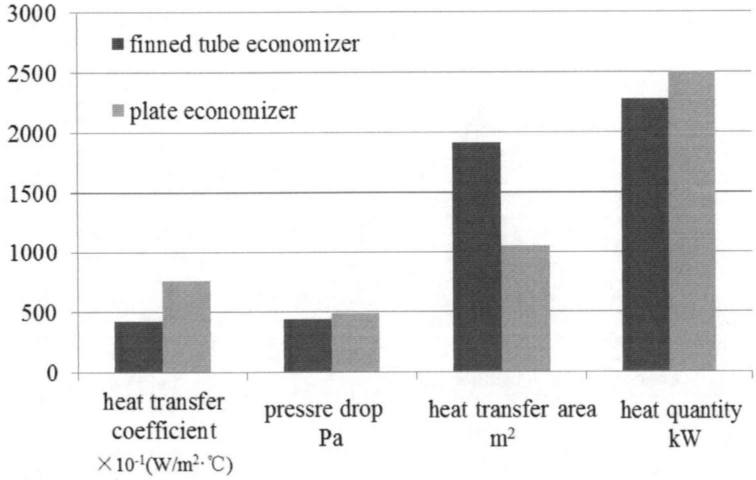

Figure 5. The data comparison of the two kinds of low pressure economizers

5. Conclusion

In this paper, the welded plate heat exchanger is applied to the field of low pressure economizer. The thermal economy of plate economizer and finned tube economizer is analyzed by the equivalent enthalpy drop method. The results show that,

(1) If the low pressure economizer system is connected in series with the original slag cooler system, the condensate state can be further improved, and the thermal economy of the unit can be greatly improved compared with the separate slag cooler system. The low pressure economizer system can make full use of the waste heat of boiler flue gas.

(2) Under the same pressure drop, the heat transfer coefficient of the new plate economizer is increased by more than 50% compared with the finned tube economizer, and the volume of economizer is reduced by about 2/3.

(3) After adding the slag cooler, the flue gas temperature increases, which is beneficial to avoid dew point corrosion of economizer. The heat consumption rate decreases by 119.7 KJ/kWh, and the coal consumption decreases by 4.68 g/kWh.

(4) Compared with the traditional finned tube low pressure economizer, the pressure drop of the plate lowpressure economizer is only about 10% higher. The influence of the equipment on the flue fan is completely acceptable.

Acknowledgments
This research subject is funded by Nantong Science and Technology Project(GY12016009).

References

[1] Ma L Y, Yang L and Lu J 2016 *Energy* **116** 812-23
[2] Wang X, Li Y X and Li X 2018 *J. Combust. Sci. Technol.* **1** 15-20
[3] Han S, Wang F, Zhang P F, Wu S L and Xu Y G 2015 *Therm. Power Gen.* **12** 133-35

IOP Conf. Series: Earth and Environmental Science **291** (2019) 012025 doi:10.1088/1755-1315/291/1/012025

[4] Wang Y, Chen W X and Yu F 2017 *J. Eng. Thermophys.* **38** 129-33
[5] An E K and Ma J Y 2010 *J. Tongji Univ: Nat. Sci. Ed.* **38** 1492-95
[6] Li Q D and Liu Z Z 2008 *Calculation and analysis of thermal economy of power plant* (Beijing: Chin. Electron. Power Press)
[7] Lin W C 1994 *Energy saving theory of thermal system in power plant* (Xi'an: Xi'an Jiaotong Univ. Press)

Application of Building Integrated Photovoltaic in Hot Humid Climate. Case Study: Office Building in Indonesia

N A Ardiani[1,2], Suhendri[1,2], M D Koerniawan[1,2] and D Y N Naimah[2]

[1] Department of Architecture, Institut Teknologi Bandung, Bandung, Indonesia, 40132
[2] Centre for Development of Sustainable Region (CDSR), Yogyakarta, Indonesia, 55281

*Corresponding email: nissaardiani@gmail.com

Abstract. This study investigated application of photovoltaic (PV) panels for office building. Bandung, Indonesia is chosen as case study due to its hot and humid climate. To conduct this study, first, site selection was evaluated, followed by load evaluation using standard requirement for energy-efficient office building. Then, design and size of the system was defined. Feasibility of this study was evaluated from technical and economic indicators, i.e. energy delivered, energy export and import to the grid, yearly savings, and payback period. Simulation of this study was performed with various scenarios on the design electricity load. This study shows that solar PV panels are feasible to be installed in flat roofs and western façade. With 365 m^2 of solar panel in total, it can generate 66 MWh electricity per year. If PV provide electricity for lighting inside the building, the payback period is 12 years. This payback period become faster, i.e. 6.24 years, if all the electricity is sold to the grid.

1. Introduction

Renewable energy systems in buildings are now more widely used because more people aware about decreasing fossil energy and greenhouse gas emission. They try to find and explore how to implement sustainable building concept into buildings. The active system of renewable energy has been studied and practically applied in the buildings with additional benefit to reduce the electrical bills from the grid. For building application, sustainable energy systems become slightly affordable to be applied in the small to large scale building.

Building Integrated Photovoltaics (BIPV) technology and its application in the façade construction enhance aesthetic of the building, improve economic, and technical issues [1]. In the past decade, BIPV as a renewable power generation source has been contribute partially in meeting electrical and thermal loads of the building [2].

BIPV application could give a noticeably different result depends on its location. Different selected location implies in deviating climate, built environment, structure of electricity industry, market stimulation mechanisms, government policies, the offering of local product, end-user demand, current industrial performances and the forms of tariff calculation for grid-connected [3]. Norton et. al. also stated that grid-connected BIPV is the simplest system consist of PV array and inverter that connect directly to the conventional electricity network or source so that batteries are not mandatory. Therefore, performance of the system depends on local climate, PV efficiency, array's orientation and inclination, energy load characteristics and the inverter [3]. Luthander et. al. reviewed papers about PV self-consumption and examine the alternatives to enhance the produced electricity load that can be consumed by the building directly, which are battery storage and demand-side management (DSM) [4].

In the tropical regions where solar radiation and temperature are relatively high, PV has a negative impact on the building's thermal performance. Previous studies showed that PV installation might lead to high indoor temperature due to its non-responsive climate material (low thermal mass) [5]. The utilization of BIPV has not been widely applied in developing countries such as Indonesia. People are still thinking that installing PV system is expensive so that subsidized electricity are cheaper. However, PV as renewable energy sources would be beneficial to meet the electricity needs without relying on the electricity supply from the grid. Nevertheless, technical issues with its application that might occur in the operation of the buildings must be handled carefully.

This study aims to examine the feasibility of solar PV application and its performance in the specific location by considering weather data e.g. solar radiation and ambient temperature. The design would incorporate standard requirements of office buildings while bring comfort to the occupants.

2. Methodology and Strategies

In this paper, renewable energy systems i.e. a small-scale building integrated photovoltaic (BIPV) system were calculated and simulated to examine the feasibility of it for an office building, in Bandung, Indonesia. The BIPV would be studied in the ongoing real project of Institut Teknologi Bandung's (ITB) new office buildings named ITB Innovation Park Ganeca (IIP Ganeca). It consists of the 4-story building with floor area of 1,540 m^2 and a flat rooftop (See figure 1). As can be seen in figure 2, this office building is located on Ganeca Street, just across the ITB Ganeca campus. However, the longest façades of the building are facing East and West due to limitation of available site location.

Figure 1. The application of the BIPV in the ITB Innovation Park Ganeca Building

This study conducted in four steps, i.e. site analysis, load evaluation, calculation of size and design of the BIPV, and the last is energy and economic analysis. Detail of all the four steps are described below.

2.1. Site analysis

Several simulations were run using Bandung, Indonesia as location of the building. Geographically, Bandung is located in 6.91° South and 107.60° East. Bandung is categorized as a hot and humid climate. The average amount of solar radiation (HT) in Bandung, Indonesia is 483.3 Wh/m^2. This measurement starts from 6 am until 6 pm in the horizontal plane 0^0. Because its location near the equator line, the average of solar radiation is quite high, and it is possible to utilize photovoltaic panel to produce electricity. The ambient temperature of this location was recorded with a minimum of 17^{0}C, maximum of 29.2^{0}C, and average 23.3^{0}C according to climate-data.org [6].

Figure 2. The site location of ITB Innovation Park Ganeca Building

2.2. Load Evaluation
In load evaluation, number of occupants and standard requirements in an office building in Bandung were considered. According to Energy Use Intensity (EUI) by ASEAN-USAID study in 1987, it is proposed that the EUI for an office building in Indonesia is 240 kWh/m^2/year [7]. However, Minister of Energy and Mineral Resources regulations in 2012 stated that an efficient office building using air conditioning should consume 8.5-14 kWh/m^2/month or about 102-168 kWh/m^2/year [8]. The total daily load was then dispersed into 24 hours depending on the electricity use and scheduling in the office building. Energy load for lighting is 15% of the total energy consumption of the building [9].

2.3. Determine the size and design of PV systems
For the simulations, the PV type used is the 250 WP Greentek Polycrystalline with Inverter On-Grid ICASolar 20 kW. The cost of the package consists of 88 PV Panels 250 WP and Inverter 20kW was IDR 522.500.000 or USD 35,761.21 and cost of the package consists of 48 PV Panels 250 WP and Inverter 10kW was IDR 285.000.000 or USD 19,734.79 [10].

2.4. Analyze the energy produced by the systems and the simple payback period of PV application

3. Calculation and result
The PV system used in the building are using these formulae:

Energy delivered = η_a x (A module x PV array) x Solar radiation (1)

Energy grid (import/export) = (Energy delivered x η_{inv}) – energy load DC (2)

Savings per year = (Energy load DC x electricity tariff/kwh x 365 days x 0.85) + (Energy export to grid x feed-in tariff/kwh x 365 days x 0.85) (3)

$$\text{Simple payback period} = \frac{initial\ cost\ (Ic)}{first\ year\ savings\ (\Delta\ Ec)} \quad (4)$$

In this calculation, if the Energy grid is in a positive number, it means that the PV is exporting electricity to the grid. The total exported energy was then multiplied by fit-in tariff/kWh. In contrast, if the Energy grid is in a negative number, it means that the PV is importing electricity from the grid. The total imported energy was then multiplied by tariff/kWh. In total, there are 224 PV arrays which is applied in to the roof and the western façade.

If the PV covered the whole building with energy load of 240 kWh/m^2/year, it could deliver a total energy of 65,957 kWh/year and still need to import 368,867 kWh from the grid. After calculating the savings per year, the payback period of investing 224 PV panels is 15 years.

If the PV only covered the energy load for lighting which is only 15% from total building energy load (36 kWh/m^2/year), it could deliver the same total energy of 65,957 kWh/year and it will export of 733 kWh/year to the grid. The payback period would reduce to 12 years.

IOP Conf. Series: Earth and Environmental Science **291** (2019) 012026 doi:10.1088/1755-1315/291/1/012026

A calculation was performed to see the difference if the total energy load reduces to 150 kWh/m^2/year and if the PV arrays reduced to 176 panels. The initial cost for investing 224 PV panels was IDR 1,330,000,000 or USD 91,997.43 while for 176 PV panels was IDR 1,045,000,000 or USD 72,283.69. The result of the calculation can be seen in Table 1.

Table 1. Energy delivered in one day and yearly savings

Energy load baseline	Electrical Energy Load DC / year	Total PV Panels	Energy delivered /year	Import/ Export	Savings /year	Payback Period
kwh/m2/year	kWh	Pcs	kWh	kWh	IDR	years
240 (total)	434,824	176	58,075	-376,749	78,981,898	14.00
	434,824	224	65,957	-368,867	89,701,156	15.00
36 (lighting)	65,224	176	58,075	-7,149	93,800,121	12.00
	65,224	224	65,957	733	112,051,051	12.00
150 (total)	271,765	176	58,075	-213,690	78,981,898	14.00
	271,765	224	65,957	-205,808	89,701,156	15.00
22.5 (lighting)	40,765	176	58,075	17,310	125,035,309	9.00
	40,765	224	65,957	25,192	148,755,661	9.00

After having the result of previous calculation, the payback period of investing renewable energy application in IIP Ganeca could be faster if the energy load that covered by PV panels are lower. In Table 1, reducing energy load result in higher exported energy to the grid which also followed by the increment of yearly savings. However, when the PV covered 240 kWh/m^2/year and changed into 150 kWh/m^2/year to meet the standard of an efficient office building, the amount of yearly savings does not change. The calculation with the same method was then continued to unveil how many loads covered by PV that could increase yearly savings. It is revealed that when the PV only covered less than 80 kWh/m^2/year, the yearly savings increased, and the payback period decreased. When the load covered is 0 kWh/m^2/year, which means all the energy delivered by PV was exported to the grid, then the income from fit-in tariff is the maximum with payback period reduce to 6.24 years. The result can be seen in Table 2 below.

Table 2. Simulation of load covered by PV to calculate yearly savings and payback period

Load covered by PV	Savings and Income from PV	Payback Period
KWh/m2/year	IDR	years
240	Rp 89,701,156.21	14.83
220	Rp 89,701,156.21	14.83
200	Rp 89,701,156.21	14.83
180	Rp 89,701,156.21	14.83
160	Rp 89,701,156.21	14.83
140	Rp 89,701,156.21	14.83
120	Rp 89,701,156.21	14.83
100	Rp 89,701,156.21	14.83
80	Rp 89,701,156.21	14.83
60	Rp 90,368,889.94	14.72
40	Rp 104,620,440.6	12.71
20	Rp 155,701,061.4	8.54
0	Rp 213,040,246	6.24

4. Conclusion

To sum up, the application of renewable energy resources such as PV panels in an office building in Bandung, Indonesia could reduce electricity bills. 224 modules of PV panels applied in the flat roof and western façade of IIP Ganeca with a total area of 365 m2. The study shows that the system could generate electricity almost 66 MWh per year. The PV panels should be placed in 00 tilted roofs and the 90o tilted western facade where the solar radiation higher than the other side to obtain maximum result. This study also revealed that the lower energy load covered by the PV, the yearly savings will be increased and lessen the payback period. The fastest payback period is gained when electricity from PV panels are sold to the grid.

However, it is important to maintain the PV panel's temperature remains stable to avoid overheating which could resulting inefficiency. Surely, the advance calculation to lessen the payback period which considering the summation such as calculation of value per building component, discount rate, incentives, etc. The increased value that could be used by building owner because the building has fulfilled the Green Building certification, will also be a benefit for marketing. From this calculation and simulation, we could also convince that if Indonesian government would give more incentives for every renewable energy system applied, the impact would be greater than business as usual and the energy could be utilized for the other needs such as public services.

5. References

[1] Tripathy M, Sadhu P K, Panda S K 2016 A critical review on building integrated photovoltaic products and their applications *Renewable and Sustainable Energy Reviews* **61** 451-465
[2] Biyik E, Araz M, Hepbasli A, Shahrestani M, Yao R, Shao L, Essah E, Oliveira A C, Caño T D, Rico E, Lechoñ J L, Andrade L, Mendes A, Atli Y B 2017 A key review of building integrated photovoltaic (BIPV) systems *Engineering Science and Technology, an International Journal* **20** 833-858
[3] Norton B, Eames P C, Mallick T K, Huang M J, McCormack S J, Mondol J D, Yohanis Y G 2011 Enhancing the performance of building integrated photovoltaics *Solar Energy* **85** 1629-1664
[4] Luthander R, Widén J, Nilsson D, Palm J 2015 Photovoltaic self-consumption in buildings: A review *Applied Energy* **142** 80-94
[5] Aaditya G, Mani M 2016 BIPV: a real-time building performance study for a roof-integrated facility *International Journal of Sustainable Energy* **37** 249-267
[6] Climate-data.org 2018 Climate of Bandung (accessed November 26, 2018). URL https://en.climate-data.org/asia/indonesia/west-java/bandung-3246/
[7] USAID Indonesia 2015 Panduan Penghematan Energi di Gedung Pemerintah Sesuai Amanat Peraturan Menteri ESDM No. 13 tahun 2012 tentang Penghematan Pemakaian Listrik URL http://www.iced.or.id/wp-content/uploads/2015/11/Panduan-Penghematan-Energi-di-Gedung-Pemerintah.pdf
[8] Ministry of Energy and Mineral Resources (MEMR) 2012 Peraturan Menteri ESDM No. 13 Tahun 2012 tentang Penghematan Pemakaian Tenaga Listrik [Ministry of Energy and Mineral Resource Regulation No 13 Year 2012 about Electricity Usage Savings] (accessed November 21, 2018)
[9] Autodesk Knowledge Network 2018 Equipment and Lighting Loads URL https://knowledge.autodesk.com/support/revit-products/getting-started/caas/simplecontent/content/equipment-and-lighting-loads.html (accessed November 21, 2018)
[10] Greentek Indonesia 2018 Paket PLTS Rooftop On-Grid URL http://greentek.co.id/en/89-paket-plts-rooftop-on-grid (accessed November 21, 2018)

Acknowledgments

Authors wishing to acknowledge assistance or encouragement from our colleagues in Department of Architecture, Institut Teknologi Bandung. We gratefully acknowledge the funding from USAID through the SHERA program - Centre for Development of Sustainable Region (CDSR).

IOP Conf. Series: Earth and Environmental Science **291** (2019) 012027 doi:10.1088/1755-1315/291/1/012027

Multi-criteria analysis for renewable energy generation: A case of Tokyo

Dinna Fitriyana[1], Mutia Prabawati[2], Frank Tanuwijaya[2], and Dintani Y N Naimah[3]

[1] Institut Teknologi Bandung (ITB)
[2] Eindhoven University of Technology
[3] Centre for Development of Sustainable Region (CDSR), Universitas Gadjah Mada (UGM)

E-mail: dinnafitriyana.df@gmail.com

Abstract. A multi-criteria analysis to determine the most suitable renewable energy technology in Tokyo, Japan, is presented in this study. Several criteria have been considered for the analysis including technical, social, economic, and environmental. Sensitivity analysis has been done to check the solidity of the data used in the analysis. The results show that the solar photovoltaic technology is preferred in Tokyo, followed by wind and biomass energy.

1. Introduction

Nowadays, most countries have realized the urgency of environmental issues, such as the unfavorable impact of fuel on the environment. These issues might reconsider their preferences within the energy sector, and accidents that happened at the Chernobyl and Fukushima nuclear power plant have reestablished global attention to renewable energy. A particular country, such as Japan, has abandoned nuclear technologies and substituted them with coal-fired power plants of which leads to greenhouse-gas emissions increase in the year 2014 [1, 2]. To encounter negative public opinion on nuclear technologies while also attempting to reduce the greenhouse-gas emissions, Japan has focused on renewable energy. Especially in Tokyo, the largest city in Japan, of which the energy-derived greenhouse gasses emissions are approximately equivalent to the amount of Austria [3]. Also, Japan aims to increase its renewable penetration to 9% in 2020 [4].

In order to determine technology development within the energy sector, economic, environmental, social, and technological criteria should be analyzed. Several studies assessed energy sector development within a scope of country or city using various multicriteria decision-making methods [5-11]. As an effort to reduce greenhouse gasses emissions in Tokyo, this paper is aimed to determine the most suitable renewable energy technology by assessing economic, environmental, social, and technological criteria.

2. Alternatives

Renewable energy sources use natural resources, such as sunlight, wind, rain, tides, waves, and geothermal energy [5]. This section provides a brief description of the renewable energy alternatives for the city of Tokyo.

Solar energy is a promising alternative since Japan is one of the leading solar power manufacturers in the world. The Japanese government encourages the implementation of solar energy in various scale throughout the country [12]. Additionally, the government offers subsidies for solar photovoltaic (PV) installation in households. Solar PV systems are flexible and can be installed on the buildings thus the land use could be minimal. However, the energy produced by solar PV is far more expensive than from other renewable energy sources [2].

In 2013, Tokyo was produced around 1.8 million tons of combustible waste as a fuel for biomass plant [13]. One kind of those waste is food waste of which 6000 tons a day can be produced in Tokyo alone [14]. The best technology to handle the food waste is methane fermentation which is also powerful to cope with global warming. This technology is under the objective of Tokyo Super-Eco Town Project to achieve more efficient disposal waste and encourage the development of environmental-related industries [15]. However, the technology requires high investment costs at commercial scale and might cause an odor nuisance [16].

Wind power technology has been considered as an alternative due to a severe nuclear crisis in Japan that now is pushing the technology to the front to meet the country's future energy demand [17]. Wind power is considered cost-effective, the power price of land-based utility-scale is one of the lowest among other energy sources [18]. Also, it does not emit greenhouse gas emission which can cause environmental damage. However, its initial investment is higher even though the cost has decreased dramatically in the past ten years. Its turbine might also cause noise and aesthetic pollution.

Other renewable energy technologies such as hydropower and geothermal have been excluded from this paper. Even though Japan has a considerable hydropower potential, the large-scale installation requires an extensive land-use, while the small-scale one has been reported as less efficient [19]. Geothermal is also worth to consider due to its tremendous potential in Japan. However, further studies showed that the drawbacks of geothermal including long construction time which might take four to 7 years to complete [20]. Thus, it might sound unrealistic to have geothermal as an alternative to achieve Tokyo's goal to reduce its emissions in the near future.

3. Methodology

The procedural steps for multi-criteria analysis (MCA) may include: (i) determining the decision context; (ii) identifying the alternatives; (iii) identifying the objectives and criteria; (iv) specifying the performance of each alternative against the criteria by creating evaluation matrix; (v) assign weighting to each of the criteria; (vi) calculating dominance scores including quantitative and qualitative data; (vii) examining the results; and (viii) conducting sensitivity analysis [21]. Several MCA procedures are distinguished each other by the techniques to process the necessary information in the evaluation matrix. Direct analysis is used in this paper of which the initial step can be to check if any of the alternatives are dominated by others. To do this, dominance scores are calculated separately for quantitative and qualitative data. Then, the quantitative and qualitative dominance scores are calculated to generate the overall dominance scores.

Due to the project constraints, the weighting process, of which usually involves reasoned inputs by stakeholders, was done by reasoned judgment and literature review by the authors. While the criteria weights are extracted subjectively, the sensitivity analysis is then conducted to analyze how variation of criteria weights might affect the results [9]. In this paper, the analysis was conducted using two techniques: (i) criterion exclusion; and (ii) turning points. Criterion exclusion was undertaken by excluding specific criteria and re-examining the decision output [6]. Meanwhile, the turning points technique was obtained through increasing or decreasing weight of specific criteria and lowering the others evenly until the first ranking changes.

3.1 MCA decision model

After analyzing the energy generation potential and conducting extensive literature, an MCA-based decision model has been formulated. The goal of the decision model is selection, evaluation, and ranking of renewable energy technologies for electricity generation in Tokyo to contribute in

IOP Conf. Series: Earth and Environmental Science **291** (2019) 012027 doi:10.1088/1755-1315/291/1/012027

achieving the 2020 Japan's goal. Four criteria and seven sub-criteria were identified that have a direct impact on the goal of the decision model. Description of criteria from social, economic, technical, and environmental aspects is presented in Table 1.

Table 1. Criteria and sub-criteria of decision model

Criteria	Sub-criteria	Description	Reference
Social	Job creation	Renewable energy projects generate employment potential especially for the local communities *Alternative that creates more job opportunities is considered better (+)*	[22]
	Quality of life	Renewable energy projects contribute to noise, traffic, odors, and visual to the local communities *Alternative that less creates bad impacts on the quality of life is considered better (+)*	[22]
Economic	Capital cost	Capital cost consists of total expense occurred in establishing a power plant. *Alternative that has less capital cost is considered better (-)*	[22]
	Electricity cost	Expected cost of the electricity generated by power plant. *Alternative that can generate electricity at a lower cost is considered better (-)*	[23]
Technical	Energy generation potential	Amount of energy sources potential which can be generated. *Alternative that has more resource available is considered better (+)*	[22]
	Land-use	Total available energy generated per area used. *Alternative that occupies less area is considered better (-)*	[22]
Environmental	Greenhouse gasses emissions	Greenhouse gasses released by power plant operations. *Alternative that emits less emission is considered better (-)*	[22]

Table 2_presents quantitative and qualitative data related to job creation, quality of life, capital cost, cost of electricity, energy generation potential, land use, and greenhouse gasses emission. It should be noted that some data used in the table are adapted from other countries since the country-specific data are not available. However, the same technology is likely to have similar relative requirements when implemented in a different country [7].

In social criterion, job creation sub-criterion was assessed through the data from California Clean Energy Future [24]. While the quality of life data sub-criterion was identified based on each alternative's noise, traffic, odors, and visual pollution [15, 25- 29]

In economic criterion, electricity cost is based on Japan data in euro/kWh [30]. The capital of-of various technologies is based on US data in euro/MWh [31].

The energy generation potential in technical criterion was calculated on an annual basis [32]. Meanwhile, the land use data were collected from US data in km^2/GWh [33].

In the environmental criterion, the greenhouse gasses emission includes carbon dioxide, methane, nitrous oxide, and hydrofluorocarbons. The emission data are median numbers expressed in carbon dioxide equivalent [34] and represent overall emissions during the technology lifecycle.

Table 2. Evaluation matrix of alternatives to generate renewable energy in Tokyo. W is weighting

Criteria	Sub-criteria	Solar PV	Biomass	Wind	+/-	Dimension	W
Social	Job creation	2.1	10.55	0.47	+	jobs/MWa	6
	Quality of life	++	+	+++	+	qualitative	9
Economic	Capital cost	+++	+	++	-	qualitative	17
	Electricity cost	0.275	0.175	0.215	-	euro/kWh	15
Technical	Energy generation potential	3878	478	52	+	GWh/year	20
	Land-use	35	1000	100	-	km^2	8
Environmental	Greenhouse gasses emissions	48	230	11	-	gCO_2	25

3.2 Weighting
Weighting is determined by the authors based on the objective of the decision model. The total weights distributed is 100. Social criterion, which is not directly linked to the primary objective, is allocated 15 weights. Meanwhile, the economic criterion is given 32 weights since both the capital and electricity costs are assumed to be the most important one. The environmental criterion is assigned 25 weights since the objective is to reduce the greenhouse gasses emissions. Lastly, the technical objective is given 28 weights due to its importance in producing energy.

4. Result and discussion
Based on the overall dominance score, results indicate that solar PV as the preferred alternatives, followed by wind and biomass energy. The result might be attributed to the abundance of solar potential and low land requirement. Despite having the lowest scores in the economic criterion, low greenhouse gasses emissions of solar PV have outweighed such economic considerations for this particular MCA objective.
Wind energy alternative takes the second place with the best scores in the environmental criterion. However, wind energy scores in energy generation potential, job creation, and land-use are lower than solar PV alternative. These factors have reduced the overall dominance score of wind energy.
Lastly, biomass, which placed in the third position, has been scored as the best in job creation and electricity cost aspects. However, its low score in greenhouse gasses emissions and energy generation potential might put it as the last preferred alternative.
Based upon the turning points technique in the sensitivity analysis, the most sensitive sub-criterion is quality of life while the most robust is capital cost aspect.
Criterion exclusion technique is conducted for job creation and quality of life sub-criterion. Job creation assessment is considered to be relatively unreliable due to the lack of data for the Japan-specific location. The quality of life score is also found to be less reliable since it is consisted of several different data to determine the qualitative score. The result from this technique shows that the criterion exclusion provides a slight change to the preferred alternatives. Solar and wind energy change the rank position due to the exclusion of quality of life aspect. It shows that the aspect is essential and a reliable data source is needed to provide a better consideration between wind and solar alternative. Meanwhile, the exclusion of job creation aspect does not give any change to the rank.

5. Conclusion

Determining the most appropriate alternative to reduce greenhouse gasses emissions while coping with Japan's 2020 goal is a complicated process. MCA might help to answer the objective; however, further research is needed to generate a more reliable result. Future work includes examining other renewable energy system alternatives and adding more aspects to the decision model. Other techniques to process the evaluation matrix are also worth to be conducted. Lastly, the weighting process could be done by involving inputs from various stakeholders. Thus, more objective results can be obtained.

6. References

[1] G. A, "Energy futures: Diversity and the need for assessment," Futures, pp. 43:820-830, 2011.
[2] D. Diakoulaki and F. Karangelis, "Multi-criteria decision analysis and cost benefit analysis of alternative scenarios for the power generation sector in Greece," Renewable and Sustainable Energy Reviews, pp. 716-727, 2007.
[3] Bureau of Environment Tokyo Metropolitan Government, "Final Energy Consumption and Greenhouse Gas Emissions in Tokyo," [Online]. Available: http://www.kankyo.metro.tokyo.jp/en/climate/index.files/b0548c2a69e7883f1945aca50f606a92.pdf.
[4] N. Shinjuku, "Tokyo Renewable Energy Strategy," 2006. [Online]. Available: http://planning.cityenergy.org.za/Pdf_files/world cities/asia/city of tokyo/Tokyo%20Renewable%20Energy%20Strategy.pdf.
[5] I. Vardopoulos, ""Multi-criteria Analysis for Energy Independence from Renewable Energy Sources Case Study Zakynthos Island, Greece," International Journal of Environmental Science and Development, pp. 460-465, 2017.
[6] D. Browne, B. O'Regan and R. Moles, "Use of multi-criteria decision analysis to explore alternative domestic energy and electricity policy scenarios in an Irish city-region," Energy, pp. 518-528, 2010.
[7] M. Amer and T. U. Daim, "Selection of renewable energy technologies for a developing county: a case of Pakistan," Energy for Sustainable Development, pp. 420-435, 2011.
[8] C. Wimmler, G. Hejazi, E. D. O. Fernandes, C. Moreira and S. Connors, "Multi-criteria decision support methods for renewable energy systems on islands," 2015.
[9] H. Ren, W. Gao, W. Zhou and K. I. Nakagami, "Multi-criteria evaluation for the optimal adoption of distributed residential energy systems in Japan," Energy Policy, pp. 5484-5493, 2009.
[10] S. D. Pohekar and M. Ramachandran, "Application of multi-criteria decision making to sustainable energy planning: a review," Renewable and sustainable energy reviews, pp. 365-381, 2004.
[11] J. Sliogeriene, Z. Turskis and D. Streimikiene, "Analysis and choice of energy generation technologies: The multiple criteria assessment on the case study of Lithuania," Energy Procedia, pp. 11-20, 2013.
[12] Global Energy Network Institute, "Solar Energy in Japan - Summary," [Online]. Available: http://www.geni.org/globalenergy/library/energytrends/currentusage/renewable/solar/japan/summary.shtml.
[13] Clean Authority of Tokyo 23 Cities, "Quantity and contents of waste," [Online]. Available: http://www.union.tokyo23seisou.lg.jp.e.de.hp.transer.com/somu/koho/toke/nakami/index.html.
[14] Japan Times, "An appalling waste of food," 2013. [Online]. Available: http://www.japantimes.co.jp/opinion/2013/01/21/editorials/an-appalling-waste-of-food/.VDkLuPmSx8E.
[15] Asia Biomass Energy Cooperation Promotion Office, "Methane Generation by Fermenting Biomass is Flourishing in Japan," [Online]. Available: http://www.asiabiomass.jp/english/topics/0910 03.html.

[16] Anaerobic Digestion, "The Advantages and Disadvantages of Anaerobic Digestion vs Composting," [Online]. Available: https://blog.anaerobic-digestion.com/anaerobic-digestion-vs-composting/.
[17] Associated Press The Guardian, "Japan to scrap nuclear power in favour of renewable," [Online]. Available: http://www.theguardian.com/environment/2011/may/10/japan-nuclear-renewables.
[18] Office of Energy Efficiency and Renewable Energy, "Advantages of Wind Power," [Online]. Available: https://www.energy.gov/eere/wind/advantages-and-challenges-wind-energy.
[19] K. Takemata, "Hydro and geothermal development in Japan. Workshop Scaling Up Financing to Expand the Renewables Portfolio, OECD Conference Centre," [Online]. Available: http://www.iea.org/media/workshops/2013/scalingupfinancingtoexpandrenewables/7JPOWER.pdf.
[20] B. Matek, "Annual U.S. Global Geothermal Power Production Report Geothermal Energy Association," [Online]. Available: http://geo-energy.org/events/2014%20Annual%20US%20%20Global%20Geothermal%20Power%20Production%20Report%20Final.pdf.
[21] J. S. Dodgson, M. Spackman, A. Pearman and L. D. Phillips, Multi-criteria analysis: a manual, 2009.
[22] J. J. Wang, Y. Y. Jing, C. Zhang and J. H. Zhao, "Review on multi-criteria decision analysis aid in sustainable energy decision-making," Renewable Sustainable Energy Rev, 2009.
[23] P. MA, P. PA and C. AI, "Evaluation of liquid bio-fuels using the Analytic Hierarchy Process," Process Saf Environ Prot, 2008.
[24] Californias Clean Energy Future, Preliminary Estimates of Job Creation, 2012.
[25] Union of Concerned Scientists, "Environmental Impacts of Biomass for Electricity," [Online]. Available: http://www.ucsusa.org/clean energy/our-energy-choices/renewable-energy/environmental-impacts-biomass-for-electricity.html.VDm TvmSx8E.
[26] Tribal Energy and Environmental Information Clearing House (TEEIC), "Biomass Energy Operations Impacts," [Online]. Available: http://teeic.indianaffairs.gov/er/biomass/impact/op/index.htm.
[27] G. C. Ledec, "Addressing The Environmental and Social Impacts of Wind Power," [Online]. Available: https://www.climateinvestmentfunds.org/cif/sites/climateinvestmentfunds.org/files/Social%20Impacts.pdf.
[28] Technical Assistance Services for Communities (TASC), "Biomass Power Plant Informational Summary," [Online]. Available: http://www.epa.gov/superfund/community/tasc/pdfs/biomass-report.pdf.
[29] Energy Sector Management Assistance Program (ESMAP), "Addressing The Social Impacts of Wind Power Development," [Online].
[30] K. Komatsubara, Wind Energy Japan, 2012.
[31] Energy Information Administration (EIA), "Annual energy outlook 2011," Energy Information Administration (EIA), 2011.
[32] W. Tatsuya and E. Sachio, "Estimation of Renewable Energy Potential and Use- A Case Study of Hokkaido, Northern-Tohoku Area and Tokyo Metropolitan, Japan," [Online]. Available: http://www.ep.liu.se/ecp/057/vol12/012/ecp57vol12 012.pdf.
[33] U.S Department of Energy, [Online]. Available: http://www.nrel.gov/docs/fy04osti/35097.pdf.
[34] Intergovernmental Panel for Climate Change (Working Group III Mitigation of Climate Change), "Annex III: Technology-specific cost and performance parameters," [Online].

Acknowledgments
We gratefully acknowledge the funding from USAID through the SHERA program - Centre for Development of Sustainable Region (CDSR).

House-scale Renewable Energy System Sizing in Liverpool

Suhendri[1] and Tri Pratiwi Handayani[2]

[1]Institut Teknologi Bandung, Indonesia
[2]Universitas Muhammadiyah Gorontalo, Indonesia

suhendri.aritb@gmail.com

Abstract. Renewable energy technology such as solar thermal, PV, and wind have been developed to be building integrated. This exercise is aimed to sizing those three renewable energy systems for a typical house in the UK, specifically Liverpool. The methodologies used in this exercise are load evaluation, resource assessment, and finally sizing the systems. The results from all of sizing calculation show that there will be 6 evacuated tube collectors with 6m2 area, 14 PV modules with 20.7 m2 area, and 1 wind turbine with 2.5 rotor diameter. The renewable energy systems for electricity sized are predicted to cover 78.47% of the electricity load with PV contribution as many as 54.5% and the rest 24% is covered by wind turbine.

1. Introduction

Household sector shares a significant amount of energy consumption to the world annual energy consumption [1]. Indeed, it also contributes directly or indirectly to CO2 emission to the atmosphere. Changing the basis of energy source for household sector from fossil fuel to renewable energy can reduce its contribution to world CO2 emissions [2]. Nowadays, using renewable energy resources for domestic demand has been a phenomenon at least in developed country since the issue of climate change emerge [3].

Renewable energy technology such as solar thermal, PV, and wind have been developed to be building integrated. However, these building integrated renewable energy technologies are not always be a good solution. Under-sizing, bad performance, or misplaced are some of the problems that may happen. In order to work in optimum condition, they need to be assessed and planned correctly. Solar thermal collector, PV, and wind turbine as the common building integrated renewable energy technology are being used in this study. Specifically, PV and wind turbine are used to generate electricity and solar collector is used to domestic hot water. Location of the study is in Liverpool, the United Kingdom.

2. Objectives

This study is aimed at sizing three renewable energy systems for a typical house in the UK. Three renewable energy systems to be designed are solar thermal, photovoltaic, and wind turbine technology. The resulting system size from the research should be reasonable for typical household in the UK in terms of cost, and dimensions as well. Analytical calculation was used to assess the renewable energy sources and sizing the systems. The data from each method supports each other.

3. Methods

As mentioned in the introduction part, using renewable energy technology systems are not always be good especially for the user. Oversized systems make the investment cost high and the systems works inefficiently. Undersized systems lead to unfulfilled energy demand from the user. Therefore, sizing the systems in the right methods is important [4].

3.1 Load evaluation

The purpose of installing renewable energy technology system is to supply the energy demand in more sustainable ways. To supply energy demand is still the main purpose of it, therefore before sizing the system, how much energy load to be covered should be clear. A typical UK house (Figure 1) for a family of 4 people is taken as a case study. More details about the electricity and hot water loads in this case will be elaborate in Part 4.

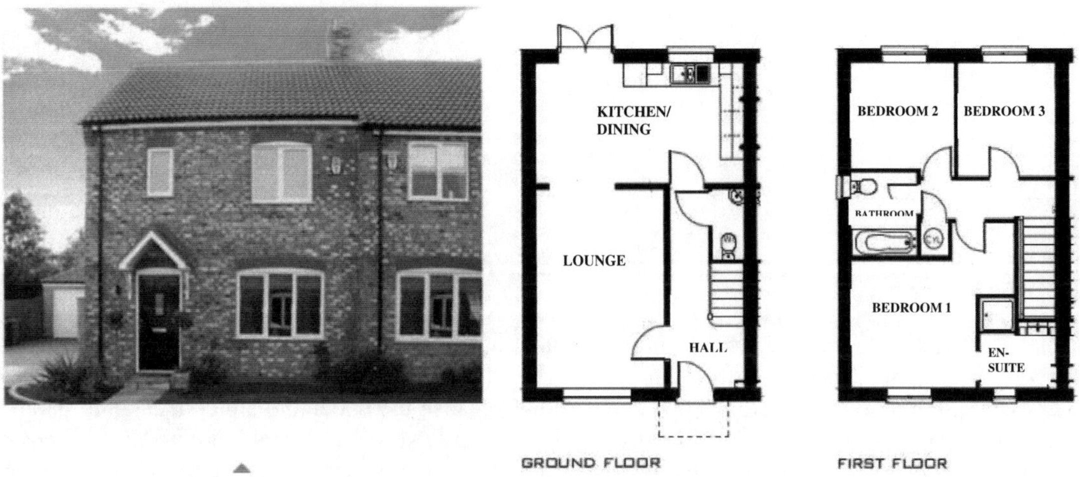

Figure 1. Case house [5]

3.2 Resource assessment

After location and case study for the load have been set, resource assessment is now being more contextual and relevant. Resources assessment is key information whether it is reliable or not to have any building integrated renewable energy system in order to supply energy demand. Solar radiation data and wind speed data are two resources to be assessed. Solar radiation can be taken from NASA's website. Wind speed data is provided by weather data file of EnergyPlus.

3.3 System sizing

Determining renewable energy system specification and its size can be done if the energy load and resource has been known. Roughly, the system size is a product of energy demand divided by resource available. It is common to slightly oversize the system due to the assumption done in resource assessment, because usually the real condition is not as high as the calculation. In the end, the important information from system sizing is to know the cost of the system.

4. Results and discussions

4.1 Solar thermal sizing

Solar thermal system sizing is based on winter conditions where hot water requirement is high and solar radiation is low. From the calculation it is known that number of solar collector needed is 6. With storage tank volume of 65 litres per collector, the storage tank is sized to be 390 litres. That volume is provided by a 1 m height tank with diameter of 70.4 cm.

Temperature profile of water in the tank during winter period can be seen in Figure 2. Energy from solar collector can maintain the temperature above 30°C, but it can only increase the water temperature to about 43°C. Energy from auxiliary system may be added in this condition to get required temperature (60°C).

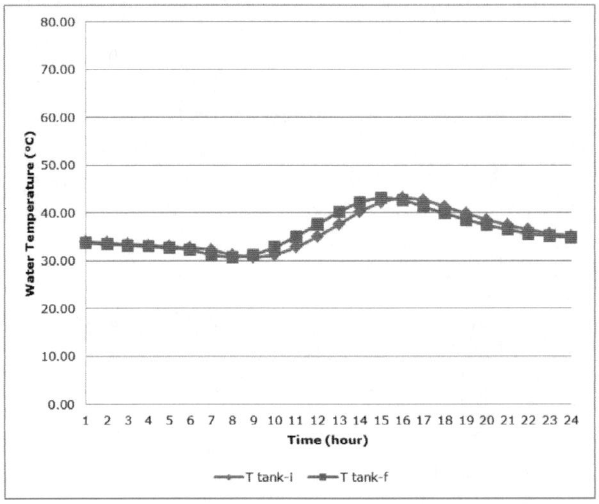

Figure 2. Water tank temperature profile in winter

During summer, to avoid the water to be overheated above 90°C, 4 collectors are preferred than 6. This variation on number of solar collectors might be overcome by using active valve for the pipeline that passing through the collectors [6]. It can be done because the collectors are connected in parallel [7]. This concept is illustrated schematically by Figure 3. By using 4 collectors, hot water demands in summer can fully covered by solar energy. As it shown by Figure 4, water temperature is maintained to be above 60°C and also not exceed 90°C.

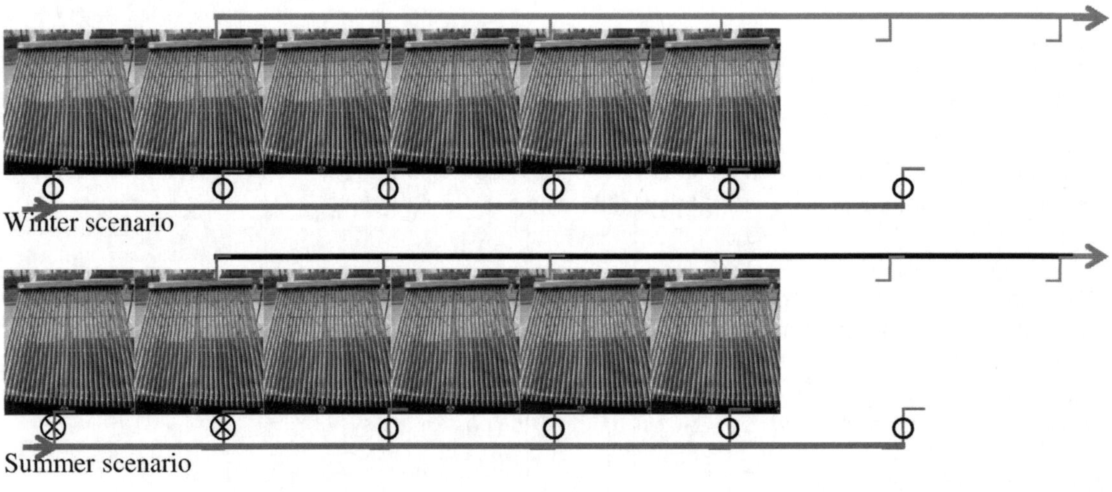

Figure 3. Solar collectors connection scheme

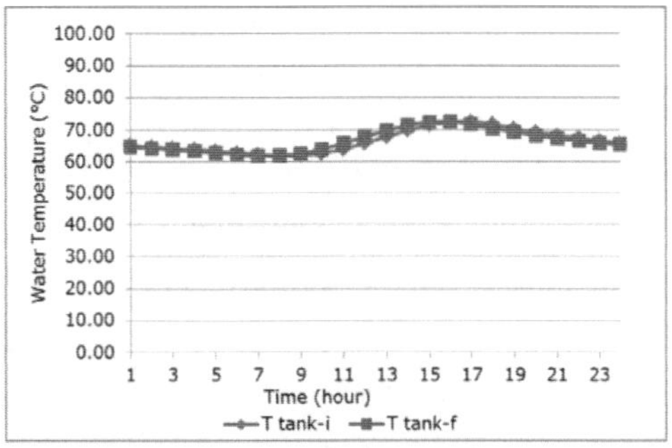

Figure 4. Water tank temperature profile in summer

4.2 Pv sizing

For the same reason as solar collectors, PV system sizing is based on winter conditions. From the calculation, it is known that number of PV needed is 14 PV modules, with PV specifications as written in the Table 1. With such number of PV, the energy balance between electricity delivered by PV, the grid, and the electrical load is shown by Figure 5 (winter) and Figure 6 (summer).

Table 1. PV Specifications

Pmax(W)	250	Vmpp(V)	31	Impp(A)	8.07
Temp. coeff	-0.463%	Voc (V)	37.9	Ioc	8.76
Temp. coeff	-0.330%	Temp. coeff	0.030%	λa	10%
Pmax (W)	179.9	Vmpp(V)	27.9	A module (m2)	1.48
Voc(V)	37.1	Isc(A)	7.07	NOCT (°C)	47.5
Vinv(V)	90-550	max Iinv	20	ηinv	97%

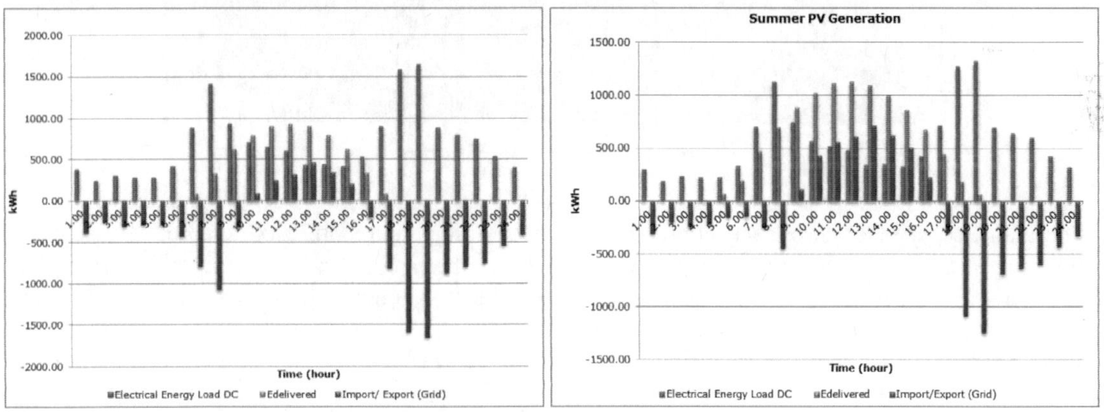

Figure 5. PV electricity energy generation in winter and summer

4.3 Wind turbine sizing

Analytical calculation uses 2.5 rotor diameter wind turbine to cover 75% electricity demand from renewable energy resources. Wind local wind speed condition, this wind turbine is predicted to generate 1,296.66kWh electricity per year.

IOP Conf. Series: Earth and Environmental Science **291** (2019) 012028 doi:10.1088/1755-1315/291/1/012028

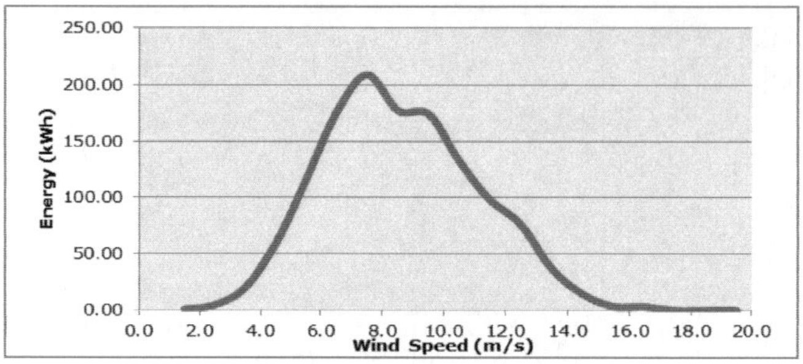

Figure 6. Wind turbine's annual generation

5. Conclusions

Sizing of a house-scale renewable energy system in Liverpool follows three general steps, which are load calculations, resource assessment, and system sizing. It is concluded that a typical house for 4 family members in Liverpool would need at least 6 evacuated tube collectors with area of 6m^2, 14 PV modules with area of 20.7 m^2, and 1 wind turbine with 2.5 m rotor diameter. The collectors and the PVs will be roof-mounted, hence positioning of these modules should be planned. The position of wind turbine also should be located in the right side of the building. Figure 7 shows the positioning of all renewable energy technologies.

Figure 7. Systems positioning

Calculation is based on two conditions, summer and winter, and optimised according to the two different conditions in the two seasons. PV and solar thermal are affected significantly by the condition in summer and winter. The challenge for the two renewable systems is that in winter condition the load increases, but energy generated decreases. Nevertheless, in average, renewable energy contributions compare to electricity load can be seen in Figure 8. All of the renewable energy systems for electricity sized are predicted to cover 78.47% of the electricity load. PV contributes as many as 54.5% and the rest 24% is covered by wind turbine.

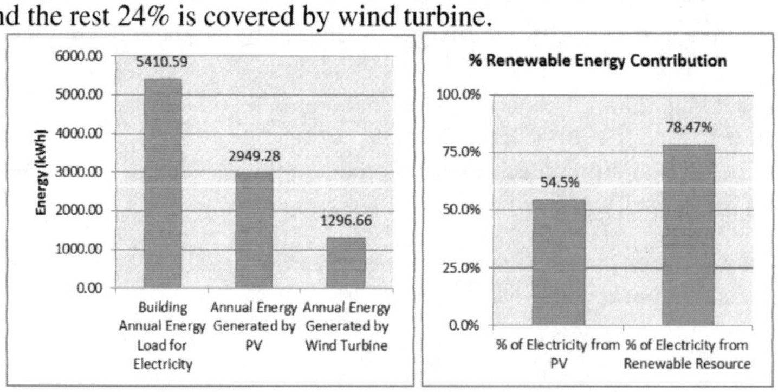

Figure 8. Renewable Energy's Contribution

6. References

[1] Heijs, Wim J. M. 2006 *Household* Energy *Consumption.* User Behavior and Technology Development p.149-157

[2] Brandon, G., Lewis, A. 1999 *Reducing household energy consumption: a qualitative and quantitative field study*. Journal of Environmental Psychology. 19. 1. p.75-85

[3] Shafiei, S., Salim, R. A. 2014 *Non-renewable and renewable energy consumption and CO2 emissions in OECD countries: A comparative analysis*. 66. P.547-556

[4] R. Burzynski, M. C. (2010). A review of domestic hot water demand calculation methodologies and their suitability for estimation of the demand for Zero Carbon houses. Reading: TSBE Centre, University of Reading.

[5] Cyden Homes. (n.d.). Retrieved February 2015, from Cyden Homes: www.cydenhomes.co.uk

[6] Solar Tube Company. (2009). Solar Tube Company. Retrieved May 11, 2015, from http://www.solartubecompany.co.uk/

[7] Sunshine Solar. (2015). Sunshine Solar. Retrieved May 11, 2015, from http://www.sunshinesolar.co.uk/

Acknowledgments

We gratefully acknowledge the funding from USAID through the SHERA program - Centre for Development of Sustainable Region (CDSR)

IOP Conf. Series: Earth and Environmental Science **291** (2019) 012029 doi:10.1088/1755-1315/291/1/012029

Hydrogen production possibility using Mongolian renewable energy

A Dorjgotov[1], B Ulziidelger[1], J Yunseong[2], K Oh Chan[2], J Ok Sung[2] and S Yong-Gun[2]

[1]Department of Chemical Engineering, School of Applied Sciences, Mongolian University of Science and Technology, 8th khoroo, Baga toiruu 34, Sukhbaatar district Ulaanbaatar, Mongolia 14191
[2]Inorganic Material Laboratoty, Department of Chemical and Biomolecular Engineering, College of Engineering, Yonsei University, 50 Yonsei-ro, Seodaemun-gu, Seoul, South Korea

Email: gotovmust@yahoo.com

Abstract. There is widespread popular support for using renewable energy, particularly solar and wind energy, which provide electricity without giving rise to any carbon dioxide emissions. Harnessing these for electricity depends on the cost and efficiency of the technology, which is constantly improving, thus reducing costs per peak kilowatt, and per kWh. Utilizing solar and wind-generated electricity in a stand-alone system requires corresponding battery or other storage capacity. The possibility of large-scale use of hydrogen in the future as a transport fuel increases the potential for both renewables and base-load electricity supply.

1. Introduction.

Hydrogen fuel cell vehicles are considered to be an alternative potential replacement of conventional fossil fueled internal combustion engine vehicles. There are several obstacles for the widespread application of hydrogen fuel cell vehicles, for example high capital and operating costs and lack of hydrogen infrastructure. However, several automobile manufacturers have committed to develop vehicles using hydrogen. The attraction of using hydrogen as an energy carrier is that, if hydrogen is prepared without using fossil fuel inputs, vehicle propulsion would not contribute to CO_2 emissions, along with its high energy concentration per unit weight. The drawbacks of hydrogen use are low energy content per unit volume, high tank weights, very high storage vessel pressures, the storage, transportation and filling of gaseous or liquid hydrogen in vehicles, the large investment in infrastructure that would be required to fuel vehicles, and the inefficiency of production processes [1]. Construction of infrastructure for hydrogen fuel is of interest as it can increase economic feasibility of clean energy as well as it contributes to further technical advancements. Production of hydrogen using renewable energy based on wind and solar energy is one of choice of best technologies [2]. Not only does Mongolia have abundant sources of renewable energy, the cost for lowering emissions is extremely low compared to most countries, and the world's most energy hungry market is right next door. On the other hand, the wide regions distribution of renewable energy resources in Mongolia is giving a on-site production possibility of hydrogen [3].

2. Resent application of renewable energy in Mongolia and future

In 2016 Mongolia generated 5.8 billion kWh in total. The power generated by the thermal power plants accounts for 95.8%, power generated by diesel stations is 0.2%, and 4.2% is generated by renewable energy sources. Also 1.9 billion kWh of electricity is imported. The interest of investors in Mongolian energy sector is considerably high which can be clearly seen from the fact that the total sum of capacity of special permission granted to renewable energy projects has reached 879.2 MW by the September of 2017 [4].

At present, close to 90,000 independent solar PV systems are reportedly in use by herders for operating lights, radios, TVs and satellite dishes. Government is also working with JICA /Japan/, GTZ /Germany/ and World Bank in the development of a comprehensive programs for off-grid aimags and soums. Under these programs already replaced diesel set in 198 soum centers, placed diesel + photovoltaic units in 15 soum centers. The capacity of these units vary from 5 kW to 200kW. The first ever large-scale solar power plant with capacity of 10MW has been constructed in Darkhan City in 2017 [5-6].

At present time, over 4000 portable, small wind generators with capacity from 50 W to 150 W are used by herders for operating lights, radios, TVs in rural area. In last years, number of feasibility studies were made for the construction of wind power plants with capacity of 50 MW at "Salkhit uul",near capital city Ulaanbaatar and 100 MW in "Umnigobi" province. Wind power utilization in Mongolia is still negligible [7].

The result of discussions held between Mongolia and Chinese officials on the high political level, the technical and economic assessment has been developed for exporting electricity to the Chinese market in 2011. The initiative to build the Asian Super Grid, in which Mongolians southern Gobi's solar and wind power resource is to be utilized, and research work on this regard has been carried out with Asian Development Bank's financing. The idea of the Asian Super Grid, also known as Gobitec is a development vision promoted by Softbank's Masayoshi Son that envisions the creation and development of a transmission grid infrastructure (High Voltage Direct Current HVDC) connecting Russia, Mongolia, China, Japan and Korea with the intent of providing the connected countries with renewable energy from the Gobi Desert. If this idea comes to fruition, development on a massive scale can be expected to take place in the Gobi Desert (figure 1).

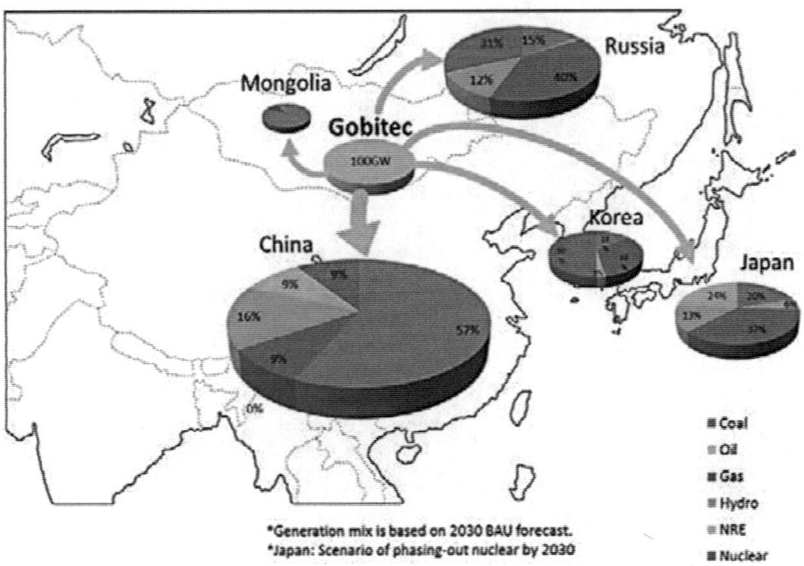

Figure 1. Asian super grid

Based on calculations to build 50 GW solar and 50 GW wind, it is estimated that 900'000 new jobs will be created in Mongolia alone from 2015-2030. Participating countries will reduce 187 Gt CO_2 per year. A total of 7'530km of transmission line will be needed to be built. The total investment cost is calculated at $293 billion with additional yearly maintenance and service cost of $7.3 billion [8].

3. Capacity of solar energy

Mongolia has substantial solar energy potential. The high intensity of solar radiation, long periods of sunshine, high air transparency, and constant presence of moisture in the air favor the development of solar energy in Mongolia. The annual number of sunshine hours in the same place is within the range of 2450-3300, which is higher than the average duration of sunshine in other countries located along the same latitude (figure 2). Approximately 70% of the total land area receives solar insolation at the rate of 3.4-6.0 kWh/m^2 per day. According to long-term expertise estimates, the solar radiation by square units increased, at the point of direction, from North to South and, at the range, from 2500 kWh/m^2 to 3550 kWh/m^2 per year [9]. The average solar radiation received by Mongolia comes to 3000 kWh/m^2 per year (figure 3).

Southern province Umnugobi aimag have abundant solar energy potential. In the region, solar radiation is 5.4 kWh/m^2 per day and the radiation area is 5542 km^2. If we assume that the area is covered by solar panel, it will generate 1.092.000 GWh electric energy per year, which is equal to 1% of world total energy consumption in 2017. The minimum potential resource of solar energy is 1500 GW, which can be generate 4.774.000GWh electric energy per year by the calculation of DOE (table 1).

Table 1. Solar radiation, potential area and electric energy

Calculated, kWh/m^2·day	Potential area, km^2	Electric energy, GWh/year
3.4	5269.5	654000
3.8	3924.7	544000
4.1	4210.6	630000
4.5	4514.8	742000
5.4	5541.9	1092000
Total	23461.5	4774000

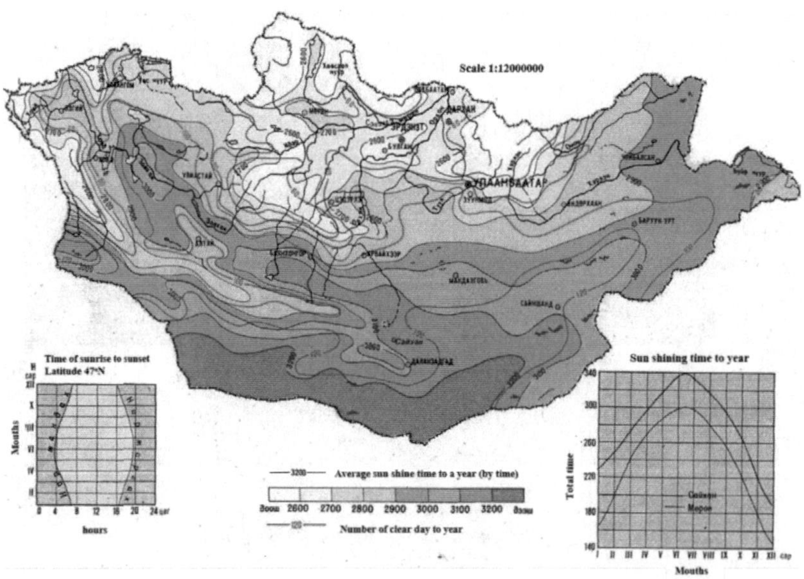

Figure 2. Duration of sunshine in Mongolia

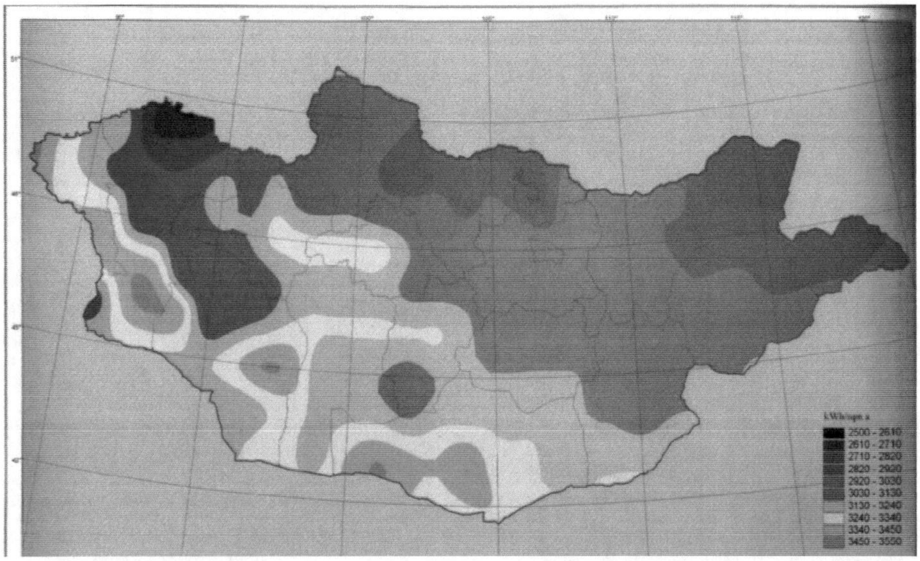

Figure 3. Direct normal solar radiation per year

4. Capacity of wind energy

In year 2000, The National Renewable Energy Laboratory of USA developed a wind energy resource map for Mongolia in cooperation with The Renewable Energy Center of Mongolia and The National Institute of Meteorology of Mongolia (figure 4). The map shown six categories considering wind power density regions.

All territory of the country is under influence of westerly mid-latitude jet stream, a high-speed ribbon of air several kilometers above sea level. The windy land mainly placed in southern and western part of the nation in Gobi and steppe region of the country. More than 10% of the total territory has been estimated to have good-to-excellent wind energy potential for utility scale wind energy applications. Regions in Mongolia with more than 160.000 km^2 area have possibility and are convenient to use high capacity wind power stations [10]. By the assumption, wind power density is about 7 MW/km^2, this amount of windy land could support over 1.100 GW of installed capacity, and deliver over 2.550.100 GWh electric energy per year (table 2). More than 40% of total territory of the country has been estimated to have moderate-to-good wind energy potential, which is suitable for rural power applications. This amount of windy land could support over 4.300 GW of installed capacity, and deliver over 8 trillion kWh per year by conservative estimations. All aimags of Mongolia have at least 6000 MW wind energy resource. Of which, 9 aimags have more than 50000 MW wind energy resource, southern province Umnugobi aimag have 300000 MW wind energy resource.

Table 2. Parameters of good-to-excellent windy lands at 30 m

Categories	At 30m level from land surface		Windy area		Capability, MW	Power per year GWh
	Wind power density, W/m^2	Wind speed, m/s	Area, km^2	%		
1	300-400	6.4-7.1	130.665	81.3	905500	1975500
2	400-600	7.1-8.1	27.165	16.9	188300	511000
3	600-800	8.1-8.9	2.669	1.7	18500	60200
4	800-1000	8.9-9.6	0.142	0.1	1000	3400
	Total		160.641	100	1113300	2550100

IOP Conf. Series: Earth and Environmental Science **291** (2019) 012029 doi:10.1088/1755-1315/291/1/012029

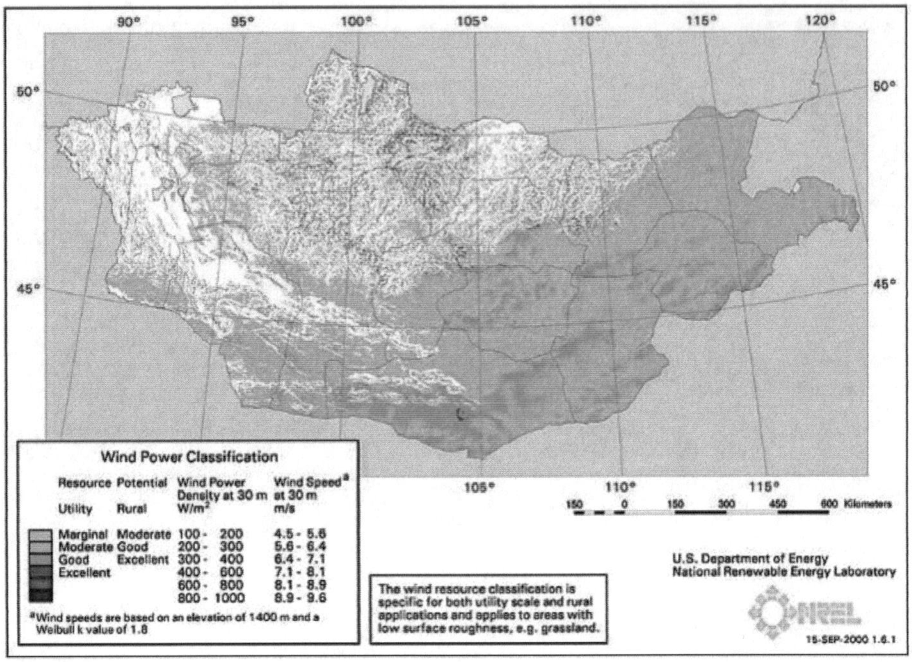

Figure 4. Mongolian wind resource map

5. Hydrogen production possibility

Hydrogen gas production through solar energy which is abundant, clean and renewable is one of the promising hydrogen production approaches.

The average efficiency of photovoltaic cells (PV) is about 20%, the average efficiency of alkaline electrolyser is 80% and the total efficiency of the system will be 16%. PV-electrolyser system with 16% total efficient requires 250 kWh of solar energy to produce 1 kg of hydrogen. Potential solar energy in the Mongolia is estimated to be more than 4.774.000GWh, enough to produce over 19 billion tonn/year of hydrogen. If the PV-electrolyser system set in Southern Umnugobi province, which have more solar energy potential, it will produce 4.35 billion tonn/year of hydrogen from the 5542 km^2 area. It is possible to produce 784 tonn/year hydrogen in 1 km^2 square.

In the case wind energy, efficiency of wind power plants is about 35%, which is higher than PV, the average efficiency of alkaline electrolyser is 80% and the total efficiency of system will be 28%. Potential wind energy in the Mongolia is approximately 2.550.100 GWh from 160641 km^2 good-to-excellent windy lands. It is possible to produce 18 billion tonn/year of hydrogen using total wind energy source because the total efficiency of system is 28%. If the wind power plants-electrolyser system set in excellent windy location with 29834 km^2 area, it can be produce 574600 GWh energy and 4 billion tonn/year of hydrogen. It is possible to produce 136 tonn/year hydrogen in 1 km^2 square of the excellent windy location. There is impossible to produce large amount hydrogen using hydro energy due to low capacity.

In the case power plant electricity is used to hydrogen gas production, electricity cost is approximately 3 $ per 1 kg hydrogen gas because 1kWh electricity price is 0.072 $ in Mongolia. If electricity from renewable resources is used to hydrogen gas production, electricity cost is approximately 1.37 $ per 1 kg hydrogen gas, it is about 2 times cheaper than power plant electricity.

6. Conclusion

5542 km^2 area of Southern Umnugobi province have huge solar energy potential, which is only 3.36% of the province and 0.35% of Mongolian territory. In the region, solar radiation is 5.4 kWh/m^2 per day.

If we assume that the area is covered by PV panel, it will generate 1.092.000 GWh electric energy per year.
Solar radiated hydrogen productivity is 5.76 times higher than wind powered hydrogen in unit area due to dense solar radiation density and long sunshine durability. It is possible to produce large amount of hydrogen in Gobi and steppe region of the country using solar and wind energy. The total calculated hydrogen productivity using sum of solar and wind potential energy is 37 billion tonn/year. The hydrogen productivity in unit area is 920 tonn/year·km^2 by the combine of sum of solar and wind energy. On-site production of hydrogen is also high possibility because 40% of total territory have moderate-to-good wind energy potential and 70% of total territory have 3.4-6.0 kWh/m^2 per day direct solar radiation. North and western regions of the country, which is mountains area, has low solar and wind potential energy. In this regions, on-site production of hydrogen is possibility using solar, wind and/or hydro energies. Possibility to large amount hydrogen production is low in the regions.

7. References

[1] Karim N and Jonathan S, 2011 *How Stuff Works* "How Fuel Cells Work: Polymer Exchange Membrane Fuel Cells"
[2] Wipke, Keith, Sprik S, Kurtz J and Ramsden T 2001 *National Renewable Energy Laboratory Report* "National FCEV Learning Demonstration"
[3] Bossel U 2006 *J. Proceedings of the IEEE* "Does a Hydrogen Economy Make Sense?" **94** 1826
[4] Gordon F 2017 Investors Keen to Support Mongolia's Renewable Energy Goals *Renewable Energy World*
[5] Bavuudorj O 2018, Ulaanbaatar, Mongolia Present Status and Challenges of Energy Sector *9th Mongolian Renewable Energy Forum Report*
[6] Batbold B 2016 The State of Renewable Energy in Mongolia
[7] Eugene G 2018 Mongolia sets out annual wind plan *Wind Power*
[8] Mano S, Ovgor B, Samadov Z, Pudlik M, Jülich V, Sokolov D, and Yoon J Y 2014, Gobitec and Asian Super Grid for Renewable Energies in Northeast Asia report prepared by Energy Charter Secretariat, Energy Economics Institute of the Republic of Korea, Energy Systems Institute of the Russian Federation, Ministry of Energy of Mongolia, Japan Renewable Energy Foundation (2014)
[9] Li Xia 2018 Mongolia to increase renewable energy capacity by 120 MW this year *Asia Pacific Edition*
[10] Batbayar Ch 2015 The role of Mongolia for development of northeast Asian super grid *The 3rd Northeast Asia Energy Security Forum*

A simulation methodology on grid displaced photovoltaic (PV) prospective for existing terrace house supply in Malaysian peninsular.

Abdul-Razak, AHN; Ahmad, N.A; Ahmad, S.S.; Haron, S. N; A.K Ermeey.
Universiti Teknologi MARA, Malaysia

haqqi250@perak.uitm.edu.my

Abstract. Accounting for the government's intention towards sustainable energy development and the deployment of Malaysia's first smart grid system, inclusive participation of residential units will transform their role from consumers to prosumers, enabling them to become "positive nodes" of the smart grid system. This paper evaluates the simulation output of a grid displaced electricity demand of a significant terrace-based housing cohort in Malaysia. Utilising Integrated Environmental Solution Virtual Environment (IESVE) as the simulation tool, solar panel retrofitted to the existing terrace houses in urban centres across peninsular Malaysia is investigated in order to identify its potential to displace current grid electricity demand, consequently shaving peak demand and reduced household's 's electricity bill.

1. Introduction

Worldwide residential energy demand shares a substantial amount of the overall national demand, where Malaysian residential energy demand equated to 19% that fell within the range of 15% to 50% of global energy consumption [1]. According to the Malaysian final energy demand data, a steady increase across various sectors were recorded (21,883 ktoe in 1995 to 51,806 ktoe in 2015) [2] . This trend is expected to continue, driven by national economic and population growth, reflected by 3% to 5% annual GDP growth [3]. UN Habitat predicts that Malaysia's urban population will reach 78% in 2030 and lifestyle shift will have a consequential impact on energy demand. Given the rising demand yet limited supply, Malaysia has set its own sustainable energy development through diversification of the energy trend mix, via Five Fuel Diversification strategy [4]. Despite this, the majority of Malaysian energy source are still fossil fuel based, which is widely known to be the main cause of environmental issue causing global warming and is exposed to volatile global energy prices [5]. Nevertheless, effort by the Malaysian government together with its leading utility company, Tenaga Nasional Berhad (TNB) are showing inclination towards renewable energy (RE) sources. Current Energy, Green Technology, Science and Climate Change Ministry aims to increase renewable energy from 2% to 20% by 2025. Malaysian residential sector final energy consumption is divided into five aggregated categories; appliances, cooking, space cooling, lighting and water heating [6]. Meanwhile, typical terrace household energy use is segregated into cooking (45%), cooling (29%), others (11%), lighting (8%) and laundry/cleaning (7%) [7]. Under the cooling category, 17% is used up for air conditioning, 10% for ceiling fan and 2% for other types of cooling devices. Increasing use of electricity to substitute Liquid Petroleum Gas (LPG) has been recorded as modern society value electricity as indispensable form of energy.

Compelled by the above issues, factors of future energy security and changes to government policy, TNB has initiated the smart grid system transformation plan [8]. Alongside the aim to achieve Sustainable Development Goals (SDG) 2030 similar to other developed countries, the transformation plan is expected to reduce carbon emission and facilitate renewable and sustainable energy production. Furthermore, smart grid implementation empowers and encourages consumers to actively participate in reducing electricity use from the grid [9]. Adopting demand side management, critical information and data linked to pricing mechanism and total energy use are relayed to the consumers thus helping them to make sustainable decision [10] Plus, the smart grid system enables consumers to become prosumers as they can visibly identify and control their daily energy consumption aside from the ability to generate electricity on-site [11]. The activation of houses as positive energy generation nodes in the smart grid may have a significant impact on the Malaysian electricity sector, thus future terraced houses must be designed and constructed according to the requirement of the smart grid, taking full advantage of its potentials.

Considering energy related issues in this sector, many studies were engaged aiming to reduce energy demand [12] increase energy efficiency [13] and optimize residential design to adopt passive design strategies [14,15] . Nevertheless, concerns of shifting lifestyle status quo have impeded such efforts [16]. As air-conditioning (AC) system becomes affordable and energy efficient, modern society have become accustomed to this new indoor climate, leading to higher electricity use [17]. In 2016, Berkeley National Lab reported that global AC installation will reach 700 million by 2030 and 1.6 billion by 2050. Accordingly, several local studies on the terrace house dwellers revealed more than half of the survey respondents owns AC unit, with at least two units installed in each households [7,18]. These indicators show AC will remain significant and more may be installed in the future. Thus, local electricity generation to offset grid demand and transform houses as positive node is imperative.

2. Terrace house electricity household consumption and the way forward

Malaysian residential electricity end use can be categorised by refrigeration (21.1%), cooking equipment (14.7%), air-conditioning (11.9%), washing machines (10.5%), illumination (5.1%) and others (36.7%) [19]. A typical monthly terrace house electricity consumption were 455kWh [7]. Under TNB's revised domestic tariff (Tariff A) [20], consumers pay RM 0.218/kWh for the first 200kWh (1kWh-200kWh), RM 0.334/kWh for the next 100 kWh (201kWh-300kWh), followed by a substantially higher rate of RM 0.516/kWh (301kWh-600kWh), RM 0.546 (601-900kWh) and RM 0.571 for the subsequent kWh used. Therefore, a consumption of 300kWh will cost RM 77.00, but typical consumption of 455 kWh will be substantially higher; i.e. RM 156.98. Such tariff system is expected to encourage sustainable and sensible use of electricity but as many researches have pointed out, current lifestyle shift disrupts this concept.

Abundance of studies have seen solar photovoltaic (PV) as a significant device to support increasing electricity demand, but this study will show that retrofitting PV on existing terrace house population will have a significant impact on grid demand. Following the falling and competitive solar PV price with increased efficiency (system cost for installing PV system in Malaysia fell around RM 7.5/watt [21]), it's integration into the current and future terrace housing stock will be significant, transforming homes into positive nodes of the future smart grid. Furthermore, there is an inversely proportionate relationship between on-site energy generation, occupancy rate and commercial peak demand [2,18,22].

3. Simulation setup

The simulation for identifying amount of grid-displaced electricity is carried out via Integrated Environmental Solutions Virtual Environment (IESVE) simulation tool. IESVE is known and widely accepted as a reliable simulation tool to carry out dynamic energy analysis. Two variables investigated in this simulation process are orientation and tilt angle. The prototype for the model house for this simulation were developed based on findings of an earlier study [18]. Four models were created and

IOP Conf. Series: Earth and Environmental Science **291** (2019) 012030 doi:10.1088/1755-1315/291/1/012030

orientated along (i) Northwest-Southeast axis, (ii) North-South axis, (iii) Northeast-Southwest axis and (iv) East-West axis (See Figure 1). The population of terrace houses based on its orientation were discussed elaborately in another study [18].

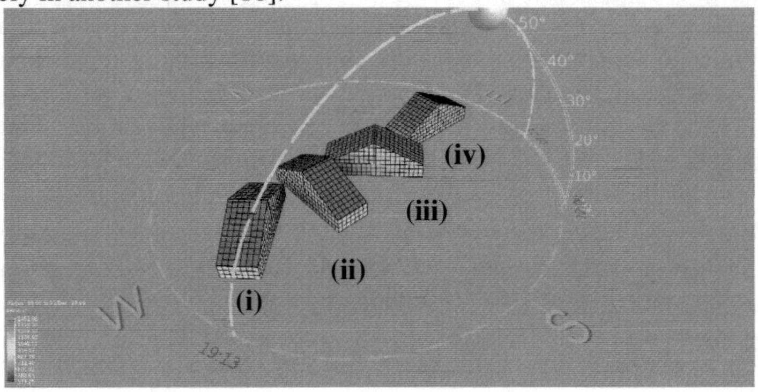

Figure 1. Simulation setup of the study model.

Despite the variety of PV array currently available, thin film is selected for this simulation as it yields higher energy conversion efficiency [21,23]. Despite 24.7% efficiency have been reported using heterojunction with intrinsic thin-layer (HIT) [24], this simulation study will use solar panel model CSUN250-60P (1.64m x 0.99m x 0.04m) with 15.4% efficiency manufactured by ChinaSunEnergy (CSUN) as it is readily available. The solar arrays were placed on the roof of the study model as shown in Figure 1, with varying tilting angle. The tilt angle was set at 0°, 15° and 23°, following findings for the best tilt angle to effectively capture solar energy [25]. The simulation analysis was carried out using SunCast and Apache modules in IESVE software.

4. Simulation findings and analysis

Findings from this simulation is read together with the orientation setup as shown in Figure 1.

Table 1. Grid displaced electricity (MWh) by varied tilt angle and orientations.

	Panel Tilt: 23°				Panel Tilt: 15°				Panel Tilt: 0°			
Orientation	**(iii)**	**(iv)**	**(i)**	**(ii)**	**(iii)**	**(iv)**	**(i)**	**(ii)**	**(iii)**	**(iv)**	**(i)**	**(ii)**
Jan 01-31	-1.8817	-1.8425	-1.8783	-1.8997	-1.6828	-1.6436	-1.6819	-1.7119	-1.4975	-1.461	-1.5208	-1.5075
Feb 01-28	-1.782	-1.7662	-1.7766	-1.8245	-1.6088	-1.5596	-1.6104	-1.6699	-1.4384	-1.3858	-1.5151	-1.4446
Mar 01-31	-2.0501	-1.9983	-2.0414	-2.0987	-1.8466	-1.7759	-1.8449	-1.9573	-1.6537	-1.5741	-1.8072	-1.6547
Apr 01-30	-1.9168	-1.8818	-1.9194	-1.96	-1.7195	-1.6629	-1.7278	-1.8159	-1.5458	-1.4717	-1.6734	-1.5458
May 01-31	-1.8516	-1.8294	-1.8599	-1.8896	-1.6603	-1.6157	-1.6679	-1.7121	-1.4901	-1.432	-1.5461	-1.4903
Jun 01-30	-1.8094	-1.7882	-1.809	-1.8325	-1.6227	-1.5843	-1.6255	-1.6499	-1.4532	-1.4045	-1.4758	-1.4519
Aug 01-31	-1.9299	-1.8917	-1.9336	-1.9693	-1.7318	-1.6765	-1.7398	-1.8048	-1.5569	-1.4871	-1.6525	-1.5556
Sep 01-30	-1.8856	-1.8526	-1.8751	-1.9324	-1.6917	-1.6334	-1.6912	-1.8099	-1.521	-1.441	-1.6639	-1.5212
Oct 01-31	-1.8696	-1.8494	-1.8707	-1.9131	-1.682	-1.6333	-1.6807	-1.7568	-1.5106	-1.4453	-1.6072	-1.5104
Nov 01-30	-1.754	-1.733	-1.7515	-1.7762	-1.5653	-1.5358	-1.5694	-1.6011	-1.4056	-1.3614	-1.4355	-1.4064
Dec 01-31	1.7265	-1.7194	-1.74	-1.7554	-1.5492	-1.5214	-1.5517	-1.5681	-1.3804	-1.351	-1.3861	-1.386
Total	- 22.3317	- 21.9992	- 22.3284	**-22.756**	- 20.0442	- 19.4817	- 20.0784	- 20.7795	- 17.9641	- 17.2708	-18.8304	-17.9823
Median	-1.8696	-1.8425	-1.8707	**-1.8997**	-1.682	-1.6333	-1.6807	-1.7121	-1.4975	-1.441	-1.5461	-1.5075

Referring to Table 1, the simulation findings denoted that PV arrays installed along the North-South axis (ii) with a tilt angle of 23° yields the highest converted electricity, equating to 22.756 MWh/year and a median of 1.8997 MWh/month. The lowest yielded electricity conversion occurs in months of November and December as cloud cast affect solar radiation intensity received by the solar panels. This phenomenon occurs on all tilting angles and orientations. The simulation also show that panel installed along West-East (iv) with tilting angle of 0° is the least to capture solar energy, yielding only 17.2 MWh/year with a median of 1.5461 MWh/month.

5. Discussion and conclusion

The simulation finding above indicate the potential surplus energy generated on site, if considering the monthly typical terrace house consumption of 455kWh as mentioned earlier. Accounting for the data in Table 1, average electricity generated by houses along North-South axis is 1895 kWh/month. So, each housing unit will be able to store or sell back to the grid around 1440kWh of energy surplus and transform consumers into prosumers making these houses as positive nodes of the smart grid. A study in [18] revealed that existing terrace house supply orientation can be categorized into North-South (30.12%), NorthEast-SouthWest (20.77%), East-West (24.27%) and Northwest-Southeast (22.52%). Utilising NAPIC record indicating existing terrace house supply of 2,236,640 units, potential units built along North-South axis alone may be able to generate 970 GWh of electricity. However, the Feed in Tariff (FiT) substitute, Net Energy Metering (NEM) only allow for 12kWp (72 kWp max for three-phased system) of grid-linked electricity generation. Therefore, excess electricity could only be locally stored and consumed. As future Electric Vehicle (EV) becomes more attractive and relevant, excess energy can be used to charge the EVs, offsetting the need to rely on non-sustainable resources. In addition, surplus energy can be diverted to commercial use hence potentially shave peak demand via feed-in supply.

The discussion above points out the potential of terraforming these terrace houses into a positive node of the future smart grid, but more study is required from various angle, including the electrical engineering and architectural design aspect which have not been given priority in any previous research.

6. Reference

[1] Saidur R, Masjuki H H and Jamaluddin M Y 2007 An application of energy and exergy analysis in residential sector of Malaysia Energy Policy 35 1050–63
[2] Suruhanjaya Tenaga (Energy Commission) 2017 Energy Statistics Handbook 2017
[3] ERCD E R and R C D 2010 Key Indicator for Asia and the Pacific 2017 vol II
[4] Wong J, Seng Y, Taylor P and Morris S 2011 Optimal utilisation of small-scale embedded generators in a developing country e A case study in Malaysia Renew. Energy 36 2562–72
[5] Sovacool B K 2013 Assessing energy security performance in the Asia Pacific, 1990–2010 Renew. Sustain. Energy Rev. 17 228–47
[6] Suruhanjaya Tenaga (Energy Commission) 2017 National Energy Balance 2016
[7] Kubota T, Jeong S, Toe D H C and Ossen D R 2011 Energy Consumption and Air-conditioning usage in residential buildings of Malaysia J. Int. Dev. Coop. 17 61–9
[8] TNBR 2014 Tnbr insight TNBR Insight 4
[9] Stephenson J, Ford R, Nair N K, Watson N, Wood A and Miller A 2018 Smart grid research in New Zealand – A review from the GREEN Grid research programme Renew. Sustain. Energy Rev. 82 1636–45
[10] Kakran S and Chanana S 2018 Smart operations of smart grids integrated with distributed generation: A review Renew. Sustain. Energy Rev. 81 524–35
[11] Haider H T, See O H and Elmenreich W 2016 A review of residential demand response of smart grid Renew. Sustain. Energy Rev. 59 166–78
[12] Mirrahimi S, Mohamed M F, Haw L C, Ibrahim N L N, Yusoff W F M and Aflaki A 2016 The effect of building envelope on the thermal comfort and energy saving for high-rise buildings in hot-humid climate Renew. Sustain. Energy Rev. 53 1508–19
[13] Jamaludin A A, Inangda N, Ariffin A R M and Hussein H 2011 Energy Performance: A Comparison of Four Different Multi-residential Building Designs and Forms in the Equatorial Region IEEE First Conf. Clean Energy Technol. CET 253–8
[14] Rahmah W, Zaki M, Nawawi A H and Ahmad S S 2011 Indoor Environmental Conditions in

Passive Architecture Terraced House 1–3

[15] Ibiyeye A I, Mohd F Z J and Zalina S 2015 Natural ventilation provisions in terraced-house designs in hot-humid climates: Case of Putrajaya, Malaysia Pertanika J. Soc. Sci. Humanit. 23 885–904

[16] de Dear R 2011 Revisiting an old hypothesis of human thermal perception: alliesthesia Build. Res. Inf. 39 108–17

[17] Attia S, Evrard A and Gratia E 2012 Development of benchmark models for the Egyptian residential buildings sector Appl. Energy 94 270–84

[18] Abdul-Razak A H N, Leardini P M and Nair N-K C 2015 Adapting Malaysian housing for smart grid deployment based on the first nationwide energy consumption survey of terrace houses Living and Learning: Research for a better built environment ed R H Crawford and A Stephan (Faculty of Architecture, Building and Planning, The University of Melbourne, Melbourne, Australia.) pp 311–21

[19] Chong C, Ni W, Ma L, Liu P and Li Z 2015 The use of energy in Malaysia: Tracing energy flows from primary source to end use Energies 8 2828–66

[20] TNB 2016 Pricing &b Tariff TNB

[21] Ghazali A, Salleh E I, Haw L C, Mat S and Sopian K 2017 Performance and financial evaluation of various photovoltaic vertical facades on high-rise building in Malaysia Energy Build. 134 306–18

[22] Abdul-Razak A H N 2017 Malaysian residential housing for the smart grid : identifying optimization attributes for design and energy performance improvements How to face the scientific communication today. International challenge and digital technology impact on research outputs dissemination p 109

[23] Halabi L M and Mekhilef S 2018 Performance Analysis of Multi-Photovoltaic (PV) - Grid Tied Plant in Malaysia Performance Analysis of Multi-Photovoltaic (PV) -Grid Tied Plant in Malaysia IOP Conf. Series: Earth and Environmental Science

[24] Taguchi M, Yano A, Tohoda S, Matsuyama K, Nakamura Y, Nishiwaki T, Fujita K and Maruyama E 2014 24.7% Record efficiency HIT solar cell on thin silicon wafer IEEE J. Photovoltaics 4 96–9

[25] Khatib T, Mohamed A, Mahmoud M and Sopian K 2015 Optimization of the Tilt Angle of Solar Panels for Malaysia Optimization of the Tilt Angle of Solar Panels for Malaysia

Acknowledgement

This study has been made possible with funding from Performance Based Research Fund (PBRF) The University of Auckland, Universiti Teknologi MARA and Ministry of Higher Education of Malaysia

IOP Conf. Series: Earth and Environmental Science **291** (2019) 012031 doi:10.1088/1755-1315/291/1/012031

Energy use prediction with information theory and machine learning technique

YW Tong[1], WY Yang[2], and DL Zhan[3]

[1] School of economics and management, Nanchang University, Nanchang, China
[2] College of automation, Shenyang Aerospace University, Shenyang, Liaoning, China
[3] Department of Mathematics, Sichuan University, Chengdu, Sichuan, China

E-mail: ellabc369@gmail.com

Abstract: Appliances energy consumption plays an increasingly important role in the overall building electric energy consumption and its temporal trending. However, predicting appliances energy consumption is complicated by lack of causal understanding of the appliances energy use as well as too many potential predictors that might be relevant to the appliances energy use. In this study, we apply information theory and advance machine learning neural network technique to first rank the importance of potential drivers that dominate appliances energy consumption and secondly model the temporal evolution of appliances energy consumption with a restricted set of environmental predictors. Our results showed that temperature and humidity were the two most important environmental drivers in the house appliances energy consumption modeling. Furthermore, using those environmental drivers, the machine learning model was able to accurately capture the temporal dynamics of appliances energy consumption.

1. Introduction

Appliances energy consumption takes up about 20% - 30% of the building electric energy consumption [1], thus significantly contribute to the daily dynamics as well as long term trend of regional energy consumption. For example, a 10% increase in the domestic building electricity use was resulted from appliances energy use in UK [2]. Hence, improving our understanding of the temporal changes, drivers, and effects of appliances energy consumption in buildings is critical important. Historically, a variety of approaches have been proposed to understand the appliances energy consumption. For example, EenergyPlus model simulates high frequency building energy use with the utility bills [3]. Data-driven machine learning techniques have also been widely employed. Markov Chain Monte Carlo technique was successfully used to estimate building occupancy and potential energy demand and consumption [4]. A multiple linear regression model was established to explain the variability of monthly energy consumption with environmental drivers [5] and showed that over 90% of energy consumption variability could be explained. Despite a vast amount of efforts have been allocated to predict appliances energy consumption, it is still challenging to accurately predict high-frequency dynamics of appliances energy use, given the complexity and non-linearity of the predictive variable itself.

In addition, it is still under debates about the dominant drivers of building appliances energy use. Previous work indicated that a wide spectrum of factors including socioeconomic factors, dwelling

IOP Conf. Series: Earth and Environmental Science **291** (2019) 012031 doi:10.1088/1755-1315/291/1/012031

characteristics, climate conditions, region of interest, presence of double pane windows, energy efficient light fixture, number of refrigerator and entertainment devices, and so on [6].

In this work, our objective is two-fold. We first aim to understand the causal relationship between environmental drivers and appliancess energy consumption and to verify whether exterior or interior drivers can best predict future energy consumption. Secondly, we aim to model the high frequency of appliances energy use temporal dynamics with advanced machine learning technique.

2. Methodology

2.1 Data description

We used observational data of house appliances energy use as well as detailed interior and exterior environmental drivers from an energy efficient house located in Stambruges, Belgium [7]. Appliances energy consumption (in Wh) was continuously recorded from 1/11/16 to 5/27/16 about 137 days and the temporal resolution is ten minutes. Humidity and temperature within the house were recorded by wireless monitoring system and exterior temperature, humidity and pressure, window speed, visibility, dew point were observed at nearby Chièvres Airport.

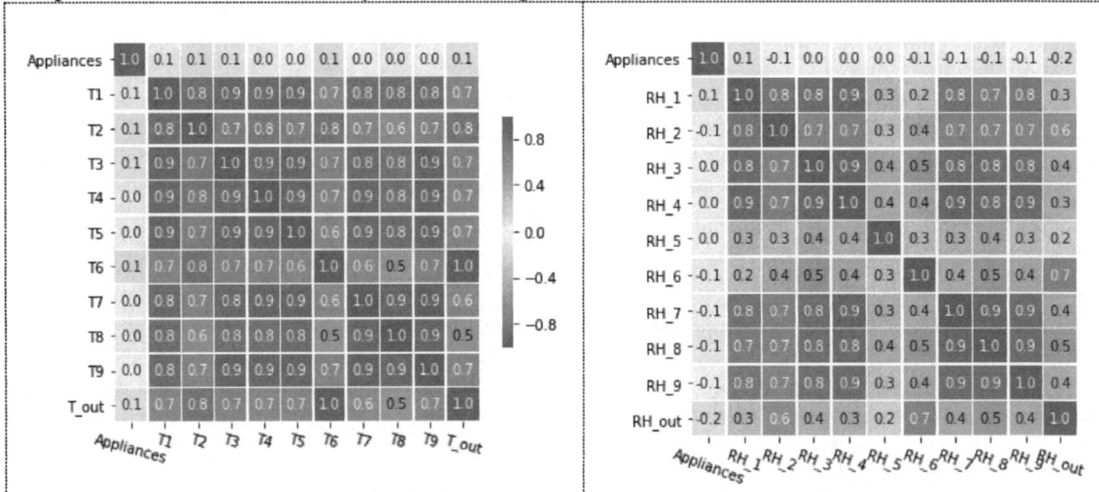

(a) Temperature (b) Humidity

Figure 1. Observed relationship between appliances energy use and environmental drivers: a) Pearson correlation of the appliances energy use with T(temperature), b) Pearson correlation of the appliances energy use with RH(humidity).

Our data preprocessing showed that the linear correlation between appliances energy use and temperature or humidity were low; while temperature or humidity from some rooms were significantly correlated with temperature or humidity from other rooms, respectively (Figure 1). The density distribution of appliances energy use showed a long tail; the mean, median and standard deviation are about 98, 60 and 102 (Figure 2)

IOP Conf. Series: Earth and Environmental Science **291** (2019) 012031 doi:10.1088/1755-1315/291/1/012031

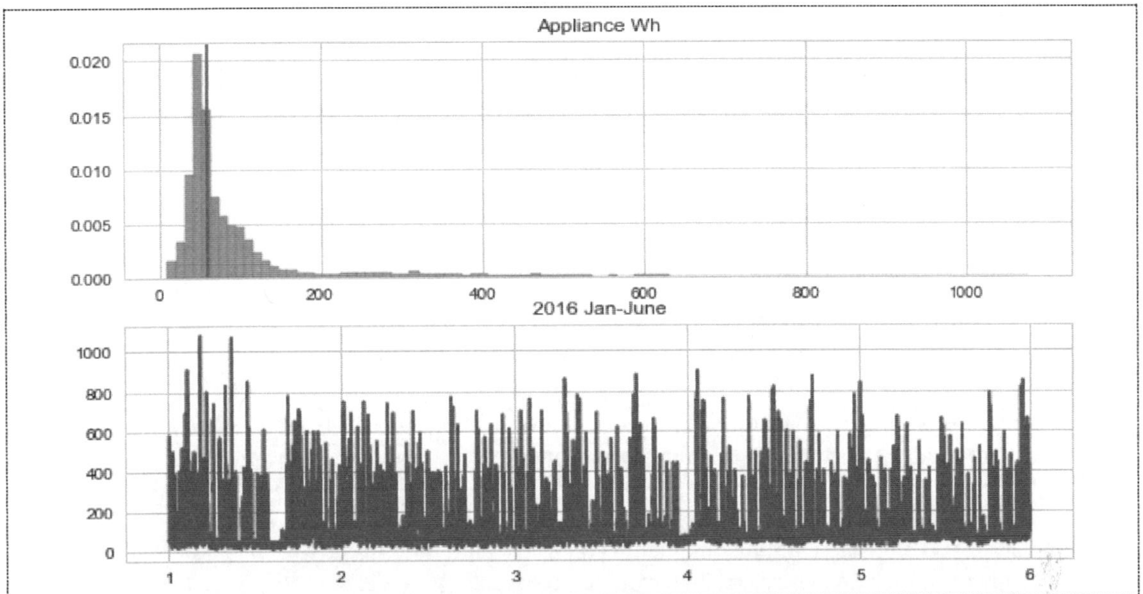

Figure 2. Upper panel shows the density distribution of appliances energy consumption(Wh). Lower panel shows the time series of appliance energy from 1/11/2016 to 5/27/2016.

2.2 House energy consumption modeling

2.2.1 Information transfer modeling. $X=\{X_t : t=1,2,...,n\}$ is a discretized time series with distribution P(x). In our data set, X presents the variables including inside temperature, inside humidity, outside temperature, outside humidity, outside visibility and dew point. The Shannon Information Entropy H describes the uncertainty in realization of X, The target Y is a time series and presents electricity consumption in this study. Transfer Entropy measure the reduction in Shannon Information Entropy of current Y, due to the acknowledgement of the past X at some time lag [8] (Figure 3), which is between current Y and past X:.

$$T(X \rightarrow Y,\tau) = H(X_{t-\tau},Y_{t-\varpi}) + H(Y_t,Y_{t-\varpi}) - H(Y_{t-\varpi}) - H(X_{t-\tau},Y_t,Y_{t-\varpi}) \quad (1)$$

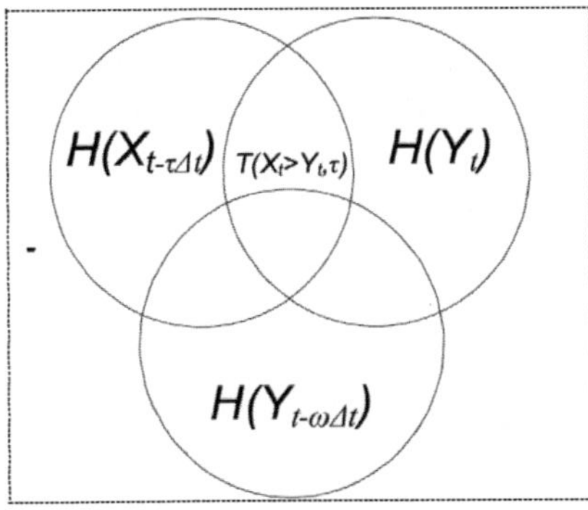

Figure 3. Venn diagram. Information flow, which transfers from current X to current Y is equal to reduction of uncertainty in variable current Y.

2.3 Recurrent Neural Networking

Recurrent Neural Networking (RNN) [9] is an artificial neural network, which could keep apart of historical information as memory to interact with the new input and generate effective outputs:

$$
\begin{aligned}
f_t &= \sigma(W_f \bullet [h_{t-1}, x_t] + b_f) \\
i_t &= \sigma(W_i \bullet [h_{t-1}, x_t] + b_C) \\
C'_t &= \tanh(Wc \bullet [h_{t-1}, x_t] + b_C) \\
C_t &= f_t * C_{t-1} + i_t * C'_t \\
o_t &= \sigma(W_o [h_{t-1}, x_t] + b_o) \\
h_t &= o_t * \tanh(C_t)
\end{aligned} \tag{2}
$$

where $\sigma()$ is the sigmoid activation function. W_f, W_i, W_o are the weight parameters, which are optimizated by fitting the model to observations during training. RNN outputs h_t given inputs x_t and h_{t-1}.

This dataset contained 25 features and the environmental data from Chièvres Airport weather station are not totally the same as the data in the building. In addition, some of them had a strong linear relationship (Figure 1, Pearson correlation > 0.7). That mean that some features were redundant in this data set. In RNN model, the filtered data set was split to 3 data sets. Dataset 1 contained all filtered variables. Dataset 2 only contained appliances and interior environmental variables, and dataset 3 only contained appliances and exterior environmental variables.

Because the features were recorded with different measurement units, it was necessary to scale them to a same numerical range. Here, we ranged all feature into the range zero and one. In this dataset, the temporal resolution was 10 mins, which mean the data vary too quickly and the random process was not stationary. To make sure that the random process was stationary, the time-series data were reorganized by averaging to1 hour temporal resolution.

3. Results & Discussion

3.1 Causality modeling

The transfer entropy quantitatively estimated the magnitude of influence from each predictor to the target variable appliances energy consumption. All environmental drivers were considered, and we grouped the results into significant exterior drivers (T_out, RH_out, Tdewpoint, visibility, press, windspeed) and interior drivers (T1, T6, RH_1, RH_8, RH_5, RH_6). It is important to notice that the Pearson correlation between appliances energy consumption and any of these identified driving factors were very small (Figure 1), which means the causality relationships confirmed by transfer entropy were high non-linear.

Among the identified exterior environmental drivers temperature was the most important driving factors that control appliances energy consumption (Figure 4). Similarly, among the identified interior environmental drivers room temperature T1 was selected to be the most influential one. Consistently, relative humidity from both inside and outside the building were identified to be the secondary important drivers that controlled the appliances energy consumption.

IOP Conf. Series: Earth and Environmental Science **291** (2019) 012031 doi:10.1088/1755-1315/291/1/012031

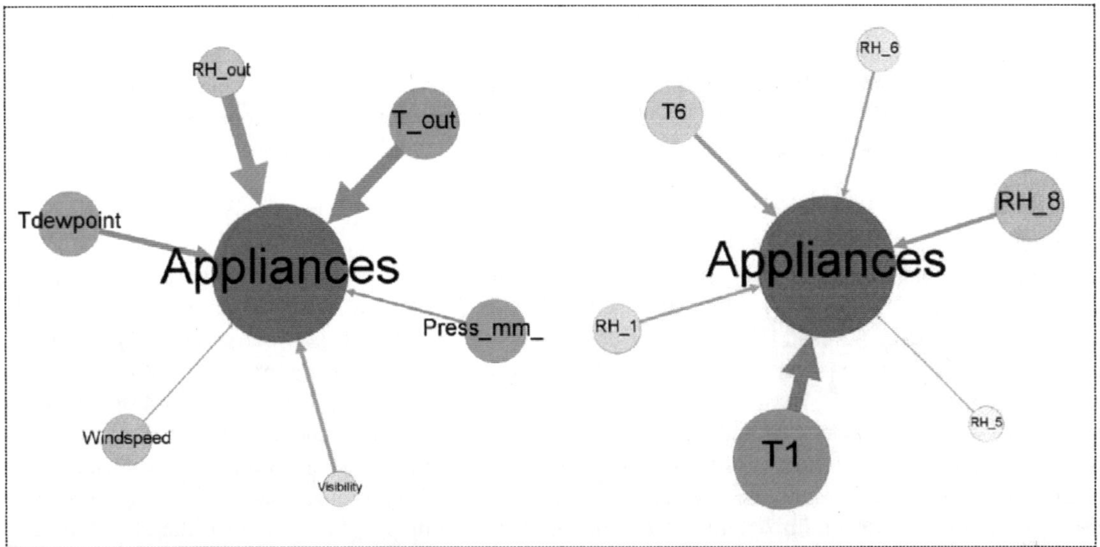

Figure 4. Causal relationship between appliances energy use (center node) and exterior environmental drivers (left panel) or interior environmental drivers (right panel) The size of node and the thickness of arrows are scaled with the amount of information transfer from one variable to Appliances energy consumption.

3.2 predictive modeling with recurrent neural network

We used 80% of observed appliances energy use to train the recurrent neural network model, and tested the trained model with the rest 20% data. Modeling experiments were carried out for three cases using (1) all environmental drivers; (2) only exterior environmental drivers (Figure 4 left panel); (3) only interior environmental drivers (Figure 4 right panel). Our modeling results showed that the memory based recurrent neural network was able to reproduce the observed temporal dynamics of appliances energy use under all three circumstances. It demonstrated that both interior and exterior environmental drivers are informative.

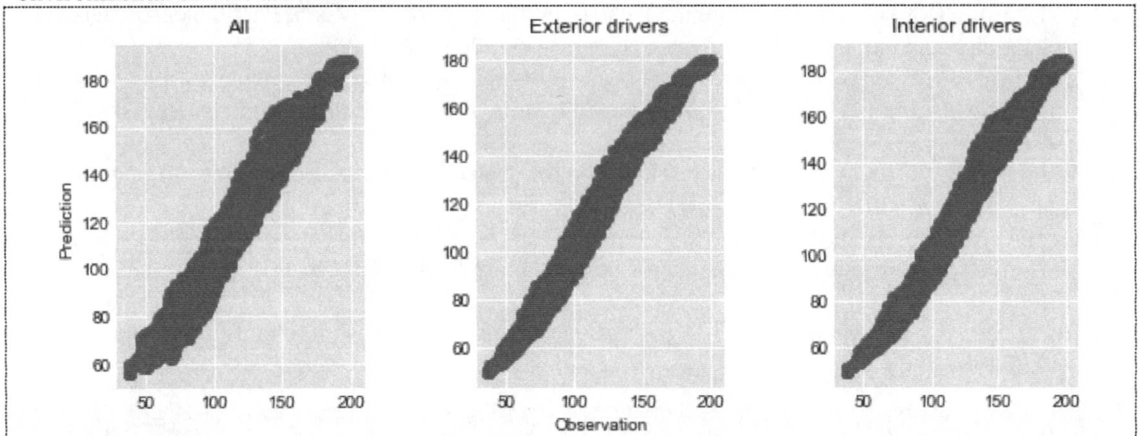

Figure 5. Observed versus predicted appliances energy use with recurrent neural network. Left, middle, and right panels correspond to using all, only exterior, and only interior environmental drivers, respectively.

3.3 Limitation and future work

This study is limited to availability of appliances energy consumption data. We focused one well developed house, but in order to build a more generic energy use predictive model, different type of house with different energy use preference are needed. In this study, the data set only contains limited features and many other potentially important features are not discussed here such as building's geometry or floor plan. Ideally external environmental driver observation is supposed to be the condition right outside the house. We used Chièvres Airport observed conditions (12 km away from the house) as a proxy of the house external conditions, which may inevitably bring in some uncertainties.

4. Conclusion

The appliances energy consumption plays an important role in the overall electric energy consumption. Understanding the relationship between appliances energy use and relative variables is critically important. This study focused on understanding the causal relationship between potentially important drivers and the house appliances energy use as well as predicting the energy use with a limited number of environmental drivers. Via statistical analysis based on information theory, we identified exterior temperature and room one temperature as the most important drivers, and exterior relative humidity and the room eight relative humidity as secondary important drivers. Our predictive modeling using recurrent neural network successfully reproduce the observed appliances energy consumption time series with those identified important environmental drivers, thus further confirmed the important and predictable relevance between appliances energy use consumption and the environmental drivers.

5. Reference

[1] A. Kavousian, R. Rajagopal, M. Fischer, Ranking appliances energy efficiency in households: Utilizing smart meter data and energy efficiency frontiers to estimate and identify the determinants of appliances energy efficiency in residential buildings, Energy Build. 99 (2015) 220–230

[2] S. Firth, K. Lomas, A. Wright, R. Wall, Identifying trends in the use of domestic appliances from household electricity consumption measurements, Energy Build. 40 (5) (2008) 926–936

[3] N. Fumo, P. Mago, R. Luck, Methodology to estimate building energy consumption using EnergyPlus Benchmark Models, Energy Build. 42 (12) (2010) 2331–2337.

[4] I. Richardson, M. Thomson, D. Infield, A high-resolution domestic building occupancy model for energy demand simulations, Energy Build. 40 (8) (2008) 1560–1566.]

[5] C.-L. Hor, S.J. Watson, S. Majithia, Analyzing the impact of weather variables on monthly electricity demand, IEEE Trans. Power Syst. 20 (4) (2005) 2078–2085.

[6] R.V. Jones, A. Fuertes, K.J. Lomas, The socio-economic, dwelling and appliances related factors affecting electricity consumption in domestic buildings, Renew. Sustain. Energy Rev. 43 (2015) 901–917.

[7] Knuth, K. H. (2005), Lattice duality: The origin of probability and entropy, Neurocomputing, 67, 245– 274, doi:10.1016/j.neucom.2004.11.039.

[8] Vicente, R., Wibral, M., Lindner, M., & Pipa, G. (2011). Transfer entropy—a model-free measure of effective connectivity for the neurosciences. Journal of computational neuroscience, 30(1), 45-67.

[9] Hochreiter, S., & Schmidhuber, J. (1997). Long short-term memory. Neural computation, 9(8), 1735-1780.

IOP Conf. Series: Earth and Environmental Science **291** (2019) 012032 doi:10.1088/1755-1315/291/1/012032

The Impact of Economic Growth, Energy Consumption And Demographic Dynamics On Carbon Emissions: A Cross Country Analysis In 133 Nations

P T T Trang[1], N T X Hanh[1] and B P P Thanh[1]

[1]Resource Management and Environmental Department, Thu Dau Mot University, Tran Van On Str., Thu Dau Mot City, Viet Nam

trangpham20_8@yahoo.com

Abstract. The growth of economic has been a determinant factor increasing the volume of greenhouse gases, particularly CO^2 emissions. To provide appropriate measures to control CO^2 emission, it is necessary to address how such factors as population and economic growth impact the emission of carbon dioxide in any developing country. However, the answer to what are the key factors or whether these factors have jointly impacted on CO^2 emissions has been debated so far. This paper chooses five main factors, including economic growth, population growth, energy consumption, age structure and urbanization as independent variables for multiple regression model to analyse the relationship between these variables to CO^2 emission variable in 133 countries. The result shows that all these factors jointly impacted on CO^2 emissions energy use and demographic factors should be controlled due to the highest contribution to the level of CO^2. The research findings are expected to shed a light on those countries policy making in coping with climate change.

1. Literature review

The relationship between carbon emissions and economic growth has been debated for many decades [1]. Some of the empirical studies support the EKC hypothesis, which states that the level of carbon emission and economic growth follow an inverted U-shaped curve [2]. Some find that there is positive linear relationship between two variables [3]. After performing analysis a panel of 88 countries, [4] also conclude that a positive relationship exists between CO^2 emissions and economic growth. In general, there is many existing theories suggested that GDP growth and carbon dioxide emissions are strongly positive interrelated [5]. While many studies have primarily focused on the impacts of affluence on carbon dioxide emissions, other empirical studies suggested that population change also significantly increase the volume of greenhouse gas emissions for the last decades [6]. Yet, identifying the relationship between population growth and CO^2 emissions continue to be debated because most studies focus on the direct effect of population on the level of carbon dioxide emissions through only population size without considering the demographic dynamics [7]. Meanwhile, in fact, energy demand and greenhouse gases emissions can also be affected by a range of demographic dynamics, such as urbanisation rate, age structure and household size. Even the impact of distribution of population may be more important than population size and thus it needs to be considered, for instance the more population living in urban, the more energy are consumed, resulting in high level of CO^2 [8]. Age structure of the population should also be taken into account in explaining the effect of population

IOP Conf. Series: Earth and Environmental Science **291** (2019) 012032 doi:10.1088/1755-1315/291/1/012032

change on CO^2 emissions because those people in working age have higher involvement of economic activities, resulting in more energy consumption than those in dependent group [9]. Many recent studies have employed the demographic factors as significantly additional explanatory variables for the increase of CO^2 [10]. According to [7], shrinking household size significantly influence the growth of carbon emissions in China. However, adverse environmental impacts are more a direct function of energy consumption than of indirect via the impact of population or economic growth. Thus, energy consumption should be employed to explain the relationship between CO^2 level and economic growth or population growth [11]. In general, all above empirical studies base on basic literature on environmental degradation from IPAT model in which three driving forces expected to influence the level CO^2 emissions includes economic growth or the affluence, population and technological change or specifically, the energy efficiency of economic activities [12]. They are interrelated in explaining the increase of CO^2 emissions in recent decades. The impacts of economic activity have become more severe when combined with demographic growth, as long as population increases lead to increases in energy consumption and consequently higher level of CO^2 emitted.

2. Data selection

The data of carbon dioxide emissions taken from Carbon Dioxide Information Analysis centre of World Bank are observations of total CO^2 emissions from all activities measured in kilo tonnes. It is also similar for data of total energy consumption variable, with all forms of commercial energy excluding firewood and other traditional fuels are converted into oil equivalents. The affluence or economic growth variable is captured by real GDP in U.S dollar. The population figures are based on the estimation of national censuses taken form World Bank. All data is cross-sectional data collected in 133 countries in 2010, with the aim of maximizing possible sample size. Regarding age structure variable, based on World Bank definition the structure of population of each country are divided into two age categories: dependent group (aged less than 15 or over 64) and another one is working group (aged 15-64). According to [9], those people in working age have higher participation in economic activities, resulting in more energy consumption than those in dependent group. Thus, it is expected that countries with higher percentage of working population (% total population) might produce more CO^2. However, the data related to age structure from World Bank only provide the data of the ratio of dependent group compare to working group, which is not relevant for my analysis. So, from the data of total population and the ratio of dependent group compare to working group in every country, the authors have built own data of percentage working population compared to the total population.

$$\% \textit{ Working population} = (\textit{population} / (1 + \textit{age dependency ratio}/100)) * 100/\textit{population} \quad (1)$$

This variable is named working population. Regarding urbanization, urban population (% of total population) data is used. Summary of all variables are presented in table 1.

Table 1. A summary of variables

Variables	Obbs	Mean	Standard deviation	Min	Max
CO^2 (kt)	133	264454.3	913228.6	513.38	8286892
GDP(current US$)	133	5.83e+11	1.84e+12	2.12e+09	1.50e+13
population	133	5.39e+07	1.63e+08	318041	1.34e+09
Energy use (kt of oil equivalent)	133	105917.4	321660.7	744.529	2516731
Working Population (% total population)	133	64.99488	6.517571	49.77217	85.80555
Urban population (% total population)	133	62.47129	21.18596	13.444	100

3. Data analysis

3.1. The relationship between each independent variable and CO^2

Table 2. Correlation of each independent variable to CO^2

Variables (non-ln)	GDP	Population	Energy use	Working population	Urban population
Correlation coefficients	0.7545	0.8040	0.9842	0.1521	0.0423
Variable(ln)	LnGDP	LnP	LnE	LnA	LnU
Correlation coefficients	0.9231	0.6904	0.9459	0.4165	0.3165

All independent variables has positive linear relationship to CO^2 variable individual, shown in table 2, which is in the line with what theories suggest. Of which, GDP, population and energy consumption variable appear to strongly correlate to CO^2 variable, with coefficient of correlation 0.7545, 0.8040 and 0.9842 respectively. However, there is not enough information to conclude GDP, or population or energy use should be the main factor causing the high level of CO^2. Instead, a multiple regression needs to be conducted to demonstrate the true relationship between those variables because apart from direct effect, GDP also has indirect effects on CO^2 via other independent variables. Furthermore, visually the linear relationship between each independent variable to CO^2 is demonstrated clearer when all variable are taken logarithm.

3.2. Multiple regressions for five independent variables

The model used is a multiple regression formulation, shown in table 3:

$$CO^2 = B1 + B2GDP + B3population + B4energyuse + B5workingpopulation + B6urbanpopulation + e \quad (2)$$

Where e is the error term

Table 3. Multiple regression summaries

Variables	Coefficient	t-ratio
GDP	-1.47e-07*	-16.01
population	-.0000637***	-0.74
energy use	3.531644*	48.67
working population	393.9955***	0.30
urban population	215.1645***	0.52
Number of obs	133	
F(5, 127)	3092.35	
Prob > F	0.0000	
R-squared	0.9919	
Adj R-squared	0.9915	
Root MSE	84035	

Note: *, **, *** significant at 1%, 5% and 10% respectively

Although the high multiple coefficient of determinant is 0.9919, the signs of coefficients of GDP and population variable are not as expected. Only coefficients of GDP and energy use is individually significantly less than 1% significant level (P<0.01), while the estimated coefficient on population, working population and urban population are individually less significantly at (more than) 10% significant level and the overall F test for fit is also significant. Obviously, there is a problem of imperfect multicollinearity in which independent variables are highly linearly related. However, the paper cannot drop any of five variables because theory suggests that all of them should be included in the model. Instead, the research use logarithm model for the model specification.

3.3. Testing for heteroskedasticity

The Breusch-pagan/cook-Weisberg test has been used for homoscedasticity. The result of low p-value = 0.0000, <0.01 indicates that the heteroskedasticity is likely to be present. In this case, taking logarithm all variables might reduce the heteroscedasticity.

$$Ln\ CO^2 = B1+B2lnGDP+B3lnP+B4lnE + lnA+lnU + e \tag{3}$$

The high p-value = 0.0846, >0.05 indicates that the heteroskedasticity is not likely to be present, or the authors cannot reject the null hypothesis that errors are homoscedastic. Obviously, taking logarithm all variables is efficient to reduce problem. Furthermore, the assumption normality of residuals has been checked; the results show that the residual of linear model is quite skew compared to the normal distribution of residual of transformed model.

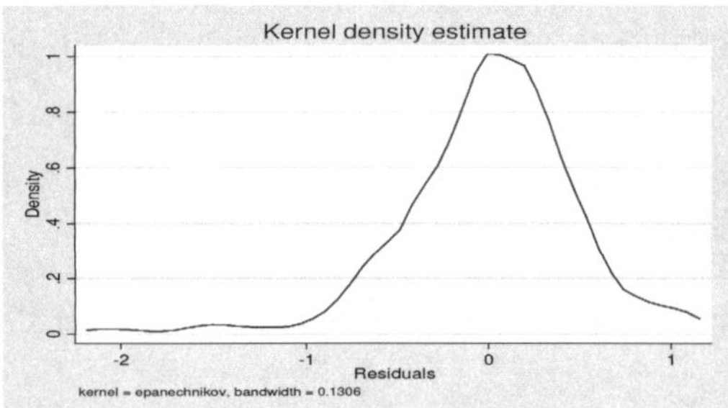

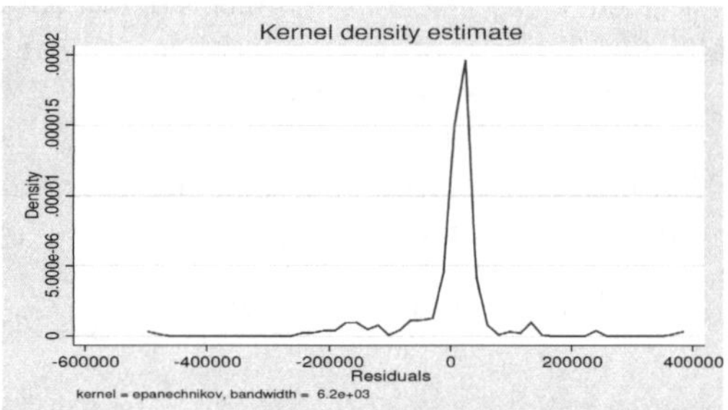

Linear model *Transformed model*

Figure 1. Distribution of residuals

As a result of these above result, a log-log model is better suitable than linear model in this case. The model used is a multiple regression formulation, shown in table 4:

$$Ln\ CO^2 = B1+B2lnGDP+B3lnpop+B4lnEnergy+lnAge+lnUrban+e \tag{4}$$

Table 4. Multiple regression summaries of transformed model

Variable (ln)	Coefficient	Standard error	t	P>t
LnGDP	.1560336	.0732798	2.13	0.035
LnP	.1507816	.0651587	2.31	0.022
LnE	.7235889	.090821	7.97	0.000
LnA	3.750907	.559135	6.71	0.000
LnU	.287917	.1474757	1.95	0.053

The t-statistic tests show that the low p-value (< 0.05) for the coefficient of GDP, population, energy use and working population implies: the effect of the (logarithm of) all four variables on the CO^2 emissions is individual significant at 5% level of significance. However, the p-value for the coefficient of the logarithm of urban population is less individual significant, 0.053. This means that we should exclude this variable due to no relationship between the level of CO^2 and the urban population in studied countries. However, the estimated result will be bias due to omitted variable because the theories have recommended that urban population should be considered as main demographic factor influencing CO^2 in recent years. Additionally, population size might not only represent its direct effect on CO^2, but also its indirect effect via urban population. Therefore, based on the F-statistic tests, with the low p value < 0.01 to reject the null hypothesis that all coefficients for the explanatory variables are simultaneously equal to zero. Obviously, urban population variable is jointly significant with other independent variables to explain the increase of CO^2 in 133 countries. Therefore, urban population should be included in the final model.

3.4. Testing for omitted variables

A Ramsay reset test has been conducted to test omitted variables with transformed model.

$$Ln\ CO^2 = B1 + B2lnGDP + B3lnpop + B4lnEnergy + B5lnA + B6lnU + e \qquad (5)$$

The authors fail to reject the null hypotheses that the model has no omitted variables because the high p-value= 0.9958>0.1. So, statistically, this double log model is correctly specified, with no omitted variables. However, the F test is not always true with 100% as there are many other factors which are important in contributing to the levels of CO^2 as well such as industry composition and household size. Due to the limit of a small essay combined with the limit data, the research have just included five relevant variables. After analysing combined with theory, a log-log model has been considered as a final useful model.

$$Ln\ CO^2 = B0 + B1lnGDP + B2LnP + B3lnE + B4lnA + B5lnU + e\ (Model\ specification) \qquad (6)$$

Table 5. Estimated results

Variable	Coefficient	t-ratio
LnGDP	0.1560336**	2.13
LnP	0.1507816**	2.31
LnE	0.7235889*	7.97
LnA	3.750907*	6.71
LnU	0.287917***	1.95
Number of obs	133	
F(5, 127)	393.91	
Prob > F	0.0000	
R-squared	0.9394	
Adj R-squared	0.9370	
Root MSE	.47955	

Note: *, **, *** significant at 1%, 5% and 10% respectively

The model has produced a high adjusted multiple coefficient of determinant of 0.93, implying that 93 % of variation in CO^2 emissions can be explained by the model, shown in table 5. The sign of all coefficients is in the line with the suggestion from the basic theory. Since both CO^2 and all explanatory variables are in logs, all slope coefficients are estimated global elasticity of CO^2 with respect to a relevant determinant, while keeping other independent variables constant. For example, the interpretation of B1 = 0.1560336 suggests that if GDP increases by 1%, the emission of CO^2 will increase proportionally by approximately 0.15% while keeping other independent variables constant. Similar interpretation is applied to the other explanatory variables. It can be seen that working population and energy use appear to have greater predictive power than the other variables, with coefficients of 3.750907 and 0.7235889 respectively.

4. Conclusion

The estimated result of the model shows that all these factors jointly impacted on CO^2 emissions. There is little association between economic growth and CO^2 emissions while energy use and demographic factors should be controlled due to the highest contribution to CO^2 level. Of demographic variables, age structure is a more reasonable explanation for the impact on carbon emissions than urbanization while keeping other independent variables constant. This finding is likely to be accepted because urbanization from rich countries can improve energy use efficiency and pollution treatment, mitigating the damage to the environment. By contrast, no matter what efficient level of energy use has been controlled, higher percentage of working age population will lead to the increase of CO^2, especially in those developed countries. The essay also suggested that using household size in explaining the impact on carbon emissions should be more effective approach for further research on this issue.

5. References

[1] Odhiambo NM 2012 Economic Growth and Carbon Emissions inn South Africa: An empirical Investigation *International Journal of Economics and Business Research* **28** p 37-46
[2] Sun JW 1999 The nature of CO^2 emission Kuznets curve *Energy Policy* **27** p 691-694.
[3] Shiafik N 1994, Economic development and environmental quality: an economic analysis, *Oxford Economic Papers* **46** p 757-773.
[4] Dinda S and Coondo D 2006 Income and emission: a panel data-based cointagration analysis *Ecological Economics* **57** p 167-81
[5] Annicchiarico B, Bennato AR & Chini EZ 2014, 150 Years of Italian CO^2 Emissions and Economic Growth, *CREATES Research Paper*
[6] Shi A 2001 Population Growth and Global Carbon Dioxide Emissions, *Development Research Group*, The World Bank.
[7] Zhu Q and Peng X 2012 The impacts of population change on carbon emissions in China during 1978–2008 *Environmental Impact Assessment Review* **36** p1-8.
[8] Zarzoso IM Morancho AB and Lage RM 2006 The Impact of Population on CO^2 Emissions: Evidence from European Countries *Environmental and Resource Economics* **38** p 497–512.
[9] Zagheni E 2011 The Leverage of Demographic Dynamics on Carbon Dioxide Emissions: does Age Structure Matter? *Demography* **48** p371–399.
[10] Cole AM and Neumayer E 2004, Examining the Impact of Demographic Factors on Air Pollution *Population and Environment* **26** p 5-21
[11] Yavuz NC 2014 CO^2 Emission, Energy Consumption, and Economic Growth for Turkey: Evidence from a Cointegration Test With a Structural Break *Energy Sources* **9** p 229-235.
[12] Dietz T 1994 and and Rosa E.A 1994 Rethinking the Environmental Impacts of Population, Affluence and Technology *Human Ecology Review* **1** p 277-300
All data from World Bank, http://data.worldbank.org/indicator.

A Review on Energy Performance in Malaysian Universities Through Building Information Modelling (BIM) Adaptation

A F Latif[1], N A Ahmad[2], M R Abdullah[3], A Ismail[3], A A A Ghani[2]

[1]Master of Green Architecture, Faculty of Architecture Planning and Surveying, Universiti Teknologi MARA, Perak Branch, Seri Iskandar, 32610, Perak, Malaysia.
[2]Department of Building Surveying, Faculty of Architecture Planning and Surveying, Universiti Teknologi MARA, Perak Branch, Seri Iskandar, 32610, Perak, Malaysia.
[3]Department of Building, Faculty of Architecture Planning and Surveying, Universiti Teknologi MARA, Perak Branch, Seri Iskandar, 32610, Perak, Malaysia.

firdaus.aflrekabina@gmail.com

ABSTRACT. Public universities in Malaysia are challenged with the issue of energy wastage due to improper energy audit on the energy performance. The wastage of energy occured due to the lack of awareness between students and staff, and limited particular guidelines for the management to improve the situation. Effort need to be made in retrofitting the buildings, to make it more energy self – sufficient, that leads to energy saving and improve the building's energy performance. The aim of this research is to improve the energy performance in an existing university building with the adaptation of Building Information Modelling (BIM). This paper will review on the energy performance process based on actual condition of an existing university building with the adoption of BIM based on previous researchers. It will identified the potential strength and weakness of applying BIM through energy performance process. The outcome of this paper will be used to integrate BIM in retrofitting the existing university building in order to improve energy performance. Furtheremore, it will help the facilities management in university to gain more accurate energy usage during the maintenance work.

1. Introduction

In Malaysia, existing universities have recorded highest replace of energy consumption in the building sector, compared to the new developed universities that have been integrated with Green Building Index (GBI) requirements. It is recorded by Suruhanjaya Tenaga Malaysia, that the existing universities in Malaysia have consumed 38% of the electricity generated in the country [1]. In addition, the energy use in these buildings is predicted to increase 5% by every year for the near future which leads to the inefficient in these universities in order to improve energy performance. However, this attempt is a complicated process with many factors needs to be considered. The factors included are the current building features, types of building materials and electrical appliances that affect the way energy has been used in the universities. A building can be retrofitted to improve its energy performance, by reviewing the energy usage of many appliances in the building, such as air conditioning system [2] and the electrical appliances in the building, such as personal computer, photocopier and lighting system [3]. The aim is to create a good energy performance in the building that meets all of the green design concepts as suggested by the Green Building Index (GBI). By accomplishing the green design objectives, the building will be less costly to operate; [4] it will increase the value, prolong its sustainability and contribute to a healthier and more productive environment for the building users [5]. In Malaysia, the guideline and standard to be referred for energy consumption is (i) Malaysian Standard MS 1525 : 2014 Code of Practice on Energy Efficiency, (ii) Uniform Building By-Law (UBBL) 1984 and (iii) Energy Efficiency and Conservation Guidelines [6]. At the moment, many researches have focused on the study of energy performance for commercial buildings. A very limited study on energy performance for universities have been carried out with regards to the adoption of Building Information Modelling (BIM) [7]. Building Information Modelling (BIM) initially provides the univerisity with more accurate perception towards the energy consumption features [8] that able to help in predicting energy wastage of a building. This paper will reviews several benefits on using BIM in improving energy performance process for universities. BIM is capable to aligning the decision-making process and allows for more effective and efficient energy audit[9].

2. The Issue

One of the main issue in public universities in Malaysia are the wastage of energy due to the lack of awareness between students and staff, and limited particular guidelines for the management of the university to improve the situation [10]. Lack of energy audits and minimal in-depth monitoring in universities also lead to higher energy consumption [10] which eventually influence the energy performance. Without proper evaluation on energy performance, universities are unaware of the energy wastage issues and keep using the energy inefficiently. Since Malaysian utility company, Tenaga Nasional Berhad (TNB) has raised the electricity tariffs in 2015, the biggest concerns for many universities are it will affect the universities energy expenditure [11]. The energy consumption in Malaysia increased rapidly every year, with an average of 2,533 GWh per year which involves universities as major energy users [12]. It is reported that energy usage in Malaysia public universities are increasing every year [13] and consume high energy due to its activities and populations. According to the Ministry of Higher Education (MOHE) (2018), based on the statistic recorded, there are 20 public universities, 53 private universities and 10 international universities in Malaysia with huge universities' population. According to the Planning and Research Unit (2018), there are 560,359 students from public higher education (IPTA) and 484,963 students from private (IPTS) enrolled in Malaysian higher education institutions. This gives greater impact on the energy demand and performance.The university's facilities especially the energy supply play an important roles in accommodating users' need in universities [14]. Appliances like computer, photocopier machine and air-conditioning system, which being used by the staffs also have influenced the energy performance, and usually the staffs tend to use these appliances excessively [15]. Usually, in many universities in Malaysia, the administration blocks which comprises with the Head of Management University and academic affairs office used cooling appliances during office hours. It is a challenge to save the energy consumption in this building since it is usually centralised and other appliances are usually switch on during the day. By using computerised solution, future energy performance can be predicted, projected and solved. This is why simulation process, like using Building Information Modelling (BIM) is essential during the retrofitting process on improving energy performance in existing building university.

3. The Review On Bim Usage For Energy Performance

Usually, in many energy performance's studies, the researchers adopts experimental and computational study methods [16]. The past researchers conducted study on the related retrofitting existing building by using Building Information Modelling (BIM) for energy performance [17]. The study is generally focusing on the daily electricity usage pattern recorded over the period of one year for non-working days and working days [18]. Table below shows the list of researchers that obtained the study related with BIM adoption for energy performance in universities (see Table 1).

Table 1 : Matrices from Past Researcheres focusing on BIM for Energy Performances Studies for University

Author / Keywords	Author 1: Alam M.J., 2014	Author 2: Reeves T., 2015	Author 3: Otuh N.F., 2016	Author 4: Hammond R.S., 2014	Author 5 : Joanna S., 2012
Building Information Modelling (BIM) in University	Building Energy Simulation, HVAC System, Human Comfort, Energy Plus	Building Information Modelling (BIM), Building Energy Modelling (BEM), Simulation, Energy Consumption, Daylighting, Natural Ventilation	BIM, Energy, Sustainability, Revit, GBS, Green Building Studio	BIM, Sustainable Design, Retrofit, Renovation	Simulation Software Tools, Energy Consumption, Buildings
BIM Tools / Software	Energy Plus	IES VE, Ecotect, Green Building Studio	Autodesk Revit, Autodesk VASARI, Autodesk Green Building Studio (GBS)	Autodesk Revit, Autodesk Green Building Studio, Autodesk Insight	Energy Plus, ESP-r, IES VE, TRNSYS
Keywords of Research	Energy Simulation to estimate building energy consumption using Energy Plus	Guidelines for Using Building information modeling for Energy Analysis of Buildings	BIM Based Energy/Sustainability Analysis For Educational Buildings – A Case Study	BIM in Sustainable Design: Strategies for Retrofitting / Renovation	Energy Simulation Software for Buildings: Review and Comparison

To achieved the main objective of this research, the experimental study need to be obtained by using instrumentational equipment on site to get the actual data. Therefore, a desktop study will be significant to be conducted to study on the energy performance of a existing building with the aid of Building Information Modelling (BIM).

Based on Table 1, the previous researchers have studied on the importance of retrofitting existing building in order to improve energy performance. It also displayed the importance of renovation works [19] and refurbishment works [20] by using technology computerized software in order to enhance energy performance. In term of environmental assessment for sustainable building, the integration of

BIM requires the software to analytically specific with information like energy related environmental profiles [21]. In BIM, it is important to has standardisation of methodologies which able to be integrated with many building standards and codes. There are several codes and standard can be applied for improving energy performance in existing building, for example, ANSI, ASHRAE, USGBC, IES Standard 189.1, Standard for the Design of High-Performance Green Buildings, the Malaysian Standard MS1525:2014, the Malaysian Green Building Index (GBI) Rating, Code of Practice on Energy Efficiency (EE) and Use of Renewable Energy for Non-Residential Building (Department of Standard Malaysia, 2014) [22].

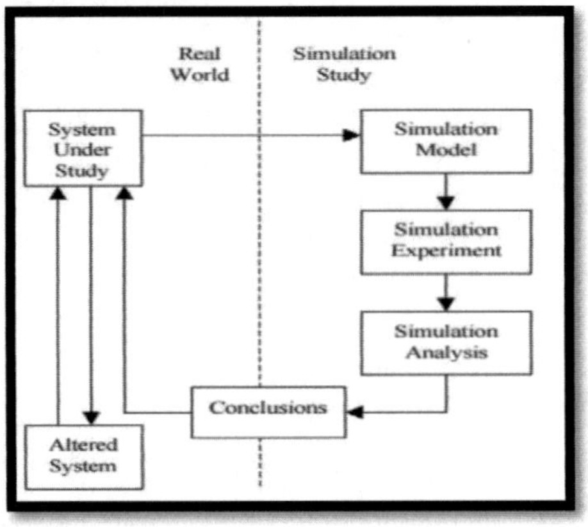

Figure 1: Simulation Study Schematic [23]

Many researchers also highlighted the importance of simulation approach in energy performance study. Simulation is a significant tool to be used to examine possible scenarios that can be followed with related building guidelines [23]. It will be used to simulate every data gained from the site. Based on figure 1 above, it will explained the process flow from the real world to simulation study process. From the stage of collecting the data from actual condition or site, the next step will be conducted on the process of 3D modelling, simulation model, simulation experiment, simulation analysis and producing result. Therefore, the energy performance study for universities needs advanced energy auditing technologies and environmental assessment frameworks for refurbishment of existing building, computer simulation software can be used to model and simulate the energy performance of retrofit measures [24].

4. Conclusion

In building and energy industry, it is familiar to used simulation with the adoption of Building Information Modelling (BIM). The software has been used in sustainable energy simulation platform such as Autodesk AutoCAD, Autodesk Revit, Autodesk Naviswork, Autodesk Ecotec, Integrated Environmental Solution (IES-VE), Autodesk Green Building Studio (GBS), Energy Plus, eQuest and others. This paper has reviewed significant component of the embodied energy used to transfer inside building component from the location to building site. Therefore, in this paper, the preparation on the desktop study will be running through Autodesk Revit for 3D model and Autodesk Green Building Studio (GBS) or Integrated Environmental Solution (IES-VE).

5. References

[1] Hassan J.S., Zin R.M., Majid M.Z.A, Balubaid S., Hainin M.R., 2014, *"Building Energy Consumption in Malaysia: An overview"*. Vol.70, Jurnal Teknologi (Science and Engineering), pp.33-38

[2] Kilpatrick R.A.R., Banfill P. F.G., 2011, *"Energy Consumption In Non-Domestic Buildings: A Review of Schools"*, World Renewable Energy Congress, pp 1008-1015

[3] Gul M.S., Partidar S., 2015. *"Understanding the energy consumption and occupancy of a multi-purpose academic building"*, Vol.87, Energy and Building, pp 155-165

[4] Bullen, P.A., 2007. *"Adaptive reuse and sustainability of commercial buildings"*, Vol. 25(1-2), Facilities Journal, pp 20-31

[5] Fiksel J., 2003. *"Designing Resilient, Sustainable Systems",* Vol. 37(23), Environment Science and Technology, pp 5330-5339

[6] Suruhanjaya Tenaga & Pusat Tenaga Malaysia, 2007, *"Energy Efficiency and Conservation Guidelines"*

[7] Ali A.S, Keong K.C, Zakaria N., Zolkafli U., Akashah F., 2013, *"The effect of design on maintenance for school buildings in Penang, Malaysia",* Vol. 31(3), Str. Srvey, pp.194-201

[8] Mohandes S. R., Marsono A. K., Omrany H., Faghirinejadfard A., Mahdiyar A., 2015, *"Comparison of Building Existing Partitions through Building Information Modeling (BIM)",* Vol.75, Jurnal Teknologi (Science and Engineering), pp.287-298

[9] Azhar S., Carlton W.A., Olsen D., Ahmad I., 2009, *"Building information modelling for sustainable design and LEED rating analysis".* Vol.20, Automation in Construction, pp.217–224

[10] Dzulkefli M.H., 2017, *"The energy audit process for universities accommodation in Malaysia: A Preliminary Study",* pp. 1-9

[11] Tenaga Nasional Berhad (TNB). *"Pricing and Tariffs".* Retrieved from https://www.tnb.com.my/residential/pricing-tariffs on 19th August 2018.

[12] Sung C.T.B., 2016, *"Electricity demand, economic growth, and sustainable energy resources in Malaysia".* Extraordinary mind discuss ideas, pp. 1-8

[13] Abd-Razak M.Z, Mustafa N.K.F., Che-Ani A.I., Abdullah N.A.G., Mohd-Nor M.F.I., 2011, *"Campus sustainability: Students perception on campus physical development planning in Malaysia".* Vol.20(2), Procedia Eng., pp. 230–237

[14] Daniel C., Barron N., Guild P., Bond P.D., Association P.S.,2008, *"The Educational Attributes of Some of the World's Top 50 Universities – a Discussion Paper".* pp.1–140,

[15] Ahmad S.A., Hassan M.Y., Abdullah H., Rahman H.A., Majid M.S., Bandi M., 2012, "*Energy Efficiency Measurements in a Malaysian Public University*" Int. Con. on Power and Energy, pp. 564-569

[16] Mylona Z., Kolokotroni M., Tassou S.A., 2017, *"Experimental and Computational Study to Improve Energy Efficiency of Frozen Food Retail Stores"*, Vol.144, Energy and Buildings, pp.129-143

[17] Khaddaj M., Srour I., 2016, "*Using BIM to Retrofit Existing Building*". Vol.145, Procedia Engineering, pp.1526-1533

[18] Amber K.P., Aslam M.W., Mahmood A., Kousar A., Younis M.Y., Akbar B., Chaudhary G.Q., Hussain S.K., 2017, *"Energy Consumption Forecasting for University Sector Buildings".* Vol.10, Energies, pp1-18

[19] Hammond R.S, Nawari N.O, Walters B, 2014, *"BIM is Sustainable Design : Strategies for Retrofitting / Renovation",* pp.1969-1977

[20] Gholami E., Sharples S., Abrishami S., Kocaturk T., 2013, *"Exploiting BIM in Energy Efficient Refurbishment: A Paradigm of future opportunities"*, Vol.2, Conf. Sustainable Architecture for a Renewable Future, pp.1-6

[21] Tarja H., Arto K., 2015, *"Sustainable Building and BIM".* World Conference SB08, Technical Research Centre of Finland, pp.679-686

[22] Use of Renewable Energy for Non-Residential Building, Department of Standard Malaysia, 2014.

[23] Hassan N.A, 2016. Green BIM Software. *Building Energy Simulation*". Retrieved from slide share https://www.slideshare.net/naghamahasan/green-bim-software-dr-nagham-ali-hasan on 19th August 2018

IOP Conf. Series: Earth and Environmental Science **291** (2019) 012033 doi:10.1088/1755-1315/291/1/012033

[24] Ma Z., Cooper P., Daly D., Ledo L., 2012, *"Existing Building Retrofits: Methodology and State-of-the-Art, Methodology and state-of-the-art"*, Vol.22(12), Energy and Buildings, pp.889-902

IOP Conf. Series: Earth and Environmental Science **291** (2019) 012034 doi:10.1088/1755-1315/291/1/012034

Biomass energy from agriculture in China: Potential and evolutionary trend

Liangzhen Zang[1,2] and Caihong Zhang[3]

[1] China Institute for Rural Studies, Tsinghua University, Beijing 100084, China
[2] School of Public Policy and Management, Tsinghua University, Beijing 100084, China
[3] School of Economics and Management, Beijing Forestry University, Beijing 100083, China

zangliangzhen@hotmail.com

Abstract. This paper aims to quantify the biomass energy potential from agriculture, analyze composition and explore evolutionary trend based on the data of 31 Chinese provinces from 1993 to 2016. The results show that: (1) China's biomass energy potential from agriculture increased from 139.42 million metric tons of coal equivalent (Mtce) in 1993 to 196.51 Mtce in 2016, with an average annual growth of 1.57% during the 24-yr period; (2) In terms of resource composition, biomass energy potential from agriculture mainly derived from the straws of rice, wheat and maize that were as high as 60.97% in all different kinds of biomass energy; (3) In terms of evolutionary trend, the biomass energy potential from agriculture displayed an increasing trend, but disparities among the provinces began widening gradually.

1. Introduction

The essence of low-carbon economy is to improve energy efficiency and increase the proportion of clean energy utilization [1]. Agricultural biomass energy, a renewable energy, is a kind of energy contained in biomass that is characterized by less pollution, easy combustion and low ash [2]. Currently, the annual biomass energy potential all over the world is around 200 EJ to 600 EJ, in which biomass energy from agriculture accounts for the highest proportion [3]. Therefore, it is urgent to develop the industry of agricultural biomass energy in order to achieve sustainable development of low-carbon economy.

Compared with developed countries, the industry of China's agricultural biomass energy is still immature. The *China's 13th Five-Year Plan for Biomass Energy Development* pointed out that China's agricultural biomass energy development is facing a number of problems, such as the difficulty of efficient collection of raw materials, the low conversion degree, and the shortage of distributed commercial utilization, which seriously prohibit the development of biomass energy industry. Therefore, it is important to know the basic and decisive information about agricultural biomass energy in China, especially on the biomass energy potential. This article aims to quantify the potential of agricultural biomass energy, analyze its composition and explore temporal and regional changes. This research could provide the corresponding theoretical guide for creating a macro-policy of agricultural biomass energy in China in the future, especially for the sustainable supply of raw materials.

2. Methodology and Data

2.1. Methodology

Crop straw and agricultural processing by-product are the main raw materials to produce agricultural biomass energy. Crop straw refers to the harvest residue of crops that mainly include grain crop, economic crop, industrial raw materials crop, feed crop and medicinal materials crop. In this paper, it quantifies the biomass energy potential from straw based on the yields of grain crop and economic crop, because the last three types of crops have not been fully added up and just account for a small proportion of all crop yields in China. Explicitly, we just focus on the agricultural biomass energy coming from the straw of cereal (including rice, wheat, maize and other cereal), beans, tubers, oil-bearing crops (including peanut, rapeseeds, sesame and other oil-bearing crops), cotton, fiber crops, sugar crops (including sugarcane and beetroots) and tobacco. Agricultural processing by-product refers to the residues produced during the primary processing. The majority of residues of agricultural processing in China are rice hull, corncob, peanut shells, bagasse, beet residue and cottonseed hull. Since there is no official statistical data about the yields of the crop straw and agricultural processing by-product in China, therefore, the potential must be calculated according to the data of crop yields.

2.1.1. Method of estimating biomass energy potential of crop straw. The biomass energy potential of crop straw usually can be estimated in four steps: the theoretical potential of biomass resources of crop straw (Q_a), the available potential of biomass resources of crop straw (Q_b), the available potential for producing biomass energy of crop straw (Q_c), and the potential of biomass energy of crop straw (Q_d):

$$\begin{cases} Q_a = \sum_{i=1}^{n} P_i \cdot RPR_i \\ Q_b = \sum_{i=1}^{n} Q_{ai} \cdot \lambda_i \\ Q_c = \sum_{i=1}^{n} Q_{bi} \cdot f_i \\ Q_d = \sum_{i=1}^{n} Q_{ci} \cdot \eta_i \end{cases} \quad (1)$$

where, i is the kind of crop, n is all the kinds of crops, P_i is crop yield, RPR_i is coefficient of grass-grain ratio, λ_i is coefficient of available potential of biomass resources of crop straw, f_i is coefficient of available potential for producing biomass energy of crop straw, η_i is the coefficient of coal equivalent. Based on the previous research [4], this paper takes the coefficient of the available potential for producing biomass energy of crop straw as 0.5. All other coefficients are shown in Table 1.

Table 1. Coefficients for estimating biomass energy potential of crop straw.

Category	*RPR* (%)	*F* (%)	*H* (kgce/kg)	Category	*RPR* (%)	*F* (%)	*H* (kgce/kg)
Rice straw	1.04	0.83	0.429	Rapeseed straw	2.9	0.85	0.61
Wheat straw	1.28	0.83	0.5	Sesame straw	1.89	0.85	0.529
Maize straw	1.07	0.83	0.529	Other oil-bearing crop straw	2.63	0.85	0.584
Other cereal straw	2.32	0.83	0.545	Fiber crop straw	1.73	0.87	0.6
Beans	1.35	0.88	0.543	Sugarcane straw	0.34	0.88	0.494
Tubers	0.53	0.8	0.432	Beetroots	0.37	0.88	0.205
Cotton stalk	2.87	0.9	0.543	Tobacco stalk	0.66	0.9	0.55
Peanut straw	0.99	0.85	0.541				

2.1.2. Method of estimating biomass energy potential of agricultural processing by-product. Similarly, the biomass energy potential of agricultural processing by-product can also be estimated in four steps, i.e. the theoretical potential of biomass resources of processing by-product (Q_a), the available potential of biomass resources of processing by-product (Q_b), the available potential for producing biomass energy of processing by-product (Q_c), and the potential of biomass energy of processing by-product (Q_d):

$$\begin{cases} Q_a = Q_a^1 + Q_a^2 = \sum_{i=1}^{n} P_i \cdot PRI_i + P' \cdot (\frac{1}{\mu} - 1) \cdot PRI' \\ Q_b = \sum_{i=1}^{n} Q_{ai} \cdot \lambda_i \\ Q_c = \sum_{i=1}^{n} Q_{bi} \cdot f_i \\ Q_d = \sum_{i=1}^{n} Q_{ci} \cdot \eta_i \end{cases} \quad (2)$$

where, Q_a^1 is the theoretical potential of biomass resources of rice hull, corncob, peanut shells, bagasse and beet residue, Q_a^2 is the theoretical potential of biomass resources of cottonseed hulls, i is the i-th by-product, n is all types of agricultural processing by-product, P_i is the yield of five kinds of crops, P' is ginned cotton yield, PRI is the coefficient of theoretical potential of biomass resources of processing by-product; PRI' is the coefficient of the theoretical potential of biomass resources of cotton husk processing, μ is the clothing fraction, i.e., the proportion of ginned cotton in seed cotton, λ_i is the coefficient of available potential of biomass resources of processing by-product, f_i is coefficient of available potential for producing biomass energy of processing by-product, η_i is the coefficient of coal equivalent. Based on the existing research [5], it is thought that μ, λ and f are 0.38 1 and 0.5, respectively. All other coefficients are shown in Table 2.

Table 2. Coefficients for estimating biomass energy potential of processing by-product.

	Rice hull	Corncob	Cottonseed hull	Peanut shells	Bagasse	Beet residue
PRI (%)	0.18	0.16	0.47	0.27	0.16	0.05
η (kgce/kg)	0.49	0.60	0.60	0.59	0.60	0.57

2.2. Data

The data of crop yield used in this paper were collected from the China Statistical Yearbook, the China Rural Statistical Yearbook and the China National Bureau of Statistics (www.stats.gov.cn). The research period covered from 1993 to 2016, with a total of 24 years.

3. Results

3.1. Temporal trend of biomass energy potential from agriculture in China

The potential of biomass energy from agriculture in China demonstrated a slowly increasing trend in the 24-yr period from 1993 to 2016. Although the resource potential was only 139.42 Mtce in 1993, it reached as high as 196.51 Mtce in 2016, with an average annual growth of 1.57%. In general, the temporal trend of biomass energy potential from agriculture in China can be divided into two stages: the inverted U-shaped stage in 1993-2003 and the rapid growth stage in 2003-2016.

During the 11-yr period from 1993 to 2003, the biomass energy potential from agriculture first exhibited a upward trend, and then entered a slow downward stage. Explicitly, it was 139.42 Mtce in 1993, reached 157.19 Mtce in 1998 (the highest point) and then began to decrease gradually, with the lowest point in 2003, 139.87 Mtce. Comparatively, the biomass energy potential from agriculture greatly increased throughout the study period, from 139.87 Mtce in 2003 to 196.51 Mtce in 2016, with an average annual growth of 2.68%.

3.2. Composition of biomass energy potential from agriculture in China

Biomass energy potential from agriculture in China mainly derived from the straw of rice, wheat and maize, which accounted for 60.97% in all kinds of agricultural biomass energy. However, these three kinds of biomass energy potential from agriculture presented a completely different changing trend. From 1993 to 2016, the biomass energy potential of rice straw and wheat straw, accounting for 40%, displayed a stable increasing trend. In comparison, the biomass energy potential of maize straw exhibited a different trend during this period, i.e., it presented a upward or downward trend before 2003 and rapidly increased from 2003, 27.21 Mtce, to 2016, 51.57 Mtce.

There were two characteristics of the changing trend of other types of biomass energy potential from agriculture: First, all of them demonstrated a stable trend. There was no obvious increase or decrease from 1993 to 2016. Second, the potential of most kinds of agricultural biomass energy increased gradually, but others decreased. Explicitly, the agricultural biomass energy potential of straw of beans, fiber crops, beetroots and beet residue gradually decreased in the 24-yr period, while other biomass energy potential from agriculture presented an upward trend.

In order to further analyze the composition of biomass energy potential from agriculture in China, this article calculated the average proportion of different types of biomass energy potential from agriculture throughout the study period. As seen in figure 1, the proportion of biomass energy potential of maize straw, up to 21.62%, was the highest, followed by the biomass energy potential of rice straw at 21.57%. The third one was the biomass energy of wheat straw, accounting for 17.78%. Overall, these three kinds of biomass energy potential accounted for 60.97% totally.

In terms of the proportion of other biomass energy potential of crop straw, there was little disparity. Specifically, according to the proportion from the highest to the lowest, they were rapeseed straw (5.37%), sugarcane straw (4.29%), beans (3.78%), other cereals straw (3.70%), cotton stalk (2.37%), tubers (1.87%), peanuts straw (1.91%), other oil-bearing crops straw (1.03%), tobacco stalk (0.2%), beetroots (0.21%), fiber crops straw (0.17%) and sesame straw (0.17%). Furthermore, the biomass energy potential of agricultural processing by-product accounted for 13.88% on average in all the biomass energy potential from agriculture. Explicitly, the biomass energy of rice hull, corncob, bagasse, peanut shells, cottonseed hull and beet residue accounted for 5.14%, 4.42%, 2.79%, 0.67%, 0.78% and 0.09%, respectively.

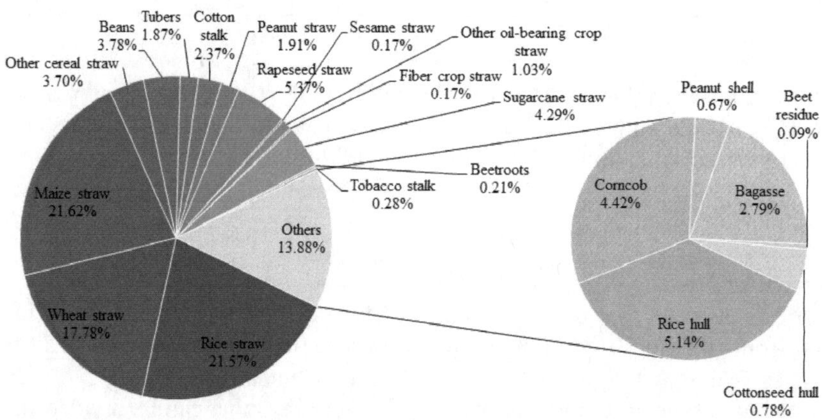

Figure 1. Proportion of biomass energy potential from agriculture in1993-2016.

3.3. Regional changes of biomass energy potential from agriculture in China

The method of kernel density was used to analyze regional changes of biomass energy potential across 31 provinces of China. The kernel density curve with five years -1993, 1998, 2004, 2010 and 2016 - was exhibited in figure 2. Based on the curve, the following characteristics were found.

First, the kernel density curve presented a trend of double-peak distribution that the first and second peaks were obvious in five curves. In terms of the kernel density, the values corresponding to the first and second main peak were very low, respectively. Furthermore, the disparity of biomass energy potential between the first and second main peak was not wide. Both of these distribution fully indicated that the biomass energy potential from agriculture in most provinces were similar each other.

Second, the kernel density curve shifted from the left to the right as the time changed. The curve in 1993, 1998, 2010, 2016 shifted to the right obviously that fully denoted that the potential of agricultural biomass energy in China was gradually increasing. Comparatively, the curve in 2004 shifted slightly to the left, but it did not move too much. The reason was that China's government adjusted the supporting policy in the field of agriculture in 2003. On the whole, China's biomass energy potential from agriculture gradually increased at all the time across all 31 provinces.

Third, the kernel density of the first peak decreased over the five study years. It indicated that the disparity of biomass energy potential from agriculture among all the provinces were gradually widening. In addition, the right parts of five curves became smooth as the time changed. Although the agricultural biomass energy potential of most provinces were slowly increasing, the resource potential in other provinces increased greatly, such as Shandong, Henan, Guangxi. Sichuan, Heilongjiang, Jiangsu. That was the reason for the increasing disparity of biomass energy potential from agriculture among all the provinces.

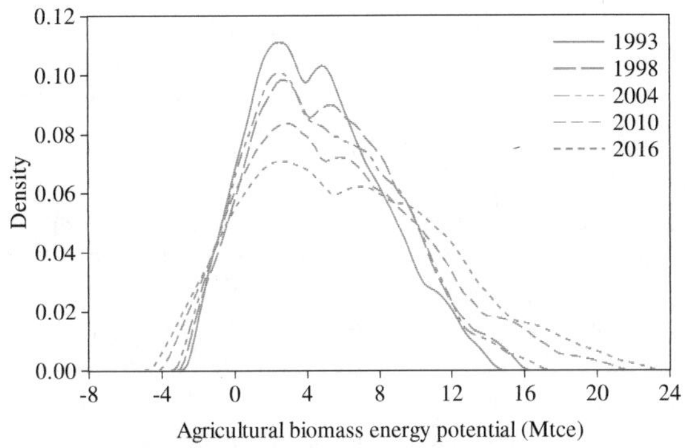

Figure 2. Kernel density curve of biomass energy potential from agriculture.

4. Conclusion

By analyzing the statistical data in the 24-yr period from 1993 to 2016 , we found that: (1) The total biomass energy potential from agriculture increased from 139.42 Mtce in 1993 to 196.51 Mtce in 2016, with an average annual growth of 1.57% and an increase of 40.95% in the past 24 years; (2) In terms of composition, biomass energy potential from agriculture mainly derived from the straw of rice, wheat and maize, accounting for 60.97% in all types of biomass energy; (3) In terms of evolutionary trend, the agricultural biomass energy potential of most provinces showed a gradual increasing trend from 1993 to 2016, while the disparity among all the provinces were gradually widening. Overall, there is a plentiful potential of agricultural biomass energy in China and the resource potential continues to increase steadily. The resources of China's agricultural biomass lay a good foundation for the development of agricultural biomass energy industry. Therefore, it is urgent to establish a long-term and stable supply mechanism according to the total resource potential in the region in order to realize the interactional development between agricultural biomass energy and fossil energy.

Acknowledgments
This research was supported by United Nations Development Programme and Ministry of Commerce of China (CPR/13/304/14/002) and China Postdoctoral Science Foundation (2017M620812)

References

[1] Chang M C 2015 *J. Clean Prod.* **99** 140-151
[2] Bilgen S, Keles S, Sarikaya I 2015 *Renew. Sust. Energ. Rev.* **48** 228-239
[3] Zang L, Zhang C 2017 *Inq. Econ. Issues* **8** 185-190
[4] Zhao X, Zhong M 2015 *J. Fujian For. Sci. Tech.* **2** 169-173
[5] Xing H, Zhao Y, Wang Y 2015 *Acta. Ecologica. Sinica* **35** 3480-89

Connecting Solar Generated Electricity during Flood Disaster in the FELDA Communities of Malaysia: The Public Perception

S J T Jamil[1], N A Ahmad[1] and J J Jamian[2]

[1]Faculty of Architecture, Planning and Surveying, Universiti Teknologi MARA, Perak Branch, Seri Iskandar Campus, Seri Iskandar, 32610 Perak MALAYSIA

[2]School of Electrical Engineering, Faculty of Engineering, Universiti Teknologi Malaysia 81310 UTM Johor Bahru Johor MALAYSA

Abstract. Solar electrification is a significant solution for supplying electricity at the evacuation centre during power shortage or blackouts in a flood disaster situation. However, it is a challenge to connect this technology to the local communities, especially for the rural people due to the limited accessibility of the technology. Strong support from the Government is needed in disseminating this technology for the locals, especially during a flood disaster where victims need the power aid relief during the power interruption. By using solar technology, the power can be generated and stored ahead of time at the flood relief centre. This paper will investigate on the perception of rural communities in Malaysia; focusing FELDA communities, in connecting the solar electrification during a flood disaster. The significant approaches for connecting FELDA communities with solar electrification need to be explored strategically. In order to investigate the acceptance of FELDA communities in connecting solar electrification in the flood evacuation centre, survey research through questionnaires has been obtained from the respondents. Interestingly, the result shows that the back-up electricity supply system is needed in order to accommodate the electricity supply at the evacuation centre when flood occurs. The research found that the local people are ready to use the solar electrification as a supportive electricity system during flood disaster. This paper has elaborated on the significance approaches towards connecting people with the technology.

1. Introduction

Malaysia has a full advantage to utilise the solar energy as an additional energy sources since this nation receives abundant solar energy all the year which allowing better benefit in energy resources perspective. Malaysia receives an average solar radiation of 1643 kWh / m^2 per year with 4500Wh/m^2 of the average solar radiation per day [1],[2]. Hence, this climate is suitable to utilise renewable energy, particularly solar energy [3],[4],[5]. By looking on the sunray data, Malaysia has the potential to apply the solar energy through PV panel generation [6],[7].

PV electrification can be used in any situation, including natural disaster. Flood disaster, for instance, can occur without signal. When these situations happen, the components of electricity system will be disrupted or shut down. In fact, most of the flood victims depend on the electricity during these emergencies moments. They needs the electricity to operate the electricity appliances such as refrigerator, lighting, radio or medical equipment [8]. During a normal practice, flood victims use gasoline or diesel generator to supplement or back up the power energy. However, this practice can link to a hazard situation since the fuels needed for generating the generator can also spark fire. In addition, the generators also produce noise and can leads to traumatic stress [8]. The body parts of the generators are also heavy and bulky which may cause hassle during the evacuation process. Therefore, it is significant to have a solution that is applicable to all parties. Using the renewable energy; through solar photovoltaic; it can acts as a back-up electricity supply during a flood disaster [9], [10]. By using this technology, the electric power can be supplied to the people immediately without delay.

During rainy season, the solar PV system is capable to generate electricity with the aid of energy storage gained during daylight time. According to [11], the benefits of solar PV systems are clean and pollution-free energy, reliable and flexible. By using the solar PV, the important system can be operated during flood seasons such (i) the multi-purpose portable system, (ii) emergency lightings for passageways, (iii) transportation, (iv) communication support and (v) additional energy in the shelters during the emergency [12]. For this reason, it is significant to investigate on the perception of communities on the solar engagement in order to measure the importance of introducing the ability of the system to meet the critical power needs during flood disaster, especially on the strategic approaches needed.

2. Strategic approaches towards public engagement for solar electrification

Strategic approaches are crucial in engaging the solar technology with the rural communities. The communities; whether the urban or rural communities, should directly involve, interact and have the opportunity to give the feedback of the technology to the Government [13].

Many form of engagement approaches can be applied to the public in order to build the relationships and trust between the Government, stakeholders and the people [13] (See Figure 2). By using the engagement approaches shown in Figure 2, the quantity and quality of communities information available can be improved especially during emergencies or natural disaster [14]. The aim of this engagement spectrum is to connect people, especially the targeted population (e.g rural communities) through various levels of strategic approaches. From the first stage of disseminating information, to the consultation, involvement, collaboration and the empowering level of the solar technology in the communities. Therefore, this research is significant in investigating the public perception towards engaging the technology through these various spectrums of engagement.

It is widely known that in other developed countries, for instance like Japan and Germany, the solar energy communities' programmes are highly supported by the technology implementers [15] [16]. However, in Malaysia, this scenario is still in growth and under gradual development [17]. The various type of community engagement approaches such as information gathering, public thinking, information dissemination, consultation, advice, counselling and etc. [18] can be implemented in order to attract the local people to involve directly within the solar energy projects. Therefore, through this study, the aim of achieving the phases of (i) information gathering and (ii) public thinking as highlighted by [18] can be gained through the survey research presented in section 4 and 5. This is important in finding a significant mechanism that can help the people in facing power interruption during flood disaster.

3. Case study: felda jengka 6 in pahang, malaysia

FELDA is the rural settlement area provided by the Government of Malaysia since the early 1960s in order to help rural people to have basic housing necessities and source of income through agricultural projects subsidised by the Government [19]. For this study, one of the FELDA settlements that have experienced flood disaster has been chosen to obtain the public perception on the solar energy usage during a natural disaster. The study area is a FELDA Jengka 6 in Bandar Tun Razak in the Pahang state, Malaysia. This settlement has own administrative building which is to cater the society and operational management. For the settlers, the FELDA management provides public amenities such as school, public hall, community hall, mosque, field, futsal court, Tomoi sport court and shoplot.

4. Methodology

The survey research was used for this study in order to gain public perception in the implementation of solar generated electricity during a flood disaster. During the survey section, the structured questionnaires were given to the FELDA residents as respondents at the study area. Door-to-door questionnaires survey was conducted through snowball sampling. Only 25 respondents were chosen to answer the questionnaires based on the specific criteria which involves i) residents that have experienced flood in the case study, ii) FELDA settlers that lived in the area for more than 5 years and iii) has chosen the FELDA hall as an evacuation shelter during the flood disaster. The survey was conducted for 2 months. The respondents were asked about (i) the experience during the flood, (ii) the impact of electricity supply system during flood, (iii) the opinion on the back-up electricity supply system provided during flood and (iv) the readiness of the people to accept the option of back-up electricity system during flood. This survey also involved questions about the engagement approach of solar electricity supply and the method to approach the solar electricity supply system to the FELDA community in order to identify the public thinking as highlighted by [18]. The data were analysed through descriptive statistics data analysis where by the percentage and frequency of the question can be identified.

5. Result

From the 25 selected respondents, 23 respondents have given effective answers for the distributed questionnaires. Therefore, the study has used the 23 respondents' opinion to have an overview and perception on the solar technology application during flood. 20 (87%) respondents out of 23 agree that the needs of back-up electric supply system to accommodate the electricity supply at the flood relief centre. Only 3 respondents or 13% disagree with the back-up electric supply system at the flood relief centre. 22% respondents not ready to use the solar generated electricity supply system as a back-up electricity system during flood. However, the rest of the respondents 78% ready to use the solar electricity system as a back-up electric energy when floods occur.

From Figure 1, it is shown that most of the respondents; 22 people (96%), agree that brochures, pamphlets and newsletters are the suitable mechanism to engage the knowledge on the solar power technology application within the local people. 17 respondents stated that the local information centre from government/NGO are the suitable mechanism to engage the solar power technology application with them. It is followed by the community service, 16 (70%) respondents. Only 1 (4%) respondent picks the focus group as a suitable mechanism to engage with the solar power technology application during flood.

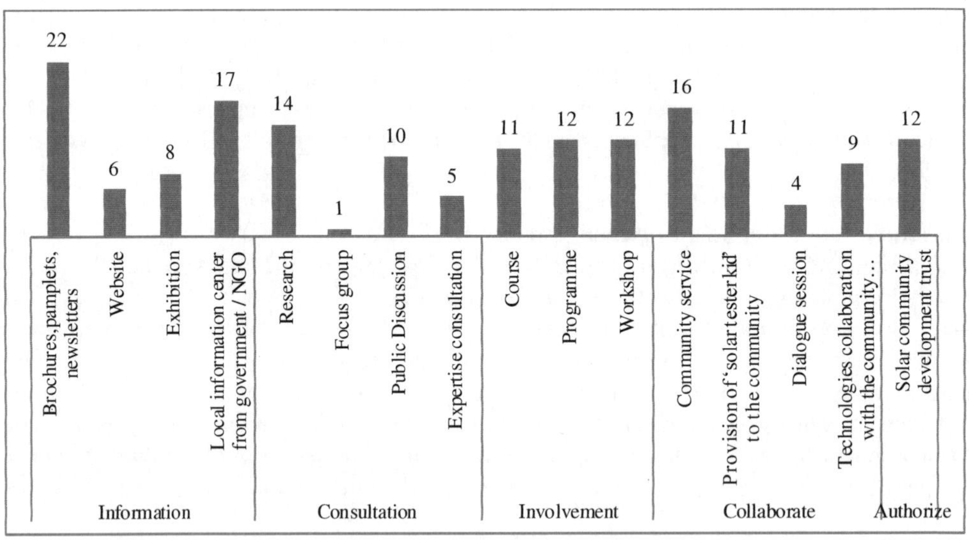

Figure 1. The Solar Engagement Approaches

Figure 2 shows data on how technologists can approach the respondents to use solar generated electricity during the occurrence of flood. Technologist can be defined as the expertise on the technical area of RE field which includes solar technology project [20]. The 21 (91%) respondents stated that the announcement through mass media from television, radio and printed media are the better mechanism where the technologist can engage the community. However, 15 (65%) respondents mentioned that the demonstration of technology is the mechanism needed to engage the respondents to use solar electrification during the flood. However, only 9 respondents agree that infografic sessions is the needed mechanism that the technologist can engage the people with solar technologies.

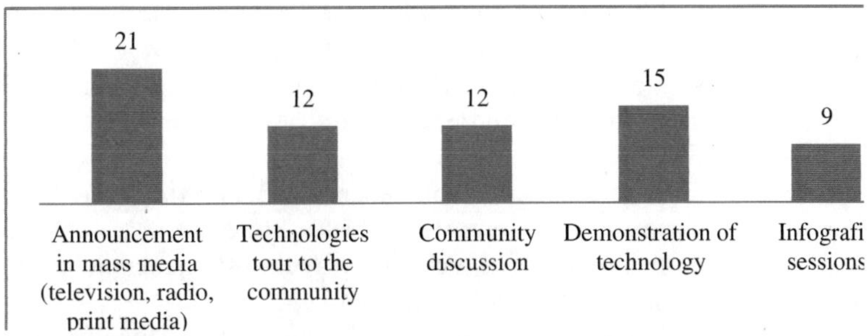

Figure 2. Solar engagement approaches through a technologist

Among the respondents, 21 (91%) respondents agree if FELDA can provide solar electricity aid programs during floods to overcome local electricity supply disruption. Only 2 (9%) respondents were disagree with the statement.

6. Discussion

The back-up electricity system is very important to enhance communities' resilience after the flood happen [10]. From the analysis, it shows 87% respondents agree that they need a back-up electricity to accommodate the electrical supply at the flood relief centre. On top of that, 78% respondents are ready to use the solar electricity supply system as back-up electricity systems during flood occurrence. Solar electricity is one of the renewable energy (RE) that meets the criteria of green technology under the Malaysia National Green Technology Policy [11].

Through the 11th Malaysia Plan, the enforcement of the Renewable Energy Act (2011) and the introduction of Fit in Tariff (FiT) have led towards the RE development [11]. It shows that the Malaysia government supports the use of solar technology. In the Budget Speech Year 2017, the Malaysian government give full supports through the national budgetary incentives to expand the program of solar technology development [11]. The mechanism of engagement is very crucial to make sure the success of technologies' usage. From the survey, 22 (96%) respondents identify that brochures, pamphlets and newsletters are the suitable mechanism to engage the solar electricity with local people. However 21 (91%) respondents mentioned that the announcement through mass media such from television, radio and printed media will give better direction to the technologies to engage the local people with the solar technology. Among the respondents, 21 (91%) respondents agree if FELDA can provide solar electricity aid programs during flood to overcome local electricity supply disruptions. This may improve the evacuation aid facilities and improve the quality of life of FELDA communities.

7. Conclusion

Connecting FELDA communities with the solar electrification technology during flood disaster is essential to help the FELDA communities to continue daily life at the flood relief center. By using the questionnaire, 23 respondants meet the criterias to collabrate in the survey. Most of the respondant agree that mass media for instance television, radio and printed media is the better public engagement approach where the technologist can engage the local people. The stakeholders from the government or non government organization should link together to help the FELDA communities to engage the solar electrification at the FELDA evacuation center. For the success of electricification in FELDA specially during flood disaster, the solar technology can implement as a relief mechanism in FELDA.

8. Reference

[1] K. H. Solangi, A. Badarudin, S. N. Kazi, and M. M. Aman, "Public acceptance of solar energy : The case of Peninsular Malaysia," pp. 540–543, 2013.
[2] H. Borhanazad, S. Mekhilef, R. Saidur, and G. Boroumandjazi, "Potential application of renewable energy for rural electrification in Malaysia," *Renew. Energy*, vol. 59, pp. 210–219, Nov. 2013.
[3] A. Y. U. W. Azhari, K. Sopian, and A. Zaharim, "A New Approach For Predicting Solar Radiation In Tropical Environment Using Satellite Images – Case Study Of Malaysia," vol. 4, no. 4, pp. 373–378, 2008.
[4] K. H. Solangi *et al.*, "Social acceptance of solar energy in Malaysia: Users' perspective," *Clean Technol. Environ. Policy*, vol. 17, no. 7, pp. 1975–1986, 2015.
[5] J. Khan and M. H. Arsalan, "Solar power technologies for sustainable electricity generation - A review," *Renew. Sustain. Energy Rev.*, vol. 55, pp. 414–425, 2016.
[6] A. M. Ismail, R. Ramirez-Iniguez, M. Asif, A. B. Munir, and F. Muhammad-Sukki, "Progress of solar photovoltaic in ASEAN countries: A review," *Renewable and Sustainable Energy Reviews*, vol. 48. 2015.
[7] C. Chong, W. Ni, L. Ma, P. Liu, and Z. Li, "The use of energy in Malaysia: Tracing energy flows from primary source to end use," *Energies*, vol. 8, no. 4, pp. 2828–2866, 2015.
[8] W. Young, "Disasters : Photovoltaics for Special Needs," 2005.
[9] H. Hassan, "Kaji bina sistem tenaga solar di pusat pemindahan banjir," *Utusan Malaysia*, 2015.
[10] S. Qazi and F. Qazi, "Green Technology for Disaster Relief and Remote Areas," *121st ASEE Anu. Conf. Expo.*, no. 10192, pp. 1–12, 2014.
[11] S. C. Chua and T. H. Oh, "Solar energy outlook in Malaysia," *Renew. Sustain. Energy Rev.*, vol. 16, no. 1, pp. 564–574, 2012.
[12] J. J. Lee, C. Laurent, and C. Becker-Birck, "Solar PV Emergency & Resilience Planning," 2013.
[13] Local Government of South Australia, "Community Engagement Handbook - A Model Framework for leading practice In Local Government in South Australia," no. March, p. i-77, 2012.
[14] Newfoundland Labrador Office of Public Engagement, "Public Engagement Guide."
[15] M. Bolinger and R. Wiser, "Case Studies of State Support for Renewable Energy," *Berkeley Lab Clean Energy Gr. CASE*, no. September, 2002.
[16] T. Wildman, "Issues of Abundance Volume 6 , Issue 12," vol. 6, no. 12, 2018.
[17] A. Ngui, "Malaysia is Third Largest Producer of PV Cells," *The Sun Daily*, no. March 2017, 2017.
[18] H. A. Marlowe, "Public engagement : Theory and Practice," 2005.
[19] Lembaga Kemajuan Tanah Persekutuan, "Mengenai FELDA," 1994. .
[20] "Sustainable Energy Technologist : Job Duties & Career Requirements Career Definition for a Sustainable Energy Technologist," *Study .com*, 2018. .

Palm empty fruit bunch gasifier characterization using Eulerian-Lagrangian CFD modelling

Priyambodo Nur Ardi Nugroho [1, 2*] and Shuichi Torii [1]

[1]Department of Advanced Mechanical System Engineering, Kumamoto University, Kurokami, 2-39-1, Kumamoto 860-8555, Japan
[2]Politeknik Perkapalan Negeri Surabaya, Jl. Teknik Kimia, Kampus ITS Sukolilo, Surabaya 60111, Indonesia.

[*]Email: priyambodo@ppns.ac.id

Abstract. Energy efficiency is crucial for many developing countries, including Indonesia. One of the solutions is to shift fossil fuels to an alternative source of energy, like biomass which is also available abundantly. Biomass conversion technique like gasification using palm empty fruit bunch is essential considering its potential in Indonesia. Advanced technology with a cost-effective and highly efficient gasification process is required to promote this method, especially in a rural area. Computational analysis for design and initial study is mandatory because gasification involves complicated chemical reactions. Biomass gasification characteristic was investigated using Computational Fluid Dynamic (CFD) program STAR CCM+. The Eulerian-Lagrangian concept was employed to model the gaseous phase (air) and solid phase (biomass particles) of biomass gasifier accordingly. Each of the gasification syngas formation - i.e., CO, N2, CH4, and H2 respectively - were analyzed. The objective of this paper is to characterize biomass gasification steps started with drying, thermal decomposition/pyrolysis, partial combustion of gases, vapor, and char, and finished with the gasification of decomposed products.

1. Introduction

Global energy demand accelerates along with the expansion of the world's economy, in 2009 about 12 billion tons of oil equivalent consumed worldwide and projected to increase by 50% by 2035 [1]. Therefore, energy supply and demand has been more and more crucial in international trade. On the other hand, carbon emission from fossil fuels is also anticipated to grow significantly, which will directly connect with climate change. The quest for renewable and environmentally friendly energy like wind, solar, hydro, and biomass had been developed intensively to answer the challenge [2].
Compared to other renewable resources, biomass possesses benefit because of its available immensely over the world [3]. After coal, petroleum and natural gas, biomass is the most important source of energy with about 10% of the total energy in the world. Along with Malaysia as the most prominent Crude Palm Oil (CPO) in the world, Indonesia is leading in the market with more than 20 million metric tons and 10 million hectares. CPO is made from oil palm tree (*Elaeis guineensis*) and produces a significant amount of biomass. About 4 kg of dry biomass produced for every 1 kg of CPO, most of them derived from oil palm empty fruit bunch and oil palm trunks [4].

Biomass naturally has a low energy density, and its bulky size makes an issue in handling, storing and transportation. Biomass needs to be converted into another form of energy. Biomass gasification seems to be a more promising technique than another conversion because of its non-oxidation condition and relatively low emission. Syngas produced could be used as fuel in an internal combustion engine, gas turbine to produce heat or power, also for feedstock in the synthesis of liquid fuels and another chemical.

Drying as the first step of gasification is necessary to eliminate moisture from the biomass. Upon entering the gasifier, biomass gains heat from the hot zone and perform the drying, thus release water from it. At a temperature higher than 100 °C, the loosely bound between water and biomass is removed irreversibly. As the temperature increase to around 200 °C, the low molecular weight volatile matter in biomass start evaporating. Pyrolysis is the second step after drying; it is the thermal crackdown of larger hydrocarbon molecules into smaller biomass molecules without chemical reaction with air or another gasification medium. Next stage is the partial combustion that has several possible paths: Gases that consist of CO, H_2, CH_4, and H_2O; or liquids in the form of tar, oil, and naphtha; and oxygenated compounds like phenols and acid, and solid-state represented by char. The last process is the char gasification and is considered the most critical reaction following the previous process [5].

Numerical analysis of gasification using commercial software STAR CCM+ version 9.02.007 was performed in this paper. At present, discussion about CFD analysis of biomass using palm empty fruit bunch is insufficient. Only a limited number of reported works on CFD modeling of palm empty fruit bunch gasification to our knowledge. For example, thermal flow and gasification of coal gasifier using applied computational model CFD. The multiphase flow was modeled with the Eulerian-Lagrangian concept, the gaseous phase was performed by the Eulerian approach, while the solid coal particle phase was performed by the Lagrangian approach [6].

2. Methodology

The meshed geometry used in this paper is presented in figure 1. The gasifier has 1350 mm in height and 290 mm in diameter based on gasifier dimension calculation [7]. There are opening for air inlet, biomass inlet, and syngas outlet. The biomass used in the simulation was palm empty fruit bunch with the proximate and ultimate analysis is given in table 1 [8]. Three meshing types used in this paper are (a) surface mesh to mesh the primary surface to specify discretized mesh appropriate for the CFD, (b) polyhedral mesh to produce a volume mesh consist of polyhedral shaped mesh and (c) the prism layer mesh to project the interior mesh back to the wall boundaries to construct prismatic shaped mesh. Mesh sensitivity analysis was performed with three different mesh to obtain the optimum meshing size and most efficient calculation time.

Table 1. Ultimate and Proximate Analysis [8]

Element	% (Weight)	Component	% (Weight)
C	61.51	Volatile	77.46
H	10.51	Fixed Carbon	17.25
O	26.00	Ash	5.29
N	1.98		

A wide variety of flow processes involve the transport of dispersed phases like solid particles, liquid droplets, and gas bubbles by continuous phase like gaseous or liquid. In principle, Eulerian-Lagrangian CFD modeling implies that continuous gas phase was solved with transport equation, while the trajectory of discrete particles was tracked with the calculated gas field. Continuous phase and discrete phase interaction were using some governing equations, by considering the exchange of mass, energy, and momentum between two phases.

This paper proposes a Lagrangian approach for biomass particle in solid and dispersed discrete phase and Eulerian method for air as a gasifying agent in gaseous and continuous phase. Governing equations for the dispersed discrete phase are particle motion, the trajectory of particle and temperature of particles. While governing equations for the continuous phase are conservation of mass (continuity equation), conservation of momentum (Newton's Second Law), and conservation of energy (1st Law of Thermodynamics). Another equation employed to model the turbulence are turbulent kinetic energy and turbulent dissipation rate, and species transport equation.

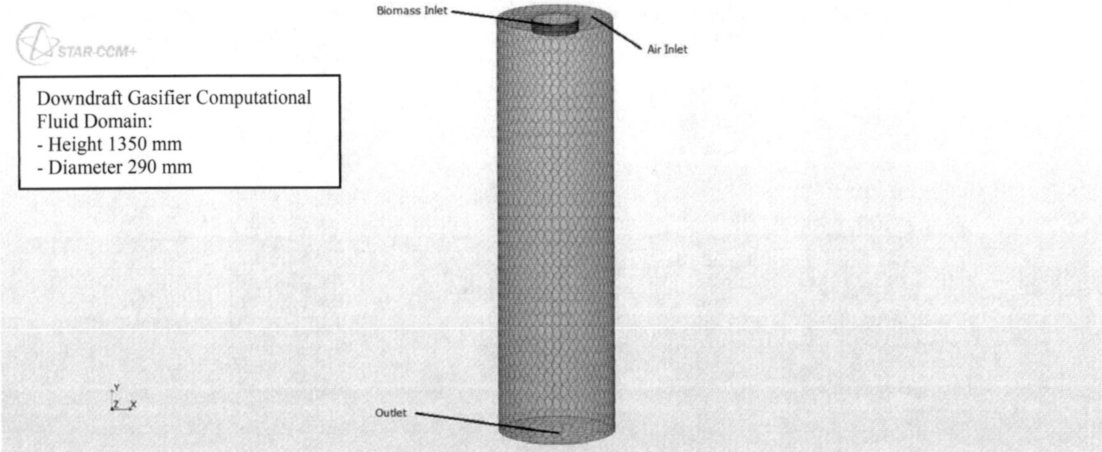

Figure 1. Meshed geometry.

Chemical kinetics equations employed are coal moisture evaporation, raw coal devolatilization, and first-order char oxidation. In coal moisture evaporation model, biomass particles are assumed to be coated by moisture particles, and moisture evaporation happened immediately just before the mass transfer process occurred. Raw coal devolatilization is defined as a process when the volatile matter is removed from the biomass particle. First order char oxidation considers remaining coal left after devolatilization. Char will be reacted with oxygen, water, and carbon dioxide to generate products [6]. Biomass char oxidations associated with several reactions between char and the gasification medium like oxygen, carbon dioxide, and steam are as follows:

$$Char + O_2 \rightarrow CO_2 + CO \quad (1)$$
$$Char + CO_2 \rightarrow CO \quad (2)$$
$$Char + H_2O \rightarrow CH_4 + CO \quad (3)$$
$$Char + H_2 \rightarrow CH_4 \quad (4)$$

Equations (1) to (4) demonstrate corresponding biomass gasification medium (oxygen, carbon dioxide, and steam) reaction with solid biomass and transform it into lower molecular weight gases such as carbon monoxide and hydrogen. All the above equations are heterogeneous because they involve different phase. Two homogeneous reactions were carried out in the biomass combustion model as follows:

$$Biomass\ Volatile + aO_2 \rightarrow CO + bH_2O + cN_2 \quad (5)$$
$$CO + H_2O \rightarrow CO_2 + H_2 \quad (6)$$

Biomass Volatile described by $CH_xO_yN_z$ in the dry ash free condition subjected to devolatilization. In STAR CCM+, Biomass Volatile was represented as CoalVolatile component, where value x, y, and z were obtained from the biomass ultimate and proximate analysis on table 1 [8]

3. Results and discussion

All biomass gasification steps (drying, thermal decomposition/pyrolysis, partial combustion and gasification of decomposed products) has been performed in this model. Biomass volatile matter has been released, and the mass fraction of the syngas product consists of N_2, CO, CH_4, and H_2 were analyzed. Heterogeneous reactions (1) – (4) occurred between char and O_2, CO_2, H_2O and H_2 as oxidizing agents resulting production of gaseous species. Homogeneous reaction (5) – (6) take place to generate CO, H_2O, N_2, CO_2, and H_2.

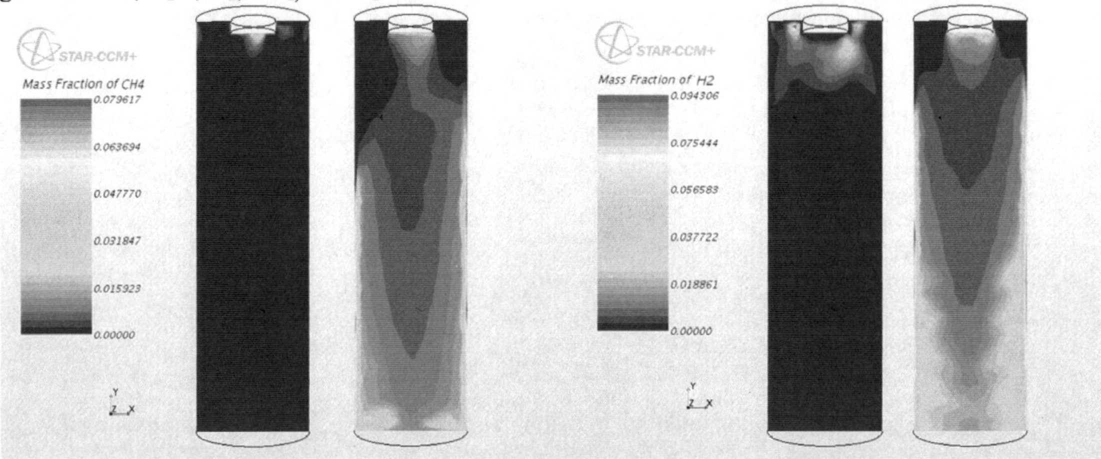

Figure 2. Mass fraction formation of CH_4 and H_2.

Formation of CH_4 could be seen in Figure 2. During early formation, the devolatilization process happened and discharged an extensive amount of volatile matter. The volatile matter then disintegrates into CH_4 and CO. As a result; there will be a more significant amount of CH_4 during this stage. As the process continued, CH_4 will be reacted with H_2O to generate CO and H_2. Therefore, in the final stage, the concentration of CH_4 will be decreased. H_2 was formed because of char oxidation, and homogeneous reaction occurred in the initial stage. As the gasification continued, H_2 will be reacted with O_2 and H_2O will be formed. H_2 consumption will lead to decreasing concentration on the later stage.

During the gasification process, there is the constant production of CO because of heterogeneous and homogeneous reactions. During the original composition, the char oxidation reaction occurred that will produce CO as a product, so in this stage, CO concentration was higher. As the gasification process continued, homogeneous and homogeneous reaction happened and generated more CO. Therefore, the concentration of CO will continue to increase because of this phenomenon as shown in figure 3. Air that consist of O_2 and N_2 enters the gasifier from the top of the gasifier. Hence, the concentration of N_2 is higher compared to the bottom as most N_2 reacted with a hydrocarbon to form tar.

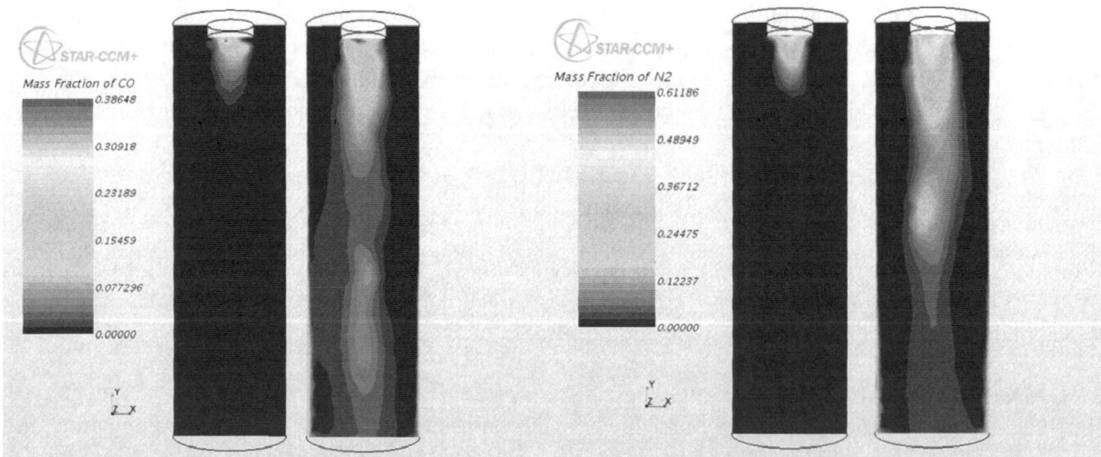

Figure 3. Mass fraction formation of CO and N_2.

IOP Conf. Series: Earth and Environmental Science **291** (2019) 012036 doi:10.1088/1755-1315/291/1/012036

Air enters the gasifier at an initial temperature of 25 °C, and temperature increases continuously as the exothermic reaction occurred. Shortly after devolatilization takes place, there will be homogeneous (5) – (6) and heterogeneous reactions (1) – (4) which are endothermic reactions. As a result, the temperature will be slightly decreasing after devolatilization. Figure 4 shows the temperature distribution along the axis of the gasifier and the contour of temperature within the gasifier. The higher temperature detected in an area where biomass contact with air. In this area, there is plenty of fresh air and unreacted biomass that will initiate exothermic devolatilization reactions. Therefore, increasing temperature could be seen at the higher part of the gasifier. Endothermic reactions occurred when we move to the lower part of the gasifier which will decrease the temperature within the gasifier. Modeling result shows compliance with references [6].

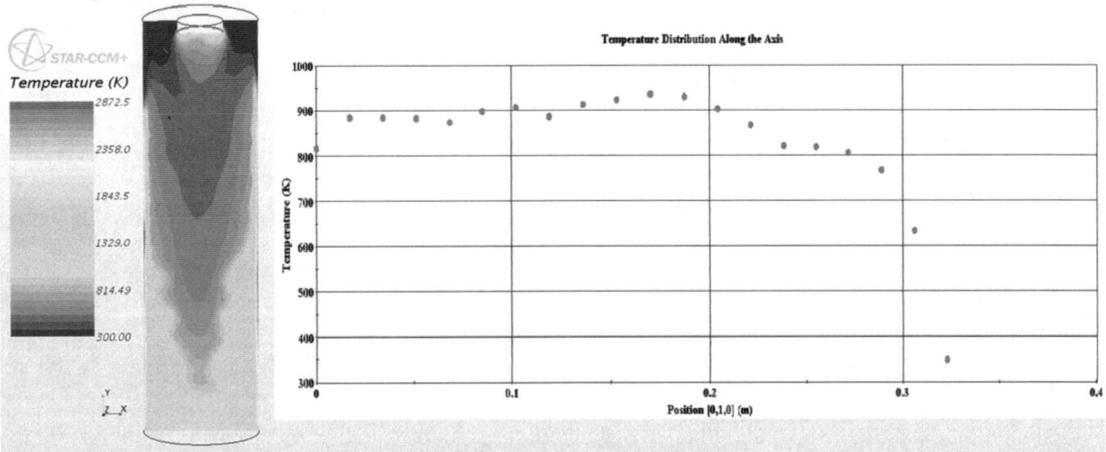

Figure 4. Temperature distribution.

4. Conclusions

Based on the results obtained from this study, the following conclusions could be made: (a) Gasification steps started with drying, followed by devolatilization or thermal decomposition partial combustion of gases, vapor, and char, and finalized with gasification of decomposition products have been successfully carried out with STAR CCM+. (b) The temperature during devolatilization was initially high caused by an exothermic reaction. Along with the end of the endothermic reaction, the temperature was decreasing. (c) Syngas produced by gasification are CO, N_2, CH_4, and H_2, all demonstrate acceptable decency with references. The future experimental investigation should be conducted to validate the modeling analysis.

5. References

[1] Tollefson J and Monastersky R 2012 The Global Energy Challenge: Awash with Carbon *Nature* **491(7426)** 654-5

[2] Pirlogea C and Cicea C 2012 Econometric Perspective of the Energy Consumption and Economic Growth Relation in European Union *Renew. and Sust. Energy Rev.* **16(8)** 5718-26

[3] Akia M, Yazdani F, Motaee E, Han D and Arandiyan H 2014 A Review on Conversion of Biomass to Biofuel by Nanocatalysts *Biofuel Res. J.* **1(1)** 16-25

[4] Saidur R, Abdelaziz E A, Demirbas A, Hossain M S and Mekhilef S 2011 A Review on Biomass as a Fuel for Boilers *Renew.and Sust.Energy Rev.* **15(5)** 2262-89

[5] Basu P 2010 *Biomass Gasification and Pyrolysis-Practical Design and Theory* Elsevier Oxford UK 662

[6] Thakare M and Nene A A 2017 Eulerian-Lagrangian CFD Modelling of Coal Gasification *Int.Eng.Res.J. Spec. Ed.* PGCON-MECH

[7] Belonio A T 2005 *Rice Husk Gas Stove Handbook* Appropriate Technology Centre Department of Agricultural Engineering and Environmental Management, College of Agriculture, Central Philippine University, Iloilo City, Philippines
[8] Salleh M A, Kisiki N H, Yusuf H M and Wan ab Karim Ghani W A 2010 Gasification of Biochar from Empty Fruit Bunch in a Fluidized Bed Reactor *Energies* **3** 1344-552

Acknowledgments
The first author would like to acknowledge financial support from the Indonesian Endowment Fund for Education (LPDP), Ministry of Finance; and the Directorate General of Higher Education (DIKTI), Ministry of Research, Technology and Higher Education, Republic of Indonesia.

IOP Conf. Series: Earth and Environmental Science **291** (2019) 012037 doi:10.1088/1755-1315/291/1/012037

Thermodynamic performance comparison between single-pressure and dual-pressure evaporation organic Rankine cycles for heat sources with outlet temperature limit

Jian Li, Jingze Yang, Zhong Ge, Yuanyuan Duan*, and Zhen Yang

Key Laboratory for Thermal Science and Power Engineering of MOE, Tsinghua University, Beijing 100084, PR China

*Email: yyduan@tsinghua.edu.cn

Abstract. Organic Rankine cycle (ORC) is a widely used technology to generate power from renewable energy and waste heat. Dual-pressure evaporation cycle holds immense potential to be used in ORC systems because it can remarkably increase the heat-work conversion efficiency and improve the adaptability of ORC to heat sources with various characteristics. This study compared the thermodynamic performance of single-pressure and dual-pressure evaporation ORCs for heat sources with outlet temperature limit, based on five organic fluids. Effects of the heat source outlet temperature limit on the characteristics of dual-pressure evaporation ORC system were also discussed. Results show that the dual-pressure evaporation cycle can substantially increase the net power output by increasing system efficiency, and is beneficial to reduce the charge volume of organic fluid in the system, compared to the single-pressure evaporation cycle. The increments in net power output of dual-pressure evaporation cycle over the single-pressure evaporation cycle can be 4.9%, 8.3%, 10.8%, 12.8%, and 14.0% at most for R227ea, R1234ze(E), R600a, R245fa, and R601a, respectively. While, compared to heat sources without outlet temperature limit, the increment in net power output of dual-pressure evaporation cycle is remarkably lower for heat sources with outlet temperature limit.

1. Introduction

Organic Rankine cycle (ORC) is a promising technology to generate power from the renewable energy and waste heat [1, 2]. The fundamental study and engineering application of ORC technology has been continuously growing in recent years [3]. The improvement of cycle type is always a hot study issue in the ORC field, because it significantly affects the overall performance of ORC system [4]. To achieve a high heat-work conversion efficiency is the primary goal for the improvement of cycle type. Moreover, a large design freedom degree of cycle structure is also crucial for ORC system to match the remarkably different heat release characteristics of various heat sources [5].

Dual-pressure evaporation cycle has two evaporation processes with different pressures and an isobaric condensation process [6]. For the dual-pressure evaporation cycle, the evaporation pressure, superheat degree, and mass flow rate of working fluid in each evaporation process can be optimized to achieve the preferable temperature match effect between heat source fluid and working fluid, and thereby substantially reducing the exergy loss in the cycle heat absorption process and also improving the adaptability of ORC to heat sources with various characteristics, compared to the conventional single-pressure evaporation cycle [6]. The superiority in thermodynamic performance of dual-pressure evaporation cycle has been proven by several studies [6] [7] [8]. Our previous work [6] compared the thermodynamic performance between single-pressure and dual-pressure evaporation ORCs based on nine pure fluids for heat sources without outlet temperature limit, and a quantitative criterion was provided to assess the optimal cycle type (single-pressure or dual-pressure) for the ORC system.

For lots of heat sources, the outlet temperature cannot be cooled to a low value due to the limit of several external factors (e.g., the scale formation for geothermal water, and dew point corrosion for industrial flue) [5]. For these heat sources with outlet temperature limit, the advantages of dual-pressure evaporation cycle in increasing the system heat absorption capacity (reducing the heat source outlet temperature) will be significantly weakened and even disappearing, compared to heat sources without outlet temperature limit. Thus, the superiority in thermodynamic performance of dual-pressure evaporation cycle over conventional single-pressure evaporation cycle should be further analyzed for heat sources with outlet temperature limit. Moreover, for the dual-pressure evaporation cycle, the selections of optimal cycle parameters, and variations in the thermodynamic performance superiority with increasing heat source temperature will be substantially different for heat sources with and without outlet temperature limit. These are also important issues to be further studied.

This study focuses on the single-pressure and dual-pressure evaporation ORC systems driven by heat sources with outlet temperature limit. The heat source temperature was selected as 100-200°C, and five common pure fluids were selected as the working fluids, to guarantee the studied heat source temperature range and working fluid are broad enough. The thermodynamic performance of single-pressure and dual-pressure evaporation ORC systems was compared, and effects of heat source temperature and working fluid thermophysical properties on the thermodynamic performance superiority of dual-pressure evaporation cycle were revealed. Effects of the heat source outlet temperature limit on the characteristics of dual-pressure evaporation ORC system were also discussed.

2. Methodology

2.1. ORC system and working fluids

The system of dual-pressure evaporation ORC and its *T-s* diagram are shown in Figure 1. The descriptions of thermodynamic processes for single-pressure and dual-pressure evaporation ORCs can refer to our previous work [6]. R227ea, R1234ze(E), R600a, R245fa, and R601a were selected as working fluids due to their preferable thermodynamic performance in the ORC system [6], and their critical temperatures are 101.75°C, 109.36°C, 134.66°C, 154.01°C, and 187.20°C, respectively [9].

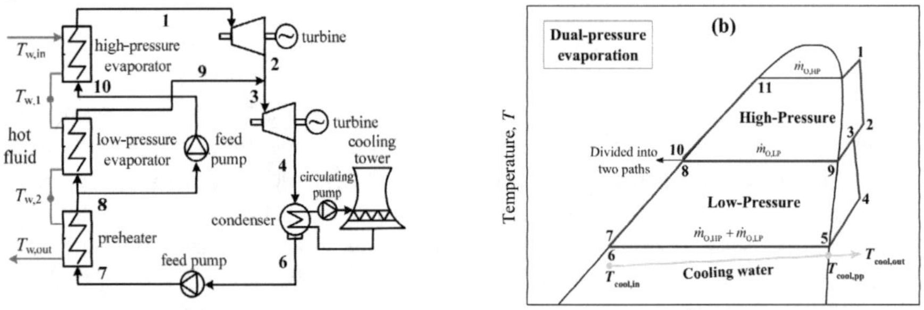

Figure 1. Schematics of dual-pressure evaporation ORC system: (a). ORC system; (b). Thermodynamic processes

2.2. System model

The inlet temperature of heat source fluid (hot water) is 100-200°C, and its outlet temperature should be not lower than 70°C. The mass flow rate of hot water is 1 $kg \cdot s^{-1}$, and its pressures are 0.5 MPa, 1.2 MPa, and 1.6 MPa for the inlet temperatures of 100-150°C, 151-180°C, and 181-200°C, respectively. The efficiencies of turbine and feed pump are 0.8 and 0.75, respectively. The cooling water inlet temperature is 20°C, and its temperature rise in the working fluid condensation process is 5°C. The pressure head of cooling system is 10 m. The ORC systems are assumed as in the steady state, and heat dissipation and pressure drop of fluids are negligible in heat exchangers and pipes.

In this study, the evaporation pressures and evaporator outlet temperatures of single-pressure and dual-pressure evaporation ORC systems were optimized to achieve the largest net power output. The net power output of an ORC system is equal to that the power output of turbines deducts the power consumed by the working fluid feed pumps and cooling system ($W_{net} = W_T - W_P - W_{cool}$). The selectable ranges of optimized cycle parameters can refer to our previous work [6], including the detailed system model equations, optimized processes and model validation. Thermophysical properties of fluids are from REFPROP 9.1 [9].

3. Results and discussion

The comparison results of optimized cycle parameters and system thermodynamic performance between single-pressure and dual-pressure evaporation ORCs are similar for five selected working fluids; thus, R245fa is selected as an example to introduce the details of comparison.

3.1. Comparisons for optimized cycle parameters

Figure 2 shows the comparisons for optimized evaporation pressures between single-pressure and dual-pressure evaporation ORCs using R245fa. With increasing heat source inlet temperature, the optimized evaporation pressure of single-pressure evaporation cycle ($p_{e\_single,opt}$) initially increases for the heat source inlet temperature below 190°C, and then it remains at the upper limit ($0.9p_c$). For the dual-pressure evaporation cycle, with increasing heat source inlet temperature, the optimized evaporation pressure in the high-pressure stage ($p_{e\_HP,opt}$) initially increases for the heat source inlet temperature below 170°C and then it remains at the upper limit ($0.9p_c$). The optimized evaporation pressure in low-pressure stage ($p_{e\_LP,opt}$) increases with increasing heat source inlet temperature. The $p_{e\_HP,opt}$ will be higher than $p_{e\_single,opt}$ unless the $p_{e\_HP,opt}$ and $p_{e\_single,opt}$ remain at the upper limits. The $p_{e\_LP,opt}$ is lower than $p_{e\_single,opt}$ and the decrement increases with increasing heat source inlet temperature; while, the $p_{e\_LP,opt}$ is nearly equal to $p_{e\_single,opt}$ when the heat source inlet temperature is lower than approximately 170°C.

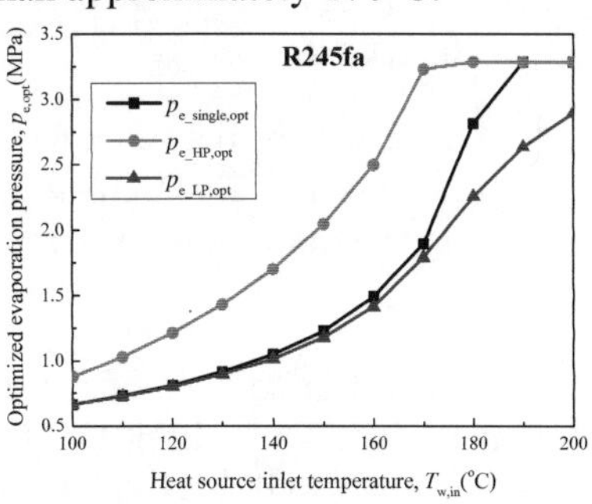

Figure 2. Comparisons for optimized evaporation pressures between single-pressure and dual-pressure evaporation ORCs.

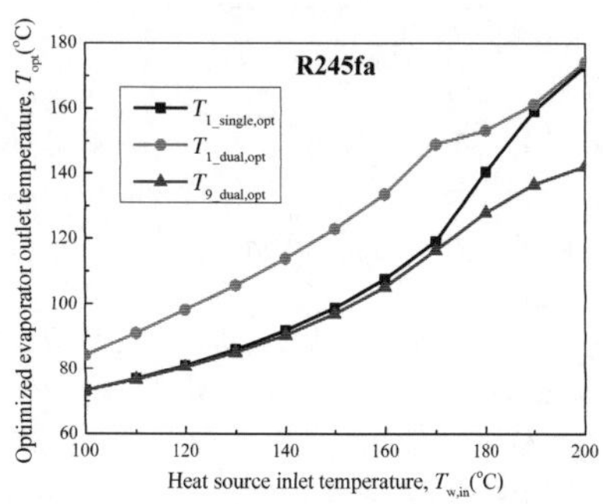

Figure 3. Comparisons for optimized evaporator outlet temperatures between single-pressure and dual-pressure evaporation ORCs.

Figure 3 shows the comparisons for optimized evaporator outlet temperatures between single-pressure and dual-pressure evaporation ORCs using R245fa. In the low-pressure stage, the optimized evaporator outlet temperature is equal to the lower limit corresponding to the optimized evaporation pressure. The optimized evaporator outlet temperatures in single-pressure evaporation cycle and high-pressure stage are equal to the lower limits corresponding to their optimized evaporation pressures for heat source inlet temperatures not exceeding 180°C and 170°C, respectively. For the single-pressure and dual-pressure evaporation ORCs, the comparison results of optimized evaporator outlet temperatures are similar to those of optimized evaporation pressures; while, the optimized evaporator outlet temperature in the high-pressure stage is slightly higher than that in the single-pressure evaporation cycle when the heat source inlet temperature exceeds 190°C.

Figure 4 presents the mass flow rates of R245fa in single-pressure and dual-pressure evaporation ORCs. In the single-pressure evaporation cycle, the mass flow rate of R245fa initially increases nearly linearly with increasing heat source inlet temperature, and then it remains nearly constant for the heat source inlet temperature of 180-200°C. In the dual-pressure evaporation cycle, the total mass flow rate of R245fa increases with a low increment as the heat source inlet temperature increases. Therein, with increasing heat source inlet temperature, the mass flow rate of R245fa in the low-pressure stage initially increases for $T_{w,in} < 170°C$ and then decreases with a low decrement. While, the mass flow rate of R245fa in the high-pressure stage initially increases nearly linearly for $T_{w,in} < 170°C$, and then increases with a low increment, with increasing heat source inlet temperature. The mass flow rate of R245fa in the high-pressure stage is significantly higher than that in the low-pressure stage, and the increment generally increases with increasing heat source inlet temperature. Therefore, the effect of high-pressure stage on the system net power output becomes more remarkable with increasing heat source inlet temperature; while, the effect of low-pressure stage on the system net power output becomes weaker due to its lower mass flow rate of R245fa. Furthermore, the working fluid mass flow rate in the dual-pressure evaporation cycle is lower than that in the single-pressure evaporation cycle, which indicates that the dual-pressure evaporation cycle is beneficial to reduce the charge volume of organic fluid in the ORC system.

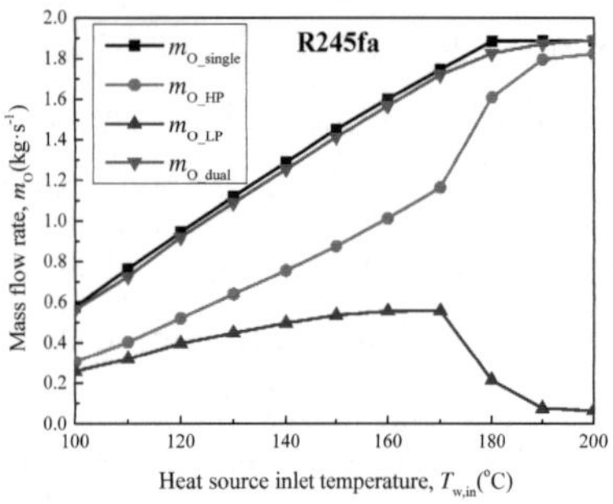

Figure 4. Mass flow rates of R245fa in single-pressure and dual-pressure evaporation ORC systems.

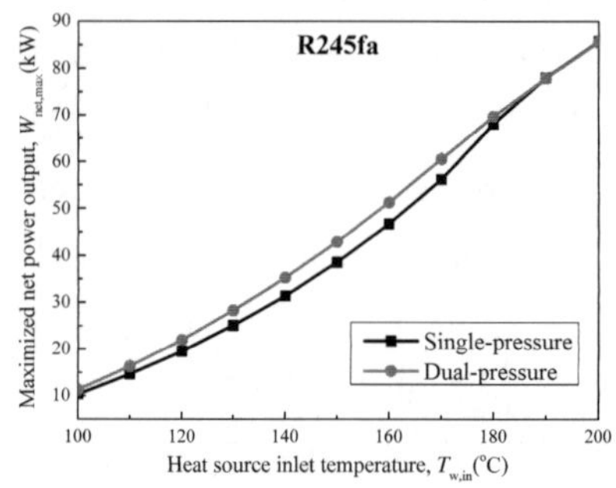

Figure 5. Comparisons for maximized net power outputs between single-pressure and dual-pressure evaporation ORC systems.

3.2. Comparisons for system thermodynamic performance

For the single-pressure and dual-pressure evaporation ORCs driven by heat sources of various inlet temperatures, the heat source outlet temperatures are always equal to the lower limit (70°C) at the optimized operating conditions. Thus, the heat absorption capacities of two ORC systems are equal, and comparisons of their maximized net power outputs depend on the system efficiency.

Figure 5 presents the comparisons for maximized net power outputs between single-pressure and dual-pressure evaporation ORC systems using R245fa. The maximized net power output of dual-pressure evaporation cycle is larger than that of single-pressure evaporation cycle when the heat source inlet temperature is below approximately 190°C, and the increment initially increases for heat source inlet temperature below 130°C and then decreases with increasing heat source inlet temperature. Compared to single-pressure evaporation cycle, the largest increment in maximized net power output of dual-pressure evaporation cycle is 12.8% which occurs at the heat source inlet temperature of 130°C. When the heat source inlet temperature exceeds 190°C, the maximized net power output of dual-pressure evaporation cycle is slightly lower than that of single-pressure evaporation cycle due to the upper limit

of evaporation pressure in the high-pressure stage ($p_{e\_HP,opt}$ cannot increase, Figure 2). In summary, with the limits of heat source outlet temperature and pinch point temperature difference in the heat absorption process ($\Delta T_{HAP,pp}$), the dual-pressure evaporation cycle can obtain a higher system efficiency due to its larger design freedom degree in the cycle heat absorption process, and thereby increases the system net power output, compared to the single-pressure evaporation cycle.
Given the evaporation pressure(s) and evaporator outlet temperature(s), the heat source outlet temperature will decrease theoretically with increasing heat source inlet temperature for single-pressure and dual-pressure evaporation ORCs. For the working fluid with a low critical temperature, the high heat source inlet temperature will be not applicable because the heat source outlet temperature is always lower than 70°C for the whole selectable ranges of evaporation pressure(s) and evaporator outlet temperature(s). Results show that R227ea and R1234ze(E) can be used for heat sources of 100-120°C and 100-140°C, respectively. R600a, R245fa, and R601a can be used for heat sources of 100-200°C. Figure 6 presents the increments in maximized net power output of dual-pressure evaporation cycle over the single-pressure evaporation cycle ($\left(W_{net\_dual} - W_{net\_single}\right)/W_{net\_single}$) for various working fluids. For R1234ze(E), R600a, R245fa, and R601a, the $\left(W_{net\_dual} - W_{net\_single}\right)/W_{net\_single}$ initially increases and then decreases with increasing heat source inlet temperature. While, for R227ea, the $\left(W_{net\_dual} - W_{net\_single}\right)/W_{net\_single}$ decreases with increasing heat source inlet temperature due to its narrow applicable heat source inlet temperature range. For R227ea, R1234ze(E), R600a, R245fa, and R601a, the $\left(W_{net\_dual} - W_{net\_single}\right)/W_{net\_single}$ can be 4.9%, 8.3%, 10.8%, 12.8%, and 14.0% at most, respectively; and the suitable heat source inlet temperature ranges of dual-pressure evaporation cycle (obtaining a larger net power output) are 100-120°C, 100-140°C, 100-160°C, 100-180°C, and 100-200°C, respectively. Thus, the critical temperature of working fluid is higher, the applicative heat source inlet temperature range is wider, and the suitable heat source inlet temperature range of dual-pressure evaporation cycle is also larger.

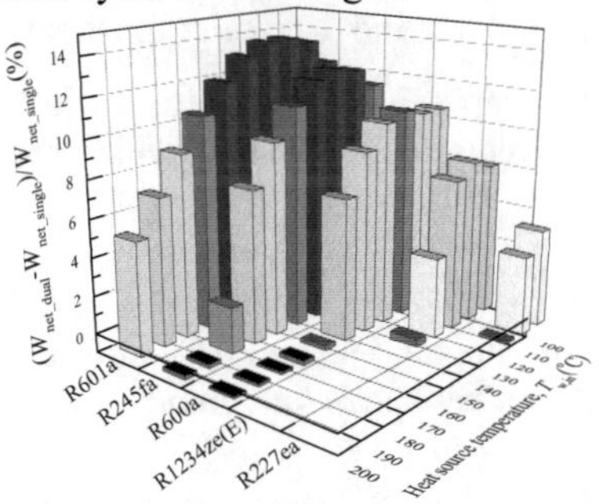

Figure 6. Increments in maximized net power output of dual-pressure evaporation cycle over the single-pressure evaporation cycle for various working fluids.

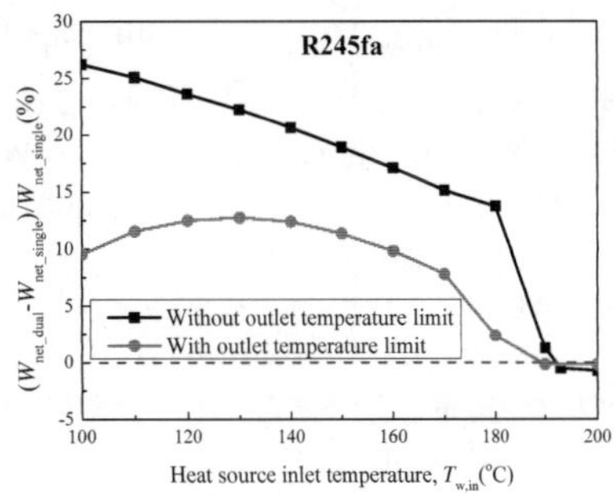

Figure 7. Comparisons for increments in maximized net power output of dual-pressure evaporation cycle for heat sources with and without outlet temperature limit.

3.3. Effects of the heat source outlet temperature limit
In this study, the operating condition will be ignored if the heat source outlet temperature below 70°C. For the single-pressure evaporation cycle, the heat source outlet temperature decreases as the evaporation pressure decreases; thus, the heat source outlet temperature may be lower than 70°C when the evaporation pressure is low. Given the heat source inlet temperature and evaporation pressure, the heat source outlet temperature may be lower than 70°C for a low evaporator outlet temperature; thus, the operating condition will be ignored. The similar situations also exist for the dual-pressure

evaporation cycle. Therefore, the applicable operating conditions of single-pressure and dual-pressure evaporation ORCs will be significantly shrunken due to the limit of heat source outlet temperature.
For heat sources with and without outlet temperature limit, the comparisons for increments in maximized net power output of dual-pressure evaporation cycle ($\left(W_{\text{net\_dual}} - W_{\text{net\_single}}\right)/W_{\text{net\_single}}$) are presented in Figure 7. The suitable heat source inlet temperature ranges of dual-pressure evaporation cycle are nearly equal for heat sources with and without outlet temperature limit. While, the $\left(W_{\text{net\_dual}} - W_{\text{net\_single}}\right)/W_{\text{net\_single}}$ for heat sources with outlet temperature limit is remarkably lower than that for heat sources without outlet temperature limit. In addition, for heat sources without outlet temperature limit, the dual-pressure evaporation cycle dramatically increases the system net power output by reducing heat source outlet temperature (increasing system heat absorption capacity), as described in our previous work [6]. While, for heat sources with outlet temperature limit, the dual-pressure evaporation cycle significantly increases the system net power output by increasing system efficiency due to its higher evaporation pressure in the high-pressure stage compared with that in the single-pressure evaporation cycle. Moreover, compared to the single-pressure evaporation cycle, the dual-pressure evaporation cycle significantly increases the working fluid mass flow rate for heat sources without outlet temperature limit; while, it will reduce the working fluid mass flow rate for heat sources with outlet temperature limit.

4. Conclusions

The thermodynamic performance between single-pressure and dual-pressure evaporation ORCs was compared for heat sources with outlet temperature limit, based on five organic fluids. Effects of the heat source outlet temperature limit on the characteristics of dual-pressure evaporation ORC system were also discussed. Main conclusions are below:
For heat sources with outlet temperature limit, the dual-pressure evaporation cycle can significantly increase the net power output by increasing system efficiency, and is also beneficial to reduce the charge volume of organic fluid in the system, compared to the single-pressure evaporation cycle.
The $\left(W_{\text{net\_dual}} - W_{\text{net\_single}}\right)/W_{\text{net\_single}}$ can be 4.9%, 8.3%, 10.8%, 12.8%, and 14.0% at most for R227ea, R1234ze(E), R600a, R245fa, and R601a, respectively. While, compared to heat sources without outlet temperature limit, the increment in net power output of dual-pressure evaporation cycle is remarkably lower for heat sources with outlet temperature limit.

5. References

[1] Tchanche B F, Lambrinos G, Frangoudakis A and Papadakis G 2011 *Low-grade heat conversion into power using organic Rankine cycles-A review of various applications*. Renew. Sust. Energy Rev. 15, 3963-79.
[2] Li J, Liu Q, Ge Z, Duan Y Y and Yang Z 2017 *Thermodynamic performance analyses and optimization of subcritical and transcritical organic Rankine cycles using R1234ze(E) for 100-200°C heat sources*. Energy Convers. Manage. 149, 140-54.
[3] Imran M, Haglind F, Asim M and Zeb Alvi J. 2018 *Recent research trends in organic Rankine cycle technology: A bibliometric approach*. Renew. Sust. Energy Rev. 81, 552-62.
[4] Lecompte S, Huisseune H, van den Broek M, Vanslambrouck B and De Paepe M 2015 *Review of organic Rankine cycle (ORC) architectures for waste heat recovery*. Renew. Sust. Energy Rev. 47, 448-61.
[5] Zhai H X, An Q S, Shi L, Lemort V and Quoilin S 2016 *Categorization and analysis of heat sources for organic Rankine cycle systems*. Renew. Sust. Energy Rev. 64, 790-805.
[6] Li J, Ge Z, Duan Y Y, Yang Z and Liu Q 2018 *Parametric optimization and thermodynamic performance comparison of single-pressure and dual-pressure evaporation organic Rankine cycles*. Appl. Energy 217, 409-21.

[7] Li T L, Zhang Z G, Lu J, Yang J N and Hu Y J 2015 *Two-stage evaporation strategy to improve system performance for organic Rankine cycle*. Appl. Energy 150, 323-34.
[8] Manente G, Lazzaretto A and Bonamico E 2017 *Design guidelines for the choice between single and dual pressure layouts in organic Rankine cycle (ORC) systems*. Energy 123, 413-31.
[9] Lemmon E W, Huber M L and Mclinden M O 2013 *NIST reference fluid thermodynamic and transport properties–REFPROP, version 9.1*. Gaithersburg: National Institute of Standards and Technology, Standard Reference Data Program.

Acknowledgments
Authors gratefully acknowledge the support provided by the National Natural Science Foundation of China (Grant Nos. 51736005 and 51621062).

IOP Conf. Series: Earth and Environmental Science **291** (2019) 012038 doi:10.1088/1755-1315/291/1/012038

Conversion of carbon dioxide into storable solar fuels using solar energy

Houda Ennaceri* **and Bernd Abel**

Chemistry Department, Leibniz-Institut für Oberflächenmodifizierung (IOM), Permoserstraße 15, 04318 Leipzig, Germany

*Email: houda.ennaceri@iom-leipzig.de

Abstract. Nowadays, there are two main energy and environmental concerns, the first is the risk of running out of fossil fuels in the next few decades, and the second is the alarming increase in the carbon dioxide concentrations in the atmosphere, causing global warming and rise of see levels. Therefore, solar-driven technologies represent a substantial solution to fossil fuels dependence, global warming and climate change. Unlike most scientific research, which aim to use solar energy to generate electricity, solar energy can also be harnessed by recycling the carbon dioxide in the atmosphere through high-tech artificial photosynthesis with the objective of producing storable and liquid solar fuels from CO_2 and water. There are two types of solar fuels, the first being hydrogen, which can be produced by mean of water splitting processes. The combustion of hydrogen generates water, which is a completely clean option for the environment. The second type of solar fuels consists of carbon-based fuels, such as methane (CH_4), carbon monoxide (CO), or alcohols such as methanol (CH_3OH) and ethanol (C_2H_5OH). The production to liquid solar fuels liquid fuels is of great interest, since they can be used in the current industrial infrastructures such as the automobiles' sector, without substantial changes in the vehicles' internal combustion engines. Therefore, guaranteeing a smooth transition from fossil fuel energy to renewable energy without radical economic consequences. Also, and most importantly, when these solar fuels are burned, they will only release the exact amount of CO_2 which was initially used, which represents an optimal process for sustainable transport.

1. Introduction

Today, the main energy concern is running out of conventional fossil fuels, which is the reason behind shifting towards renewable energies. The population growth and the aggressive consumption of energy are leading to alarming depletion of fossil fuels such as oil and gas, as well as increasing levels of pollution [1]. In fact, according to the united nation population division, the world population has reached 7 billion in 2011, and is projected to reach more than 9 billion by 2050, which will drastically increase the energy demand. According to Hubbert peak theory, there will be an extreme drop in oil resources within the next 40 years [2, 3]. Therefore, years between 2055 and 2060 will possibly be the time when the oil production will reach zero. Moreover, and considering the current energy consumption rate, the World coal institute's estimations expect that the current coal reserves will last for another 130 years, oil reserves for 42 years, and natural gas reserves for 60 years. The need to replace the traditional fossil fuels by sustainable options and new energy scenarios is therefore becoming very crucial. Another serious environmental concern is the increasing concentrations of carbon dioxide (CO_2) in the atmosphere, which is produced by burning fossil fuels and is responsible for global warming problems [4, 5]. In fact, the concentration of carbon dioxide in the atmosphere has severely increased from 280 ppm in 1750 to 400 ppm in 2014 [6]. The increase of the human-induced CO_2 has devastating consequences such as the increase of average temperatures, sea ice melting, floods and rise of sea levels, affecting the overall life, health and economy in the world. According to the Intergovernmental Panel on Climate Change (IPCC), the carbon dioxide atmospheric concentration has to be stabilized by reducing the 2006 global CO_2 emissions levels by 50% in 2050, by 100% in 2075, and above 100% by 2100 [7, 8]. Currently, the available technologies to tackle the CO_2 global warming problems is the sequestration process, also referred to as the carbon capturing and storing (CCS) process, which consist of capturing CO_2 at major outlets, compressing it in the form of a liquid for practical transportation, and storing it in underground sites or under deep sea levels [9]. Although CCS process can greatly reduce the CO_2 concentrations in the atmosphere, this procedure is extremely expensive and has different safety related uncertainties. In fact, the capital cost for a coal power plant using the CCS process is extremely higher compared to a conventional plant not equipped with CCS [10, 11]. A sustainable energy process is to produce energy from CO_2 instead of capturing and disposing it. In fact, CO_2 can be converted into value-added chemicals such as hydrogen, methanol and ethanol following some of the thermochemical routes. These fuels can be used to power cars without performing significant changes in their internal combustion engines, leading to a new route for sustainable future. In order to overcome today's drastic depletion of fossil fuels and mitigate the global warming risks related to the high concentrations of CO_2 in the atmosphere, finding sustainable options and new energy scenarios, mainly the conversion of CO_2 into value-added chemicals using sunlight is one of the top-most research priorities. In fact, the world's perception of carbon dioxide is recently changing, starting to consider CO_2 as a value-added gas, rather than a waste [12].

The purpose of our current work is the production of renewable solar fuels in liquid form such as methanol and ethanol from carbon dioxide and water. The production of solar fuels can be achieved by water splitting thermo-chemical reactions and carbon dioxide reduction under sunlight. This topic is nowadays considered one of the hottest research topics and top-most research priorities, which represents a sustainable energy source supply, and would a great impact on future energy options.

2. Background and rationale

The production of solar fuels can be achieved through artificial photosynthesis process [13, 14], which is similar to the natural photosynthesis, where green plants use the sunlight to convert water and carbon dioxide into oxygen and hydrocarbons according to the following reaction:

$$6CO_2 + H_2O \xrightarrow{light\ energy} C_6H_{12}O_6 + 6O_2 \tag{1}$$

Theoretically, the conversion efficiency of artificial photosynthesis is proven to be much higher compared to its naturally occurring counterpart. It fact, natural photosynthesis of plants do not exceed 1% conversion efficiency [15], while the artificial photosynthesis is expected to reach more than 40% conversion efficiency, which is similar to the theoretical efficiency of tandem solar cells. The artificial photosynthesis can be achieved by the use of strong catalyst materials to produce liquid solar fuels or fuel precursors such as carbon monoxide (CO) and solar hydrogen (H_2). These two fuel precursors are referred to as synthesis gas or syngas. One of the main issues regarding the CO_2 reduction to liquid fuels is the assemblage of nuclei and formation of chemicals bonds, which would allow the reduction of CO_2 molecule into more complex molecules. This can be achieved using two different methods, the first by converting CO_2 and H_2O into synthesis gas (CO and H_2), and then convert the synthesis gas into liquid fuels such as gasoline using the Fischer-Tropsch process according to Equation (2), where n is an integer, with values ranging from 10 and 20:

$$(2n+1)H_2 + n\,CO \rightarrow C_nH_{2n+2} \tag{2}$$

The second method is to attempt to produce CO_2 directly to liquid fuels through photo-catalytic or electro-catalytic processes. The first method is substantially easier to achieve since it is easy to convert carbon dioxide (CO_2) to carbon monoxide (CO), and to convert water (H_2O) to hydrogen (H_2). The second method is, on the other hand, more perplex since it involves ore kinetic challenges [16]. The key success of converting CO_2 directly to liquid fuels resides in identifying a strong catalyst which can do the complete sequence of converting CO_2 into hydrocarbons or alcohols with low kinetic barriers. The principle of artificial photosynthesis consists of sunlight absorption by a semiconducting material and the excitation energy creates an electron-hole pair. Before the photo-excited electrons ($e-$) and holes ($h+$) recombine, the holes oxides water on the catalyst's surface and provide protons for the CO_2 reduction process as shown in equations (3) to (5) [17]:

$$H_2O + h^+ \rightarrow OH^{\cdot} + H^+ \quad (3)$$

$$H_2O + 2h^+ \rightarrow H_2O_2 + 2H^+ \quad (4)$$

$$2H_2O + 2h^+ \rightarrow 4H^+ + O_2 \quad (5)$$

The electron transfer from the conduction band produces carbon dioxide anion radicals ($\cdot CO_2^-$) and hydrogen radicals ($\cdot H$) through the following equations:

$$H^+ + e^- \rightarrow \cdot H \quad (6)$$

$$CO_2 + e^- \rightarrow \cdot CO_2^- \quad (7)$$

These radicals form other stable hydrocarbons such as methane (CH_4), methanol (CH_3OH), carbon monoxide (CO), formic acid ($HCOOH$), formaldehyde ($HCHO$), or ethanol (C_2H_5OH) as shown in equations (8) to (13) [17]:

$$\cdot CO_2^- + 8 \cdot H + 8h^+ \rightarrow CH_4 + 2H_2O \quad (8)$$

$$\cdot CO_2^- + 6 \cdot H + 6h^+ \rightarrow CH_3OH + H_2O \quad (9)$$

$$\cdot CO_2^- + 2 \cdot H + 2h^+ \rightarrow CO + H_2O \quad (10)$$

$$\cdot CO_2^- + 2 \cdot H + 2h^+ \rightarrow HCOOH \quad (11)$$

$$\cdot CO_2^- + 4 \cdot H + 4h^+ \rightarrow HCHO + H_2O \quad (12)$$

$$\cdot CO_2^- + 12 \cdot H + 2h^+ \rightarrow C_2H_5OH + 3H_2O \quad (13)$$

One of the major considerations of semiconductor photocatalysis is the recombination of the photo-generated electron-hole pairs, which reduces the overall quantum efficiency [18]. The lifetime of the photo-generated electron-hole pairs is only some few nanoseconds, which is suitable for stimulating redox reactions. The reduction of CO_2 is more challenging compared to water splitting. In fact, the production of hydrocarbon fuels requires multi-electron processes, transfer of electrons and protons, and breaking the C-O bonds to create new C-H bonds [17].

The artificial photosynthesis reduction of CO_2 depends on the following central factors [19]:

- The high reduction potential of a single electron (Equation (7)): (E0 = - 1.9 V versus the normal hydrogen electrode (NHE))
- The solubility of carbon dioxide (CO_2) in water
- The competition of hydrogen formation.

In order to overcome these limitations, the choice of the candidate catalysts and the reaction conditions and parameters such as the reductants, light intensity, temperature, and pressure need to be optimized. Different metal oxides have been tested in the literature to achieve carbon dioxide thermochemical dissociation or splitting such as zinc oxide [20], titanium dioxide [21], graphene [22] and ferrites [23]. Zinc oxide and ferrite have been proven not to be good candidates for solar fuel production as zinc oxide presents volatility and quenching problems [24], and ferrites present slow kinetics [25]. Ceria

(CeO_2) has also been used for CO_2 splitting and production of solar fuels because of its high melting point and high catalytic activity for CO_2 reduction [26]. However, cerium-based catalysts have been shown to get easily deactivated by a small coverage of deposited carbon [27]. Generally, in order to achieve CO_2 reduction and water oxidation using semiconductor photo-catalysis, the position of the valence band of and conduction band of the semiconductor photo-catalyst should match the oxidation and reduction potentials of water splitting and CO_2 reduction.

3. Methods

Ethanol and methanol have high energy densities of 20.9 MJ/L and 15.6 MJ/L, respectively, compared to gasoline which has an energy density of 34.2 MJ/L [28]. Ethanol and methanol can be used in fuel cells to produce electricity, or can be directly used as alternative fuel for internal combustion in the automobiles' sector. The advantages of producing these fuels from CO_2 reside in their low toxicity, safety to handle, and the possibility to be used as car fuels without modifications in their internal combustion (IC) engines. Also, these liquid fuels do not need high pressure to be stored at room temperature. The CO_2 molecule has a high kinetic stability. Therefore, efficient catalysts are needed for a successful CO_2 hydrogenation, which is the main objective of my research.

The electrochemical reduction of CO_2 into liquid solar fuels such as methanol and ethanol requires complex multi-electron processes. In fact, the CO_2 electrochemical reduction to methanol requires 6 electrons/protons, and the CO_2 electrochemical reduction to ethanol requires 12 electrons/protons. In order to assess the catalyst's performance towards CO_2 reduction, different parameters need to be investigated such as:

- The overpotential, which is the difference between the standard thermodynamic redox potential, and the actual onset potential at which the reduction is experimentally occurring.
- The current density, which is the current divided by the geometric surface area of the working electrode
- The faradaic efficiency of each reduction product (α*n*F/Q), where α is the number of electrons transferred, n is the number of moles of a specific product, F is Faraday's constant (96485 C mol^{-1}), and Q is the charge that is passed through the electrolysis process (I*t)
- The energetic efficiency (E_0*FE/V_{in}), where E_0 and FE are the standard potential and the faradaic efficiency of the reduction product, respectively, and V_{in} is the total cell voltage
- Turnover number (TON), which is the number of moles of the formed product divided by the number of active centers.
- The Tafel slope, which relates the overpotential to the logarithm of the partial current density of the product.

4. Summary

Different metal electrodes are electrochemically active towards CO_2 reduction and can be divided in 3 groups. The first group (e.g. Sn, Hg, In, Pb, Cd, Bi) produces formic acid or formate as reduction products. On the other hand, the second group of metals (e.g. Au, Ag, Pd, Zn, Ga) generates gaseous carbon monoxide as reaction product. The last group is composed of copper only, which has been shown to be the most promising metal in the electrochemical CO_2 reduction since it is the only metal catalyst that can electrochemically convert carbon dioxide into oxygenates/hydrocarbons. However, the main drawback of using copper and copper-based catalysts is their long-term stability.

Special attention is given to the development of efficient catalysts for the (photo) electrochemical reduction of CO_2. In fact, a suitable semiconductor photo-cathode should have a high absorption in the visible range, high stability, high photo-potential efficiency, low kinetic over-potentials and good charge transport. Therefore, in order establish and scale-up a technology for solar fuel production, it is essential to develop and design new functional, low-cost, durable, highly stable and energy-efficient thin films used as photo-cathodes.

The photo-cathode performance is also sensitive to the deposition method. Hence, optimizing and controlling the film thickness, morphology, and crystal structure is of great interest. Moreover, and as explained earlier, the production of hydrocarbon fuels requires multi-electron processes, transfer of electrons and protons, and breaking the C-O bonds to create new C-H bonds. Therefore, a complete characterization of the molecular, electronic, and nuclear structure of the tested materials is very crucial to understand the bonding interactions and the chemical reactivity.

5. References

[1] van Bergen F, Gale J, Damen K J and Wildenborg A F B 2004 Energy 29 pp 1611-1621.
[2] Bentley R W 2002 Energy Policy 30 pp 189-205.
[3] Cavallo A J 2004 Natural Resources Research 13 pp 211-221.
[4] Budzianowski W M 2012 Energy 41 pp 280-97.
[5] Budzianowski WM 2012 Renewable and Sustainable Energy Reviews 16 pp 6507-6521.
[6] Dickson B, Yashayaev I, Meincke J, Turrell B, Dye S and Holfort J 2002 Nature 416 pp 832-837.
[7] Lilliestam J, Bielicki J M and Patt A G 2012 Energy Policy 47 pp 447-455.
[8] Metz B, Davidson O, Bosch P, Dave R and Meyer L 2007 Climate Change 2007: Mitigation of Climate Change (Cambridge University Press).
[9] Hicks J C, Drese J H, Fauth D J, Gray M L, Qi G and Jones C W 2008 Journal of the American Chemical Society 130 pp 2902-2903.
[10] Rubin ES, Davison J E and Herzog H J 2015 International Journal of Greenhouse Gas Control 40 pp 378-400.
[11] Nemet G F, Baker E and Jenni K E 2013 Energy 56 pp 218-228
[12] Aresta M 2010 Carbon Dioxide as Chemical Feedstock (WileyVCH).
[13] Zhu L, Lu Y and Shen S 2016 Energy 104 pp 53-63.
[14] Ampelli C, Centi G, Passalacqua R and Perathoner S 2016 Catalysis Today 259 pp 246-258.
[15] Sun L, Hammarstrom L, Åkermark B and Styring S 2001 Chemical Society Reviews 30 pp 36-49.
[16] Benson EE, Kubiak CP, Sathrum AJ and Smieja JM 2009 Chemical Society Reviews 38 pp 89-99.
[17] He H, Zapol P and Curtiss L A 2012 Energy and Environmental Science 5 pp 6196-6205.
[18] Choi W, Termin A and Hoffmann M R 1994 Journal of Physical Chemistry B 98 pp 13669-13679.
[19] Kim W, Seok T and Choi W 2012 Energy and Environmental Science 5 pp 6066-6070.
[20] Loutzenhiser P G and Steinfeld A 2011 International Journal of Hydrogen Energy 36 pp 12141-12147.
[21] Krejcikova S, Matejova L, Koci K, Obalova L, Matej Z, Capek L and Solcova O 2012 Applied Catalysis B: Environmental 111-112 pp 119-125.
[22] Xiang Q, Cheng B, Yu J 2015 Angewandte Chemie International Edition 54 pp 11350-11366.
[23] Gálvez M E, Loutzenhiser P G, Hischier I and Steinfeld A 2008 Energy Fuels 22 pp 3544-3550.
[24] Loutzenhiser P G, Meier A and Steinfeld A 2010 Materials 3 pp 4922-4938.
[25] Gokon N, Kodama T, Imaizumi N, Umeda J and Seo T 2011 International Journal of Hydrogen Energy 36 pp 2014-2028.
[26] Milanese M, Colangelo G, Iacobazzi F and de Risi A 2017 Solar Energy Materials & Solar Cells 160 pp 174-181.
[27] Wang W, Wang S, Ma X and Gong J 2011 Chemical Society Reviews 7 pp 3703-3727.
[28] Ganesh I 2014 Renewable and Sustainable Energy Reviews 31 pp 221-57.

Acknowledgments

The funding to conduct this work has been provided by the Alexander von Humboldt Foundation.

Performance analysis of a two-stage gravitational water vortex turbine

Taqi Ahmad Cheema[1†], Rizwan Ullah[1] and Abdul Samad Saleem[1]

[1]Faculty of Mechanical Engineering, GIK Institute of Engineering Sciences and Technology, Topi, 23460, Pakistan.

†Corresponding author: tacheema@giki.edu.pk

Abstract. In the present study the performance of two-stage gravitational water vortex turbine (GWVT) assembled in a conical basin with Savonius blade profile configuration has been investigated at different flow rate and vortex elevation. Two-stage GWVT with conical basin, not only increases the performance parameters with increase in flow parameters, but the lower stages also feed additional impact to the stage located above it because of forced vortex generation in the vicinity of top stage. Moreover, maximum hydraulic head drop near the orifice is the main contributor in the power generation of the bottom stage, by giving rise to strong tangential velocity. A stage starts power production as soon as the parabolic surface of the vortex approaches the corners of the blades even in the absence of a significant water-blade interaction. The developed analytical model qualitatively predicts the performance of the turbine minutely leading the experimental results quantitatively.

1. Introduction

Gravitational water vortex power plant (GWVPP) is a relatively new technology in the field of micro hydro power plants (MHPPs) for electricity generation. In a GWVPP, the power from an artificially induced water vortex under gravity is extracted through a non-conventional co-axial gravitational water vortex turbine (GWVT) that works under a head of 0.7-3m [1-3]. The past studies have mainly used a cylindrical basin for generation of gravitational vortex [4-10] with an optimum orifice diameter to basin diameter ratio of 14% - 18% [5, 7, 8], however, a similar combination for conical basins is yet to be investigated [11]. Experimental studies on GWVT performance with cylindrical basin considered the effects of various parameters such turbine position along the height of the basin, the blade size, blade numbers, testing of different blade profiles, numbers of blades and blade materials [2, 9-12]. The said experimental investigations have been missing for a GWVT with a basin of conical configuration until the recent work of Dhakal et al. [11, 13]. The vortex strength dominates in conical basin over cylindrical basin and hence, gives more output power for the same input parameters [13]. The literature has reported the use of CFD tool to trace the effect of basin diameter, notch length, canal height and cone angle on the vortex formation in conical basin [15].

The past studies have used only a single runner in a GWVT setup. However, it was suggested that a multistage turbine can be installed to extract the power of vortex reformed in a conical basin after its initial distortion by single runner [14]. Moreover, the researchers have used blades of conventional turbines which tend to cut the flow stream and is undesirable in a GWVT [5, 7, 8]. The present study addresses the aforementioned concerns by introducing the profile of Savonius type blades with conical configuration in the present study along with the performance investigation of two-stage GWVT at various flow rates and vortex elevation. Rotational speed, torque and power have been examined experimentally as well as analytically for a prototype multistage GWVT.

2. Experimental methodology

The schematic of experimental setup used in the present study for the testing of two-stage GWVT has been shown in Figure 1. The complete facility consists of a water storage tank (415 L), a centrifugal pump (3hp), an overhead water reservoir (230 L), a basin and a multistage GWVT assembly. The two valves located at the entrance of upstream channel and at the exit of basin are adjusted to ensure a proper vortex formation and maintenance of desired vortex elevation in the basin. The exiting water from the water storage tank from where the water is recirculated through a centrifugal pump. The GWVT setup has been assembled in a conical basin with 2 runners each having 4 blades. Each runner is independent in terms of power generation because of telescopic shafts arrangement. The inlet flow rates been measured with the help of volumetric (bucket) method [2, 12, 14]. Flexible arrangement for variation in flow rates is justified with the help of upstream and downstream valves. The rotational speed in revolutions per minute (rpm) was measured with photo/contact tachometer (Lotron DT-2236B). Prony brake mechanism provided the support for the measurement of torque.

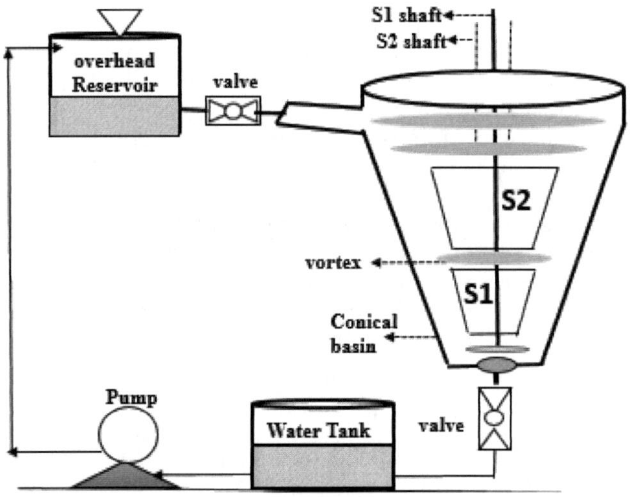

Figure 1. Schematic for the testing of GWVT

3. Results and Discussions

The performance of a two-stage GWVT has been examined experimentally along with analytical calculations by computing speed, torque and power of each stage of the turbine. Though, the aforementioned parameters are important for each individual stage; however, in the present study, main emphasis has been given to the inter-stage relationship, i.e. the effect of one stage on the other. A comparison of analytical performance parameters (APP) has been made with the experimental performance parameters (EPP) for the same conditions. This will provide validation to the analytical model as well as prediction of extended performance. The comparison has been established by investigating the effects of constant and variable inlet flow rates and vortex elevation for two-stage turbine. For the performance evaluation of both stages, ratio of rotor to basin diameter is kept constant at 0.69.

In the forthcoming cases the experimental and analytical results reveal an agreement on the same trend line of rpm, torque and shaft power (Figure 2 and Figure 3). Therefore, it can be inferred that the analytical model qualitatively predicts the correct results with some experimental limitations. N_e^i and N_a^i represent the experimental and analytical calculated rpm for stage i, where i stands for stage number. Moreover, the same technique has been followed for torque and power. It has been found that

IOP Conf. Series: Earth and Environmental Science **291** (2019) 012039 doi:10.1088/1755-1315/291/1/012039

EPP for S2 supersedes the APP, showing an additional contribution to the performance parameters which has not been reported before [2]. In case of a two-stage GWVT with conical basin, the performance parameters not only increase by the flow parameters (flow rate, vortex and hydraulic potential head) but the lower stage i.e stage one (S1) also feed additional impact to the stage located above it ; stage two (S2), to increase power generation in terms of vortex strength. This is the advantage that the two-stage GWVT with conical basin, provides better performance than a single stage GWVT with uniform basin in addition to multiple power from multiple stages.

3.1. Effect of inlet flow rate

The effect of inlet flow rate is important to determine the potential of two-stage GWVT to be installed at a particular water stream. For this purpose, performance comparison between analytical calculations and experimental results of GWVT has been presented for variable inlet flow rates in Figure 2. Initially, a relationship between S1 and S2 has been investigated because of the fact that a maximum flow rate of 5.5 L/s with downstream valve completely opened, the vortex elevation ensures to have water contact with both stages (S1 and S2) of turbine. All cases shown in Figure 2 represents a zero value of performance for S2 at a flow rate of 3.5 L/s, showing that the surface vortex is only in contact with S1. Increase in flow rate causes an increase in vortex elevation because of which runner of S2 starts power production. The variation in EPP as well as APP is negligible at all flow rates above 4L/s since S1 is completely submerged in the vortex. Rpm, torque and shaft power for S2 shows a rise in their values because of the fact that forced vortex generated by the stage below it (i.e. S1) strengthens the vortex in the vicinity of S2 (Figure 2). Another possible reason for the said increment is the gradual increase in the vortex height ensuring a rise in performance parameters with more vortex-blades interaction. In case of S1, experimental and analytical results are in good agreement both qualitatively and quantitatively.

In the absence of any stage below S1, all the power produced by it can be attributed only to the water flow rate and head drop. A small variation in effective head drop significantly affects the vortex tangential velocity. Therefore, the head drop is the main source of power generation located near the outlet. An overall assessment of Figure 2 reveals that S1 outperforms S2 in terms of rpm, torque and shaft power for all investigated inlet flow rates.

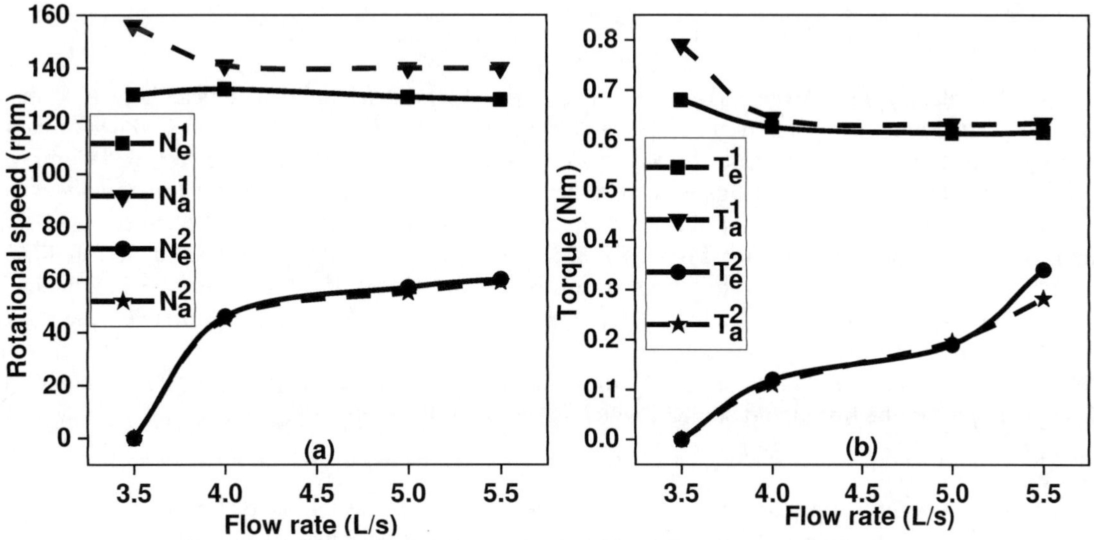

Figure 2. Effect of inlet flow rate on (a) Rotational speed (b) Torque

3.2. Effect of vortex elevation

The vortex elevation is the vertical straight-line height of the free surface of the vortex from the bottom of the basin. This is an important parameter both from the vortex formation and power generation point of view. The effect of vortex elevation on the performance parameters have been examined in context of the relationship of the stages of GWVT. At first, APP and EPP of S1 and S2 at constant inlet flow rate of 5.5 L/s have been conducted for all the two-stage of GWVT by maintaining the same flow rate of 5.5 L/s as shown in Figure 3. In all cases, outlet valve is adjusted in such a way to get fixed vortex elevation during experimentation and thus S1, may be considered completely submerged in water at all vortex heights. On the other hand, the submergence of S2 start at vortex elevation of 0.607m. An important observation need to be considered that the top stage start power production as soon as the parabolic surface of the vortex approaches the corners of the blades, though the vortex-blade contact is not significant. Therefore, the performance parameters below vortex elevation of 0.607m of S2 represent smaller magnitudes than because of the said effect.

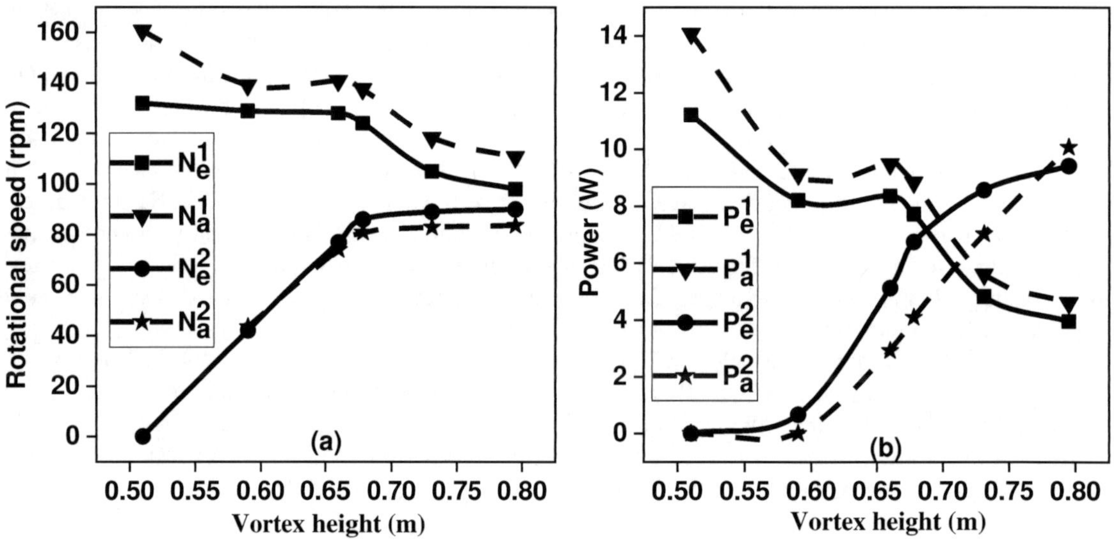

Figure 3. Effect of vortex elevation on (a) Rotational speed (b) Power

For the relationship between S1 and S2, it can be observed from Figure3 that there is a sharp decline in the rotational speed (rpm), torque and shaft power of S1. The decrease in said performance parameters may be attributed to the effect of vortex distortion because of S2. Moreover, APPs are over predicted than EPPs for all investigated vortex elevations showing that the analytical method does not account for the vortex distortion because of the neighbouring stage. On the other hand, the performance parameters for S2 increases as well as APPs are observed to have been under predicted. Since, area of contact between gravitational vortex and S2 blades increases with the increment in vortex elevation thus, there is a gradual increase in the performance parameters. Moreover, the forced vortex generated by the stage below it (i.e. S1) strengthen the vortex near the region of S2, thus the performance of S2 is enhanced. It is of worth mentioning that the analytical model is unable to accurately predict the forced vortex effect which makes it limited.

4. Conclusion

In the present study, the performance of a two-stage gravitational water vortex turbine (GWVT) in a conical basin with Savonius blade profile has been examined. An advantage of a two-stage GWVT assembled in a conical basin is that the performance parameters of top stage are fed because of forced vortex generation from the lower stage. The study concludes that the power generation of bottom stage, located near the outlet is strongly supported by hydraulic head drop. The top stage may start power production as soon as the parabolic surface of the vortex approaches the corners of the blades, though vortex-blade interaction is not significant.

5. References

[1] Rahman M M, Tan J H, Fadzlita M T and Muzammil A W K 2017 A Review on the Development of Gravitational Water Vortex Power Plant as Alternative Renewable Energy Resources In *IOP Conf. Series, Materials Science and Engineering* vol 217(1) IOP Publishing pp 012007

[2] Gheorghe-Marius, M and Tudor S 2013 Energy capture in the gravitational vortex water flow *J. of Marine Technology & Environment* vol 1

[3] Shabara H, Yaakob O, Ahmed Y and Elbatran 2015 A CFD simulation of water gravitation vortex pool flow for mini hydropower plants *J. Teknologi* vol 74(5) pp 77-81.

[4] Wanchat S and Suntivarakorn R 2012 Preliminary design of a vortex pool for electrical generation *Advanced Science Letters* 13(1) pp 173-7.

[5] Mulligan S and Hull P 2010 Design and optimisation of a water vortex hydropower plant (Undergraduate thesis) *Inst. of Tech. Sligo, Sligo, Ireland*

[6] Kueh TC, Beh SL, Rilling D and Ooi Y 2014 Numerical analysis of water vortex formation for the water vortex power plant *Int. J. of Innovation (Management and Technology)* 5(2) pp111

[7] Chattha J A, Cheema T A and Khan N H 2017 Numerical investigation of basin geometries for vortex generation in a gravitational water vortex power plant *In Int. Renewable Energy Congress (IREC) 8th* (IEEE) pp 1-5

[8] Khan N H, Cheema T A, Chattha J A and Park C W 2018 Effective Basin–Blade Configurations of a Gravitational Water Vortex Turbine for Microhydropower Generation *J. of Energy Engineering* vol 144(4) pp 04018042

[9] Power C, McNabola A and Coughlan P 2016 A parametric experimental investigation of the operating conditions of gravitational vortex hydropower (GVHP) *J. of Clean Energy Technologies* vol 4(2) pp 112-119

[10] Kueh T C, Beh S L, Ooi Y S and Rilling D G 2017 Experimental study to the influences of rotational speed and blade shape on water vortex turbine performance In *Journal of Phys. Conf. Series* vol 822(1) IOP Publishing pp 012066

[11] Dhakal S, Nakarmi S, Pun P, Thapa A B and Bajracharya T R 2014 Development and testing of runner and conical basin for gravitational water vortex power plant *Journal of the Institute of Engineering* vol 10(1) pp 140-48.

[12] Nishi Y & Inagaki T 2017 Performance and Flow Field of a Gravitation Vortex Type Water Turbine *Int. J. of Rotating Machinery*

[13] Dhakal S, Timilsina A B, Dhakal R, Fuyal D, Bajracharya T R, Pandit H P, Amatya N and Nakarmi A M 2015 Comparison of cylindrical and conical basins with optimum position of runner: Gravitational water vortex power plant *Renewable and Sustainable Energy Reviews* vol 48 pp 662-69

[14] Marian M G, Sajin T and Azzouz A 2013 Study of Micro Hydropower Plant Operating in Gravitational Vortex Flow Mode In *Applied Mechanics and Materials* (Trans Tech Publications) vol 371 pp 601-05

[15] Dhakal S, Timilsina A B, Dhakal R, Fuyal D, Bajracharya T R and Pandit H P 2014 Effect of dominant parameters for conical basin: Gravitational water vortex power plant In *Proc. of IOE Graduate Conf.* pp 381

Acknowledgments

The authors acknowledge the support of Higher Education Commission of Pakistan and GIK Institute of Engineering Sciences and Technology, Topi, 23460, KPK, Pakistan.

IOP Conf. Series: Earth and Environmental Science **291** (2019) 012040 doi:10.1088/1755-1315/291/1/012040

Device Selection of the Potential Wave Energy Site in Indonesian Seas

B Triasdian[1,2], Y S Indartono[1,2], N S Ningsih[3] and D Novitasari[4]

[1]Faculty of Mechanical and Aerospace Engineering, Institut Teknologi Bandung
[2]New and Renewable Energy Research Center, Institut Teknologi Bandung
[3]Faculty of Earth Sciences and Technology, Institut Teknologi Bandung,
[4]Center for Energy Studies, Universitas Gadjah Mada

Abstract. Ocean wave energy (OWE) is one of many promising renewable energy sources with a high energy density. By using the average wave energy data of 21-year WAVEWATCH-III (WW3) model, some areas were discovered to have a great potential for the wave energy generation. This present study was evaluating the selected Wave Energy Converters (WEC) and determining their capability to do the energy capture in the potential site of the Indonesian seas. The data of WW3 indicate the energy flux of OWE in Indonesia reached to 74.65 kW/m. The case study was conducted in the present research placed in an area of the Java Island called Cidaun, with energy flux about 42.41 kW/m. One of the Heaving Device type was used for assessing the practical energy and to analyze its development in Indonesia with certain limitations; a single round float and diameter device about two meters. The annual energy extracted from the designated WEC was approximately reached 119 GWh. Furthermore, the levelized cost of energy (LCOE) was assessed for the particular device in the selected site, resulted about 25.13 $cent/kWh or four times higher than a standard of minimum feed in tariff given by the Indonesian government.

1. Introduction

Indonesia is an archipelagic country that 70% of the area is filled with water. The area reaches to 5.8 million km^2 and elongated shoreline about 95.181 km[1]. The vast potential of the Indonesian seas can give a huge advantage to utilize ocean energy for their alternative energy resource. There are three major energy from the ocean that are potential in Indonesia; namely tidal, ocean thermal and wave.

Ocean wave energy (OWE) has a high power density and consistent propagation that made them as a promising renewable energy in Indonesian seas[2]. Wave energy presents a number of advantages with respect to other CO_2-free energy sources – resource predictability, high power density, relatively high utilization factor and low impact of environmental and visual[3]. As the tropical country, the waves propagated wind in Indonesia are not locally generate in their near-shore. However, some of places are found have an abundant energy generated by the wind waves called the swell. The west region of the Indonesian water is exposed to swells from the South Indian Ocean and affected by the predominant monsoons, especially the South West monsoon[4].

The utilization of renewable energy (RE) is not limited to the feasibility study, but also have to consider the technical requirement, environmental impact and socio-economic study. There is recorded that more than 1,000 wave energy converters (WEC) are existed worldwide that are classified by dimension, arrangement with respect to the main wave direction, distance to the shore, the principal of operation[5][6].

As an archipelagic country with more than 17,500 islands, RE utilization in Indonesia will benefit the remote. The small islands in Indonesia rarely have the local energy source that all the energy supply comes from the main island, makes the cost of energy relatively high. Based on National Energy Policy, the target of New Renewable Energy (NRE) of Indonesia is 23 % of the national energy mix by 2025[7]. Bioenergy will share 10%, hydro 3%, geothermal 7% and other NRE of 3%. Ocean renewable energy considered as one of the options of NRE to be developed in Indonesia.
In the present paper, we are emphasizing on the device-matched and financial aspects related to wave power. The device selection process will be vary by the region, due to a different society and environment. Some aspects were selected by considering on the installation process, size and the maintenance requirement of five main variants of WEC. The financial assessment of the selected design can be used to analyze the possibility of the ocean wave development in the future. These matters are also expected to become the consideration of the blue energy issue for the government related questions are likely appearing between the scientist, engineer and investors.

2. Data and methodology

The selection process of the site for the WEC model design mainly considered by the environmental effects and the energy cost assessment. Wave energy was reckoned to the result of WAVEWATCH-III (WW3) numerical model that is provided the hind-cast of wave conditions in the Indonesian seas from 1991 to 2011. The 21 years data were processed to get the seasonal data of wind and waves, therefore the pattern and character of the OWE in the Indonesian seas can be defined.
The wave energy assessment using WW3 indicated that wave energy resources in the Indonesian seas reached to 74.65 kW/m[8]. The most prominent energy is placed in the national sanctuary in Sumatra Island named Lampung. However, there are areas along the south-coast of Java that are potential to be utilized for the WEC. Thus, the research was focused to assess the utilization and recommend for one study in Java Island.
A field observation was undertaken along the south side of the West Java province to validate the selected locations from the model. There are some specifications of the area selected to be observed as follows:

- Energy above 20 kW/m
- The distance from the nearest coastline is less than 5 km
- The site is nearby the public facility
- No rocky reefs
- No sea-grass or coral reefs ecosystem

Those considerations were made to decide the most applicable site to be used as the study case. All large-scale RE conversions are an area demanding and therefore in a potential conflict with other interests[9]. Thus, all the specifications were needed to determine the optimum value between the potential sources, economics and the impact to the environment without offending any conflict in the community. One of the areas that fulfil all the specifications in Java Island, called Cidaun that shown in Figure 1. The selected area has a potential source with energy flux reached to 42.41 kW/m.

IOP Conf. Series: Earth and Environmental Science **291** (2019) 012040 doi:10.1088/1755-1315/291/1/012040

Figure 1. The selected area for the WEC development location

2.1 Wave energy converter

Babarit in 2015 specifically distinguished the WEC group and arranged all of them as five main variants; oscillating water column (OWC), Overtopping, Heaving Devices, Fixed and Floating Oscillating Wave Surge Converters (OWSC)[6]. Those specifications are extracting the energy differed one another. The OWC generates the electricity from the rise and fall of the water by a wave in a cylindrical pipe with holes[10], while Overtopping device catch the energy by elevating waves into a reservoir above sea level from which the water passes back to the sea via low head Kaplan hydro turbines[11]. Heaving device utilizes the horizontal movement, while OWSC will adapt the vertical movement to generate power. The illustration of the energy captured by five variants of WEC is presented in the Figure 2.

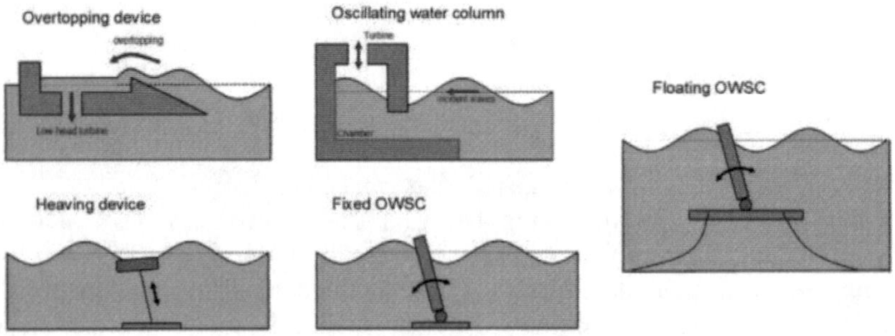

Figure 2. Illustration of the archetypal five category wave energy devices by Babarit[6]

The wave energy resource development become a priority in some countries in the world due the high energy density and minor potential impact on the environment [12]. However, the location of WEC device installations could give a significant effect to the ocean processes or environmental damage due to a different requirement of installation. The performance measure used of WEC was vary greatly from one source to another, and can be applied depending on the water depth and on the location (shoreline, nearshore, offshore)[13][14]. According to the distance of the location of installation from the shore, WEC is categorized to shoreline, near shore and offshore devices. The environmental impacts of wave energy conversion technologies as depicted in the Table 1 below.

Table 1. Environment impact related to the WEC instalment[1]

Environmental Effects	Shoreline	Nearshore	Offshore
Land use/sterilization	L		
Construction/maintenance sites	L		
Recreation	L	L	
Coastal erosion	L	L-M	L-M
Sedimentary flow patterns		L	L
Navigation hazard		L	L
Fish & marine biota	L	L	L
Acoustic noise	L		
Working fluid losses		L	L
Endangered species	L	L	
Device/mooring damage		L-M	L-M

(L: Low, M: medium)

In this research, the design of WEC is made by some deliberation on environmental issues. Due to the high cost of material and maintenance for the seawater instrument, the design is limited using no submerged utilities. The selected site will be adjusted by a particular WEC size to a specific environment.

2.2 Cost of energy

The Levelized Cost of Energy (LCOE) is a useful parameter to assess the economic feasibility of a technology. It is defined as the sum of all capital costs and lifetime operation and maintenance costs divided by the electricity generation to grid accumulated throughout the technology's lifetime [15]. The equations (1) (2) are used to calculate the *LCOE* as follows[16]:

$$LCOE = \frac{NPV(CAPEX + OPEX)}{AEP} \quad (1)$$

$$AEP = \frac{avl.\times CWR \times D \times J \times N \times 24\,h \times 365\,days}{1000} \quad (2)$$

NPV is the net present value, *CAPEX* is the capital expenditures, *OPEX* is the operational expenditures and *AEP* is the annual electricity production. The equation 2 define avl as availability, *CWR* as Capture Width Ratio, *D* as the characteristic dimension of the device and *J* as the wave resource of a site. Through a review of the existing data, the cost components for the *CAPEX* include the civil and structural costs, plant costs, electrical supply and installation to the development costs [17]. As well as the initial investment, the economic analysis must include the operation and maintenance expenditure (*OPEX*). Tables 2 and Table 3 show the summaries of the basic costs of wave energy development. The initial cost included the main sector of *CAPEX* calculation, while the operation cost related to the *OPEX*.

Table 2. Summary of initial cost [18]

Element	Cost
WEC installation	2.5 – 6.0 M€/MW
Mooring System	0.265 €/day
Mooring Installation	50,000 €/day
Underwater Cable	10% CAPEX
Cable Installation	2.07 €/m
Electrical Substation	≈ 1.2 M€

Table 3. The annual cost of operation WEC installation

Cost	€/MW h
Operation and Maintenance tasks	20 – 35
Public Services	3.5
Insurance Cost	15

OWE is an immature technology in new renewable energy. It is understandable the price of the electricity of the installation still costly compared by other advanced RE technologies. It is expected that wave energy technologies could become more competitive once approximately 2.5–10 GW of cumulative capacity has been installed[17]. World Energy Perspective in 2013 indicates that the cost of energy for existing OWE was about 286 to 1058 ($/MW h). The price was five times higher than any other NRE as general[19]. The convergence of wave energy to the given *LCOE* depends not only on resource availability but also on the ability of the sector to identify common solutions and broader design consensus, which will unlock economies of scale.

The government has a role to improve enabling condition and policy framework that could lure public and private investment in renewable energy. The Minister's Regulation No. 12/2017 insisted some renewable power purchasing prices should be referred to audited regional electricity production price[20], which the national electricity production price approximately 7.39 cents$US/kWh[21]. In order to assist the ocean energy development, especially for OWE, the price would be insufficient to attract investors in this sector. Some countries proposed a higher price to show their support. The highest price feed in tariff (FIT) of ocean energy given by the government of Italy about 34 cents $US/kWh, while the serious commitment for WEC development given by the government of Portugal with FIT price about 33 cents $US/kWh[22][23]. Another assessment was coming from the price of energy by the devices. Andres et al. in 2017 calculated the *CAPEX* of five different types of WEC from a certain number of *LCOE*. It shows that the *CAPEX* from WEC from the lowest cost are Floating OWSC, Heaving Device, Overtopping, OWC and Fixed OWSC, respectively[24].

3. Result and discussion

Cidaun is selected as the area study of the present research based on analysis of the 21-years data from WW3 model and field observation. Based on the Table 1, either the coastline or nearshore converters considered as a better option than the offshore installation since they have the least environmental impacts. The optimum design of WEC was determined from the limitation lists that is shown in the Table 4 below.

Table 4. Limitation of the WEC type to be modelled

Consideration	Type of WEC				
	OWC	Overtopping	Heaving Device	Fixed OWSC	Floating OWSC
Not disturbing the sea floor.	v	v	v		v
The maintenance could be done out of the water.	v		v		
The size of the WEC is flexible to the source and environment.			v		v

Amongst 5 different converter classifications, the Heaving Device considered as the best WEC type that fulfilled all the requirements of this research. In order to minimize the underwater maintenance and simplifying the operational, trestle was considered to be built for the full design. The overall design of the installation is shown in the Figure 3.

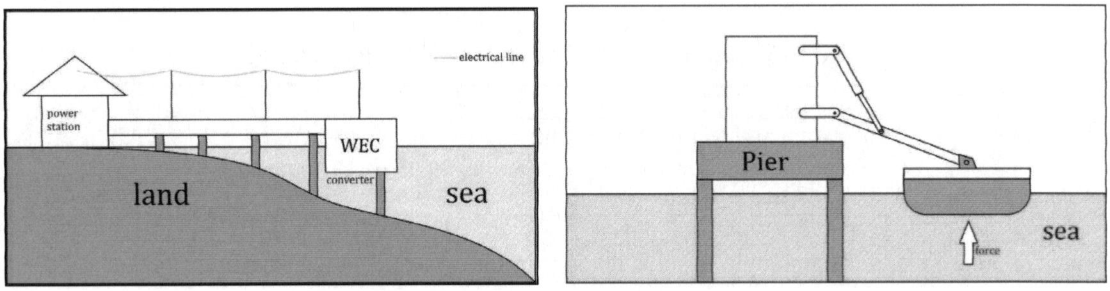

Figure 3. Design of the WEC installation

The WEC design selected is using the work principle of the heaving device with 2-meters diameter and held in a single float (buoy). The assessment of using these design in one of potential site that has an energy flux to 42.41 kW/m can extract the electricity up to 119 GWh per year.

The capital expenditure in this research was not only calculate the initial cost, but also the infrastructure requirement using the standard from the Indonesian government like trestle and the access road. The capital expenditure of the particular device in the selected site was approximately € 19,006,719.44 for 1 MW capacity with more 50% took by the infrastructure cost only. Meanwhile, the operational cost for the 15 years' instalment for the same capacity was about € 6,376,808.23. Using the *LCOE* equation, the final result of the cost of energy in the particular device in selected area (Cidaun) was approximately 25.13 $ cents /kWh. That price is four times higher than the regional electricity production price, which the standard price for the RE plant by the new energy regulation. It makes the OWE utilization become not feasible to be built in the near future.

There are several ways to cut off the cost of energy to make the development of wave energy technology in Indonesia more competitive:

- Pairing the WEC utility with other coastal project development, for instance: the jetty, harbor, or even tourism (the collaboration of such project could subside 46% from all the capital cost).
- Increase the number of WEC installation (For every additional float can suppress cost from 12% - 15%).
- Using the technology in a very remote area with a higher regional electricity price.

The technology will benefit the remote islands like Siberut and Enggano (Sumatra region) with a high energy density, 51 kW/m and 36.70 kW/m, respectively. Using the design and WEC utilities as Cidaun's case with some modifications, the Siberut Island (using two floats) and Enggano Island (using three floats) will cater the regional electricity price to 12.60 cents $/kWh and 13.68 cents $/kWh, each. For such remote area like Siberut and Enggano Islands, the development of WEC not only fulfilling the electricity demand, but also will add a significant value for the island if it combines with infrastructure project.

The FIT of wave energy in Indonesia will be more appealing if the government could increase the electricity production price for certain NRE to a reasonable price. The OWE utilization can be set up with a support as we can copy the EU Member States using push and pull mechanism. In some countries, their government is more considerate and set the higher FIT to support the OWE development. Portugal for the instance, has a full support from their government by having FIT for wave energy to 33 $/kWh. The government act can be showed with incentive, investment fund, or regulation convenience.

4. Conclusion

The present study assessed the maximum energy from one of potential sites, called Cidaun which has an average energy flux to 42.41 kW/m using one of type WEC named Heaving Device. The design included a single float with a diameter about two meters. The result depicts that annual energy can be extracted from the source by using the designated device reached approximately 119 GWh. The LCOE for the particular device is approximately 25.13 $cent/kWh or four times higher than a standard of minimum feed in tariff given by the Indonesian government. There are several ways to optimize the wave energy utilization in Indonesian seas like pairing with the other coastal development or maximizing the capacity of WEC. However, the production price cannot be suppressed without support from the government using investment relief and regulation convenience. We can also encourage the wave energy development by adapting the push and pull mechanism by EU Member States or set the higher FIT like Portugal.

5. References

[1] Centre for Renewable Energy Sources 2002 Wave energy Utilization in Europe - Current Status and Perspectives (Greece)

[2] Rusu L and Onea F 2015 Assessment of the performances of various wave energy converters along the European continental coasts Energy 82 889–904

[3] Carballo R, Castro A, Fraguela J A and Frigaard P 2009 Wave energy potential in Galicia (NW Spain) 34 2323–33

[4] Aboobacker V M 2017 Wave energy resource assessment for eastern Bay of Bengal and Malacca Strait Renew. Energy 114 72–84
[5] Langhamer O, Haikonen K and Sundberg J 2010 Wave power-Sustainable energy or environmentally costly? A review with special emphasis on linear wave energy converters Renew. Sustain. Energy Rev. 14 1329–35
[6] Babarit A 2015 A database of capture width ratio of wave energy converters Renew. Energy 80 610–28
[7] Minister of Law and Human Rights of The Republic of Indonesia 2014 Goverment Regulation of The Republic of Indonesia, Number 79 of 2014 on National Energy Policy
[8] Triasdian B, Indartono Y S and Ningsih N S 2018 Energy Capture Potential of Existing Wave Energy Converters for Indonesian Sea AIP Conf. Proc. 30002
[9] Langhamer O, Haikonen K and Sundberg J 2010 Wave power — Sustainable energy or environmentally costly ? A review with special emphasis on linear wave energy converters 14 1329–35
[10] Irhas and Suryaningsih R 2014 Study on wave energy into electricity in the south coast of yogyakarta, Indonesia Energy Procedia 47 149–55
[11] Millar D L, Smith H C M and Reeve D E 2007 Modelling analysis of the sensitivity of shoreline change to a wave farm 34 884–901
[12] Zhang Y, Lin Z and Liu Q 2014 Marine renewable energy in China : Current status and perspectives Water Sci. Eng. 7 288–305
[13] Cruz J 2008 Green Energy and Technology Ocean Wave Energy ed J Cruz (Bristol, United Kingdom: Springer)
[14] IRENA(International Renewable Energy) 2014 WAVE ENERGY
[15] OES - IEA 2015 International Levelised Cost Of Energy for Ocean Energy Technologies An Anal. Dev. Pathw. Levelised Cost Energy trajectories wave, tidal OTEC Technol. 35
[16] De Andres A, Medina-Lopez E, Crooks D, Roberts O and Jeffrey H 2017 On the reversed LCOE calculation: Design constraints for wave energy commercialization Int. J. Mar. Energy 18 88–108
[17] Magagna D and Uihlein A 2015 2014 JRC Ocean Energy Status Report: Technology, Market and Economic Aspects of Ocean Energy in Europe
[18] Astariz S and Iglesias G 2015 The economics of wave energy : A review 45 397–408
[19] Joseph Salvatore 2013 World Energy Perspective - Cost of Energy Technologies World Energy Counc. 42
[20] Menteri Energi dan Sumber Daya Mineral Republik Indonesia 2017 Peraturan Menteri Energu dan Sumber Daya Mineral Republik Indonesia Nomor 50 Tahun 2017
[21] Menteri Energi dan Sumber Daya Mineral Republik Indonesia 2016 keputusan Menteri Energi dan Sumber Daya Mineral Republik Indonesia, Nomor 1404 K/20/MEM/2017
[22] Rousseau N 2008 Ocean Energy : A European Perspective
[23] The Ocean Renewable Energy Group The Role of Feed-in Tariffs : Moving Ocean Energy Ahead in Canada
[24] de Andres A, MacGillivray A, Roberts O, Guanche R and Jeffrey H 2017 Beyond LCOE: A study of ocean energy technology development and deployment attractiveness Sustain. Energy Technol. Assessments 19 1–16

Acknowledgements
We gratefully acknowledge the funding from USAID through the SHERA program – Centre for Development of Sustainable Region (CDSR).

IOP Conf. Series: Earth and Environmental Science **291** (2019) 012041 doi:10.1088/1755-1315/291/1/012041

Design and development of integrated savonius and darrieus small scale vertical axis wind turbine for power generation

Edison E Mojica[1]*, Cian Marvince Fabay[1], Felix Kehinde[1] and John Lambert Tenorio[1]

[1] Electrical Engineering Department, College of Engineering, University of Perpetual Help System Dalta, Las Pinas City, Metro Manila, Philippines

* Corresponding author: edisonmojica@gmail.com

Abstract. The study focuses on the integration of the darrieus and savonius wind turbine to a vertical axis wind turbine for power generation. The material used for darrieus was fiber glass with three blades and diameter of 350 mm, and polyvinyl chloride (PVC) for the savonius with four blades and diameter of 100 mm. The evaluation was conducted using a controlled wind source to vary its linear wind velocity. The actual power generated was compared to the theoretical power generated to identify the efficiency of the system. With the lowest linear wind velocity of 2.63 meters per second, the actual power output of the system (combination of the darrieus and savonius power output) was 3.79 watts and the theoretical power output is 6.03 watt. This made the system to be 62.85% efficient with low wind velocity. With the highest wind velocity of 12.52 meters per second, the output of the system was 408.80 watts and the theoretical output is 650.61 watts. This made the system to be 62.83% efficient. In an overall performance of the system with varying linear wind velocity, the efficiency of the system is close to 63%.

1. Introduction

The acceptance of the recent technologies is an indication of the economic development and the measurement of the efficiency of the production. The acceptance of this technologies should take into consideration the substantial pressure to address environmental issues. The acceptance should be extensive and should provide public good. In case of environmental issues, public good leads to the absence of pollution. The absence of pollution is the adoption of the green technology. A green technology imparts benefits to consumers but additional costs to the producer [1-4].

The utilization of renewable energy has drawn much attention nowadays because of the excessive use of fossil fuel, energy depletion, and environment pollution. Alternative to fossil fuels is commonly the topic in research and development as this is evident by annual average growth rate of the consumption of the renewable energy as fuel in the generation of electricity. The capacity of wind energy has increased drastically over the other forms of renewable energy. In the context of wind energy, the horizontal axis wind turbine (HAWT) has been the center of exploration during the recent years. But, the vertical axis wind turbine (VAWT) has an essential advantages over the previous. VAWT can generate power even with relatively low cut-in wind speed, being omni-directional, and less noise. VAWT can be classified into two classes, the Darrieus (a turbine using the lift force) and Savonius (a turbine using the drag force). These two produce less noise, have a simple shape, and capable of maintaining stable performance with low wind speed [5-9].

IOP Conf. Series: Earth and Environmental Science **291** (2019) 012041 doi:10.1088/1755-1315/291/1/012041

The design of the darrieus and savonius wind turbine has different specifications. The savonius wind turbine can be applied in power generation but the self-starting mechanism might be a problem. On the other hand, the darrieus wind turbine compromises the efficiency if the blade rotates, the angle of attack is changed. The integration of the savonius and darrieus wind turbine can aid one another in addressing their flaws and help to increase their efficiencies. The proponents thought of a low cost design of a small-scale vertical axis wind turbine by integrating the savonius and darrieus wind turbine with a newer design of the blade architecture for a much higher efficiency than the conventional vertical axis wind turbine [10-15].

2. Design aspects

Figure 1 presents the design of the integrated savonius and darrieus small-scale vertical axis wind turbine. The combination of the savonius and darrieus wind turbine is intended to reduce the space for installation and thus, to increase the efficiency of the power output of a VAWT with the reduced production cost but with larger volume of electricity. The material used in the fabrication of darrieus blade is fiber glass and 80mm-diameter polyvinyl chloride (PVC) pipe for the savonius blade.

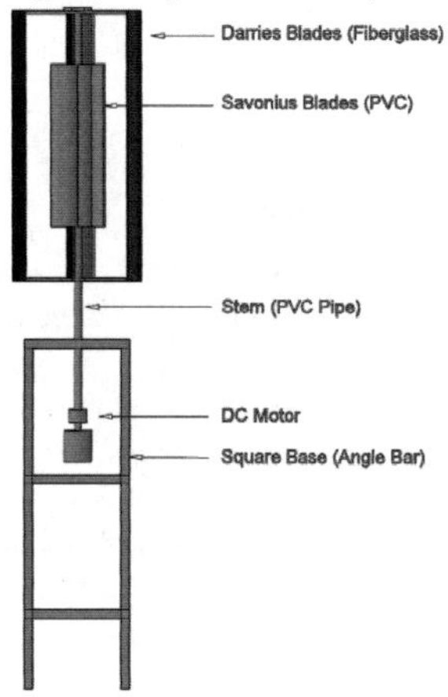

Figure 1. Design of the integrated savonius and darrieus small-scale vertical axis wind turbine

The researchers have come up with the design of the rotor blades for the darrieus with NACA 4415 because it has a better performance than the other airfoils design. The swept area, tip speed ratio, blade chord, number of blades, and, angle of attack were considered in the final design of the blade.

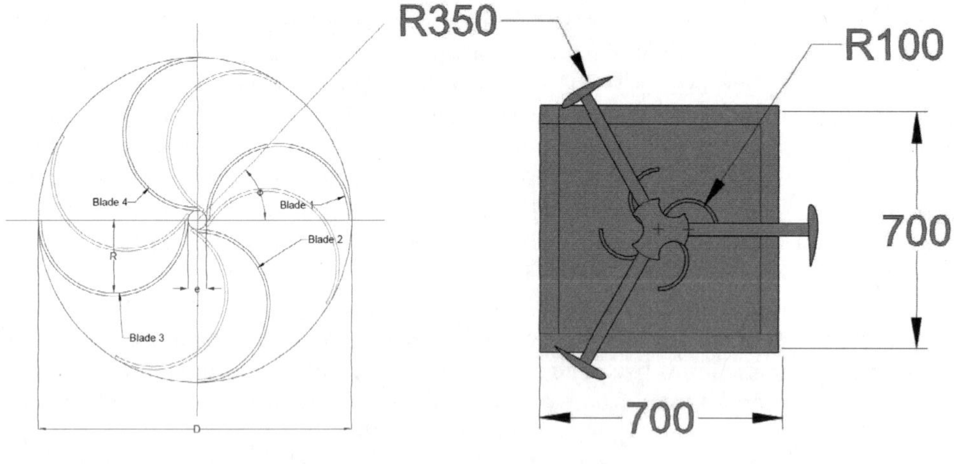

Figure 2. Design and dimensions of the savonius blade *(top view)*

The design utilized four blades of savonius (figure 2) and three blades for darrieus (figure 3) to achieve maximum power extraction since the power generation depends on how fast the blade can capture the wind and how fast these blades will rotate. The four and three blades combination increases the output efficiency with smoother rotation and most efficient design in the environment [13][14][15]. This combination of number of blades also helps in the quick start of the rotor as this is designed to rotate with higher rpm and more torque. With lower wind speed, the combination also allows the increase of the thrust and centrifugal force on the blade . The fiberglass and pvc pipe were used for the strength and stiffness of the blade. Figure 2 presents the radius of the savonius which is 100 mm and this is attached to a square base with side measuring 700 mm.

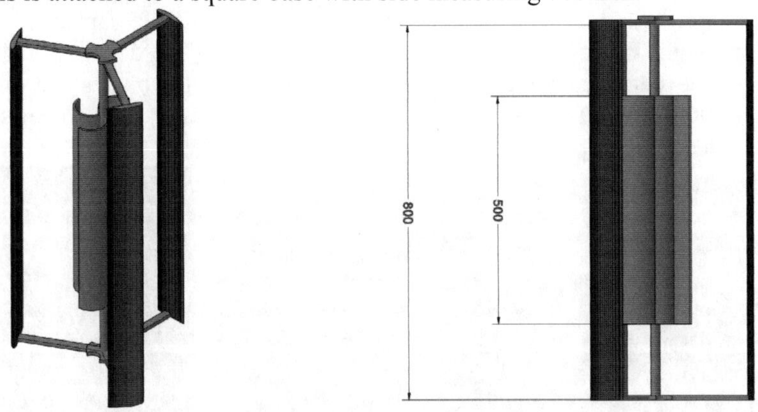

Figure 3. Design and dimensions of the integrated savonius and darrieus blade

Figure 3 shows the integration of the savonius and darrieus blade with the intention of reducing the space for installation. The length of the savonius blade is 500 mm and that of the darrieus blade is 800 mm.

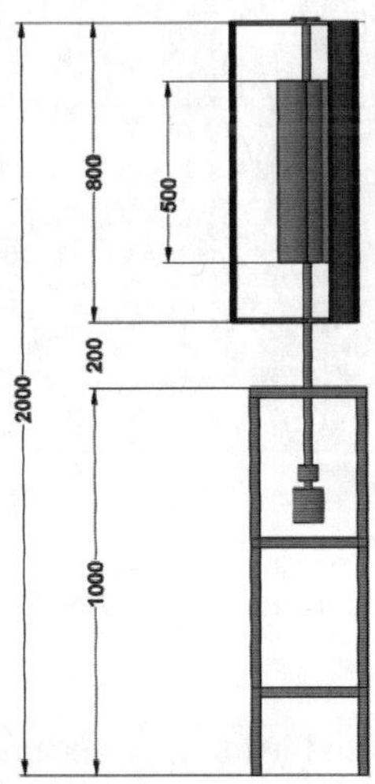

Figure 4. Dimensions of the integrated savonius and darrieus small-scale vertical axis wind turbine

Figure 4 presents the design and dimensions of the integrated savonius and darrieus small-scale vertical axis wind turbine. The height of the prototype is 2000 mm with its base height of 1000 mm.

The base (700 mm by 700 mm) was constructed using an angle steel bar to maintain its stability and it houses the dc motor.

3. Development of the system

The system combines the savonius and darrieus types of blades for a small-scale vertical axis wind turbine. The blades work in any wind direction and do not require any mechanism to change their angles. Figure 5 shows the constructed integration of the savonius and darrieus wind turbine. The design is intended to be integrated to an urban setting as this produces less noise and present no danger to anyone. As the wind blows, the mechanism starts with the rotation of the darrieus blades. With the fixed stem, clutch bearings, and the spaces between the three blades of the darrieus, the wind reaches the savonius that drives the blades to rotate also at the same time as the darrieus blades are rotating. The lower portion of the stem accumulates the generated rotation of the blades and this serves as the prime mover of the DC motor. The DC motor is connected to a battery for power storage with an inverter to match the required output voltage for consumption.

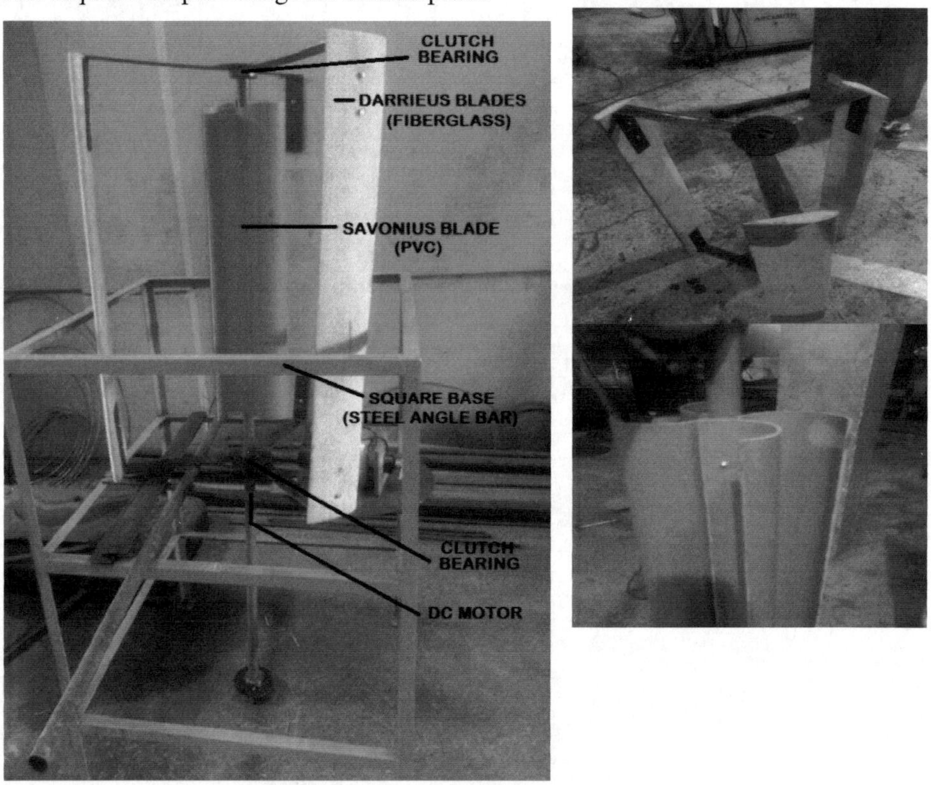

Figure 5. Development of integrated savonius and darrieus small-scale vertical axis wind turbine

4. Data and results

The evaluation of the system was conducted using a wind tunnel with varying speed as the source of wind energy. The linear velocity of the wind was varied from 2.63 meters per second to 12.52 meters per second. The result of the individual performance of the darrieus and savonius clearly shows that darrieus is more efficient in converting electrical energy from wind energy. Though, savonius is still capable of providing electrical energy from the available wind energy.

Figure 6 compares the result of the system (combining the savonius and darrieus wind turbine output) to the theoretical power output. The theoretical power output was computed using the formula $P = (\rho A V^3)/2$ where V is the linear wind velocity (in meters per second), ρ is the wind density (in

IOP Conf. Series: Earth and Environmental Science **291** (2019) 012041 doi:10.1088/1755-1315/291/1/012041

kilogram per cubic meter), and A is the swept area considering the diameter of the wind turbine. Figure 7 shows the comparison of the actual power output of the system versus the theoretical power output. Table 1 shows the performance of the individual trial with varying wind characteristics, darrieus power output, savonius power output, system power output, and theoretical power output. Considering the lowest linear velocity of the wind which is 2.63 meters per second, the power output of the system was 3.79 watts and the theoretical output is 6.03 watts. The data show that the system is 62.85% efficient in the low wind speed. With the highest linear velocity of 12.52 meters per second, the generated output of the system was 408.40 watts and the theoretical output is 650.61 watts. The result shows an efficiency of 62.77% which is almost the same as that of the efficiency with the lowest linear velocity. With all ten different trials, the efficiency of each was close to 63%. The result clearly shows that the developed system will have the same efficiency regardless of the linear wind velocity.

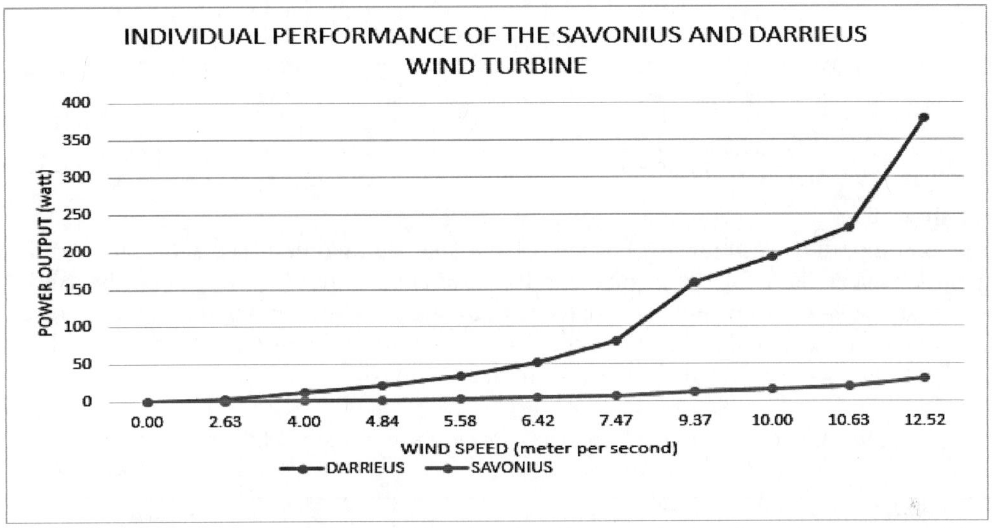

Figure 6. Performance of the savonius and darrieus wind turbine in watt versus the wind speed

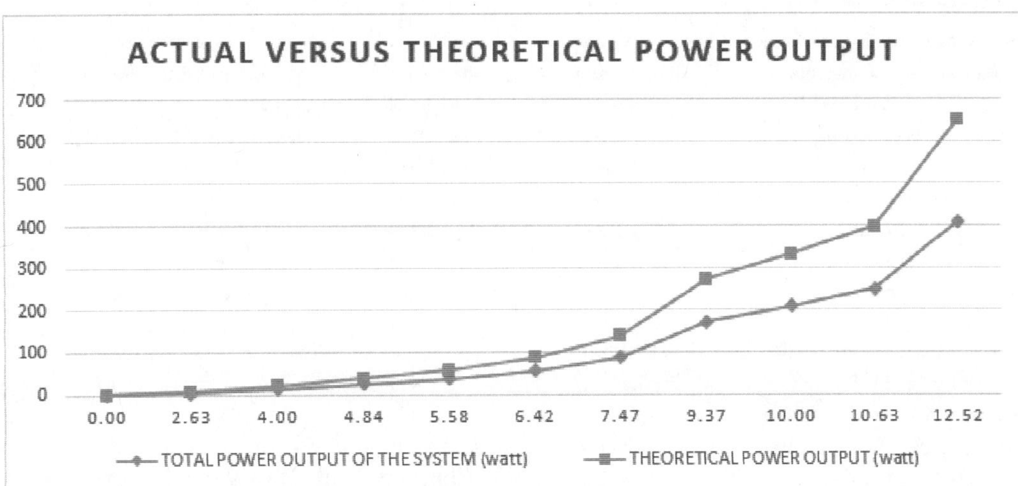

Figure 7. Actual power output of the system versus the theoretical power output in watt

IOP Conf. Series: Earth and Environmental Science **291** (2019) 012041 doi:10.1088/1755-1315/291/1/012041

Table 1. Performance of the individual trial with varying wind characteristics, darrieus power output, savonius power output, system power output, and theoretical power output.

WIND CHARACTERISTICS				DARRIEUS		SAVONIUS		SYSTEM (ACTUAL)	THEORETICAL	EVALUATION	
Linear Velocity (meter per second)	Temperature (^oC)	Density (kg/m^3)	Specific Weight (N/m^3)	Area covered by the blade	Power output (watt)	Area covered by the blade	Power output (watt)	Power output (watt)	Power output (watt)	Power output difference (watt)	Percent Efficiency (actual / theoretical)
2.63	-20	1.395	13.68	0.38	3.50	0.03	0.29	3.79	6.03	2.24	62.852%
4.00	0	1.293	12.67	0.38	12.31	0.03	1.00	13.31	21.22	7.91	62.724%
4.84	5	1.269	12.45	0.38	21.83	0.03	1.78	23.611	37.59	13.98	62.812%
5.58	10	1.247	12.23	0.38	33.39	0.03	2.73	36.12	57.60	21.48	62.708%
6.42	15	1.225	12.01	0.38	50.91	0.03	4.16	55.06	87.72	32.66	62.768%
7.47	20	1.204	11.81	0.38	80.27	0.03	6.55	86.62	138.19	51.36	62.682%
9.37	25	1.184	11.61	0.38	158.11	0.03	12.91	171.02	272.73	101.71	62.707%
10	30	1.165	11.43	0.38	192.29	0.03	15.70	207.99	331.52	123.53	62.738%
10.63	40	1.127	11.05	0.38	231.07	0.03	18.86	249.94	398.21	148.27	62.766%
12.52	50	1.109	10.88	0.38	377.95	0.03	30.85	408.8	650.61	241.81	62.833%

5. Cost analysis

The intention of the study was to develop a VAWT that may be used in urban and rural area which is low-cost and all materials should be available in the local area. With the utilization of the fiber glass for darrieus and PVC for the savonius, the total expenses in the construction of this was PhP 12,215.00 or 231USD. Other related studies reflected their total expenses ranging from 600USD to 800USD. Table 2 presents the total breakdown of expenses of the prototype.

Table 2. Breakdown of components/materials and cost for the prototype

Components/Materials	Quantity	Cost	Function
Benetech GM816 Anemometer Wind Speed	1.0 unit	Php 870.00	wind speed reading
Bosca Charge Controller DC 12V/24V 5A	1.0 unit	Php 700.00	charging controller
Bosca Power Inverter 12V-220Vac (SA-10000AF)	1.0 unit	Php 1,200.00	power output converter
Reliance Electric DC Motor/Generator 24V	1.0 unit	Php 2,300.00	power generator
Darrieus Blade Fiberglass Fabrication	3.0 units	Php 2,000.00	blade material for darrieus
Clutch Bearing	2.0 units	Php 2,500.00	rotation shaft in one-direction configuration
Ordinary Bearing	1.0 unit	Php 100.00	rotation shaft in two-direction configuration
Wire	5.0 meters	Php 75.00	conductor
Disk Brakes	2.0 units	Php 600.00	brake system
Brake Caliper	2.0 units	Php 1,000.00	brake system
2600 Ah Battery	1.0 unit	Php 1,400.00	power storage
PVC Pipe 3	1.0 unit	Php 270.00	blade material for savonius
PVC Pipe 3/4	1.0 unit	Php 100.00	stem material
	TOTAL	Php 12,215.00	

6. Conclusion

There is a great advantage in the combination of the darrieus and savonius wind turbines in one vertical axis wind turbine over the horizontal axis wind turbine. With the use of the fiber glass and PVC as materials for the blades, these are light weight and produce less noise. These were able to answer the common issues facing the wind energy technology. Also, with all materials available to any end-user, the system may be constructed and developed in rural and urban areas. The system is capable of generating electrical energy even with low wind speed and the mechanism is less noisy with minimal vibration produced. The system is 63% efficient regardless of the linear wind velocity. Thus, there is a need to further improve the system to achieve higher efficiency. The mounting height, diameter of the system, materials to used, and other parameters may be considered for further exploration of the system.

7. References

[1] Bernard, P., 2014. Turbine having optimized efficiency (France: Save Ingenierrie).
[2] Ghanegaonkar, P., Kawade, R., and Garg, S., 2014. Conceptual model of vertical axis wind turbine and cfd analysis. *International Journal of Innovative Research in Advanced Engineering. Volume 1 Issue 3.*
[3] Xisto, C., Pascoa, J., and Trancossi, M., 2016 Geometrical parameters influencing the aerodynamic efficiency of a small scale self-pitch high solidity vawt. *Journal of Solar Energy Engineering. Volume 138 Issue* pp 10.
[4] Dvorak, P., 2014. Vertical axis wind turbines: what makes them better. *Wind Power and Engineering and Development.*
[5] Islam, M., Ting, D., Fartaj, A., and Esfahanian, V., 2015. Applications of vertical axis wind turbines for remote areas. *ResearchGate.*
[6] Asr, M., Nezhad, E., Mustapha, F., and Wiriadidjaja, S., 2016. Study on start-up characteristics of darrieus vertical axis wind turbines comprising naca 4 digit series blade airfoils. *Energy Volume 112* (Elsevier). pp 528-537.
[7] Nagare, P., Nair, A., Shettigar, R., Kale, P., and Nambiar, P., 2015. Vertical axis wind turbine. *2015 International Conference on Technologies for Sustainable Development.* (Mumbai, India).
[8] Korobenko, A., Hsu, M., Akkerman, I., and Bazilevs, Y., 2013. Aerodynamic simulation of vertical axis wind turbine. *Journal of Applied Mechanics. Volume 81 Issue 2.*
[9] Tjiu, W., Marnoto, T., Mat, S., Rusian, M., and Sopian, K., 2015. Darrieus vertical axis wind turbine for power generation 2: challenges in hawt and the opportunity of multi megawatt darrieus vawt development. *Renewable Energy. Volume 75* pp 560-571.
[10] Sukanta, Roy, Ujjwal, K., Saha., 2014. Wind tunnel experiments of a newly developed two-bladed Savonius-style wind turbine. *Applied Energy 137 (2015) 117–125.*
[11] Natapol, Korprasertsaka, Thananchai, Leephakpreeda., 2015. CFD-Based Power Analysis on Low Speed Vertical Axis Wind Turbines with Wind Boosters. *International Conference on Alternative Energy in Developing Countries an Emerging Economies.*
[12] Xin, Jin, Gaoyuan, Zhao, KeJun, Gao, Wenbin, Ju, 2014. Darrieus vertical axis wind turbine: Basic research methods. *Renewable and Sustainable Energy Reviews 42(2015)212–225.*
[13] Dr. Hassan Darhmaoui, Dr. Naeem Sheikh., 2017. Savonius vertical wind turbine: design, simulation and physical testing. *Alakhawayn University.*
[14] Kok Hoe Wonga, Wen Tong Chonga, Nazatul Liana Sukimana, Sin Chew Poha, Yui-Chuin Shiahb, Chin-Tsan Wang, 2016. Performance enhancements on vertical axis wind turbines using flow augmentation systems. *Renewable and Sustainable Energy Reviews 73 (2017) 904–921.*
[15] M.Zemamoua, M.Aggoura, A.Toumib., 2017. Review of savonius wind turbine design and performance. *4th International Conference on Power and Energy Systems Engineering, CPESE 2017, 25-29 September 2017, Berlin, Germany.*

Acknowledgments
The researchers would like to take this opportunity to thank all the faculty members of the College of Engineering – UPHSD-LP for their endless support and love in providing the researchers with necessary information and facilities needed for the research and to all who directly or indirectly have lent their hand in the success of the project. Also, with the greatest gratitude to our parents for their love, prayer, encouragement, caring and sacrifices for educating and preparing us for the future and for their continuing support to complete this research work. For everything that we hope to be, we owe it to our family and our creator.

IOP Conf. Series: Earth and Environmental Science **291** (2019) 012042 doi:10.1088/1755-1315/291/1/012042

Evaluating the Sustainability Impact of Greening, Water Management, Ventilation and Interior Lighting on Car Parking in Taiwan

Shu-Hui Hsieh. Yu-Chen Chien. Yi-Chien Lee

Chaoyang University of Technology,Taiwan

ericayichien@gmail.com

Abstract. Amid increasing rates of car ownership, serious issues around parking have emerged in Taiwan in recent years. However, most debates on the construction of urban parking infrastructure have focused on assessments of its economic benefits, rather than on a more comprehensive decision-making framework that takes energy use and other sustainability issues into account. Accordingly, this paper is a first step toward creating such a framework from a green-architectural perspective.

1. Introduction

Due to rising living standards in increasingly complex societies around the globe, ownership of personal motor vehicles has increased dramatically in recent years. As such, ensuring that parking structures meet vehicle users' demand while also preserving the quality of urban life has emerged as a critical challenge for government in Taiwan, where road-widening is at best difficult, and traffic jams and accidents are commonplace. Until recently, the main factors considered when building or rebuilding parking facilities were their economic impact and the convenience of their locations, rather than any more holistic evaluation of their influence on their surroundings. Yet, with governments and the general public increasingly interested in pursuing sustainable development, such a holistic approach would appear somewhat overdue. The present research therefore suggests that the design and construction of parking facilities in Taiwan should take account of their energy use, especially with regard to interior lighting; carbon footprints; water-management characteristics; and other environmental aspects. Specifically, it will conduct three case studies, to examine:

(1) Key design differences between a traditional parking structure and a modern one, with regard to vegetation cover and water-management characteristics;

(2) A traditional underground parking facility's ventilation issues, and how these might be improved upon in a sustainable way; and

(3) The interior lighting in an above-ground parking structure and ways of reducing its lighting-related consumption of energy and other resources.

2. Literature review

2.1 Christiansen, P., Engebretsen, Ø., Fearnley, N., Usterud Hanssen, J. (2017) [1]. Parking facilities and the built environment: Impacts on travel behaviour. Transportation Research Part A: Policy and Practice, 95, 198-206.

This research analysis modern parking facilities' key features and the relevant regulations affecting them, and establishes a set of principles for designing such facilities. In the case of open-air car parks, these principles include close attention to rainwater infiltration, the percentage of the land that is excavated, the percentage of green area, and soil depth .

2.1.1 , Shiau, Tzay-An (2018) [2]. Evaluating the sustainability of constructing public off-street parking lots.
Sustainable evaluation index.

2.1.2 Sarachanga et al. (2016)[3]. Evaluation of existing sustainable infrastructure rating systems for their application in developing countries.
This study evaluated the sustainability of infrastructure in terms of four dimensions, as shown in Table 1.

Table 1. Main Criteria Considered by the Rating Systems Under Assessment

(Sarachanga et al.'s)

Economy	Environment	Society	Management
Minimising workforce travel	Principles on the use of land	Stakeholder- and community engagement	Project management (contracts, design, construction)
Increasing the community's connectivity	Flood-risk management	Assessment of impacts in neighbours	Implementing sustainable procurement
Employment growth	Maintenance, enhancement or restoration of biodiversity and habitats	Boosting local employment	Improving decision-making processes
Resource efficiency	Maintenance or enhancement of landscapes	Preserving historical, cultural and archaeological heritage	Managing risk and opportunity
Business development	Efficient water use	Increase public information	Dealing with conflicting regulations and policies
Procurement practices	Maintenance or enhancement of water quality	Promotion of community health, wellbeing and safety	
Workforce development	Reduction of greenhouse gas emissions	Promoting workforce safety	
Stimulation of sustainable growth and development	Noise management	Improving accessibility	
Improvement of the community's quality of life	Reduction of air pollutants	Enhancing public space	
	Lighting pollution management	Preservation of views and local character	
	Reduction of energy use		
	Promotion of renewable energy		
	Efficient resource management		
	Hazardous-materials management		

3. Research methods

The three cases mentioned in Section 1 will be analysed based on the three indices described in Section 2, with particular attention to vegetation cover, soil water content index, ventilation-rate index, and interior-lighting index. The results of such analysis will then be used to establish principles for sustainable parking-facility design (Figure 1).

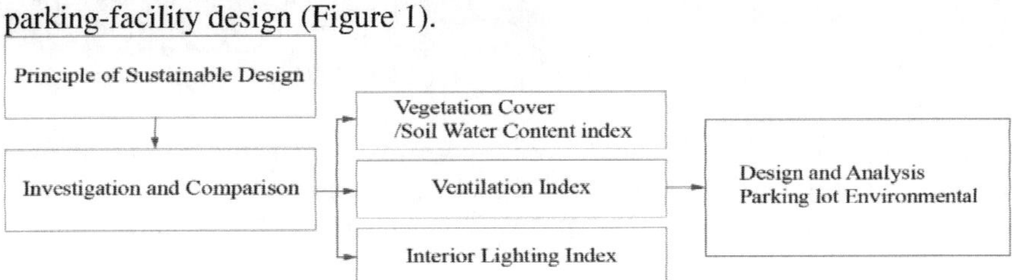

Figure1. Research Framework

IOP Conf. Series: Earth and Environmental Science **291** (2019) 012042 doi:10.1088/1755-1315/291/1/012042

4. Results and discussion

4.1 Building coverage ratio analysis

From the brown-shaded area of the left panel of Figure 2, it can be seen that the construction of Parking Facility "A" is typical from B1F to 4F, with the building taking up more than two-thirds of the site's total area. Facility "B", shown in the same figure's middle panel, comprises underground parking arranged over two floors, raising questions related to the depth of the soil covering it. Lastly, the modern structure shown in the right panel from B2F to 4Fincludes a higher proportion of green space, in part as a means of achieving better soil water content.

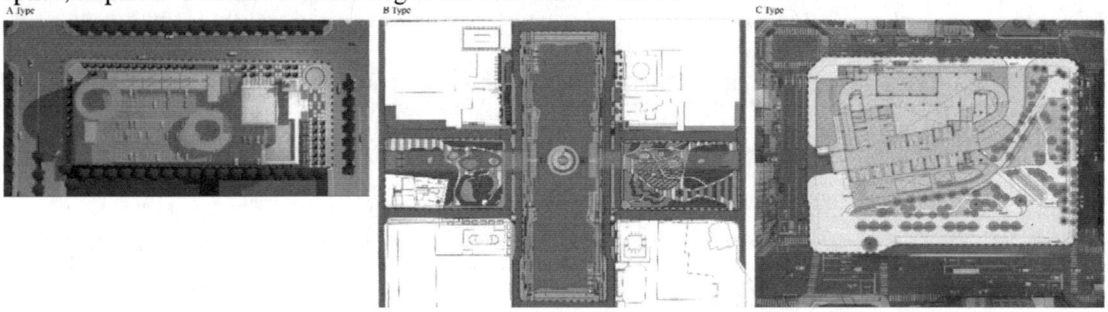

Figure 2. Plan Views of the Three Studied Parking Facilities (Building Cover Shaded in Brown)

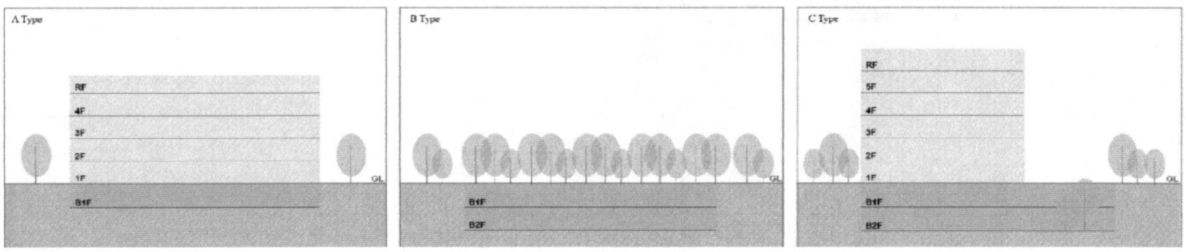

Figure 3. Sectional Views of the Three Studied Parking Facilities

4.2 Vegetation cover and soil water content indices

The soil water content index refers to natural and artificial soil layer's ability to trap and store water inside construction base Vegetation cover is computed as a percentage of the site area in which plants grow and statutory vacant space (Figure 4).

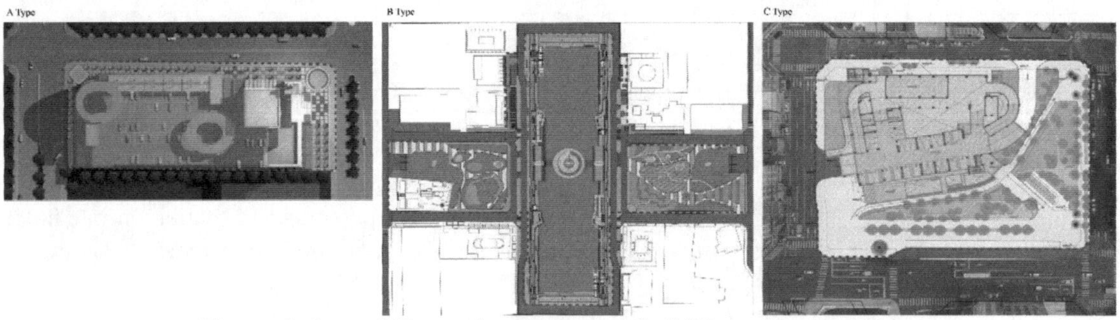

Figure 4. Area of Vegetation Cover and Soil Water Content Index

According to the statistics shown in Table 2, environmental re-design of sites "B" and "C" could reduce their building-coverage ratios, and thus raise vegetation cover to between 67% and 73%. It would also increase soil water content significantly, while providing new open space for public recreation.

Table 2. Percentage of Vegetation Cover and Soil Water Content Index

Dimensions	Traditional parking structure (Type "A")	Underground parking facility (Type "B")	Modern parking facility (Type "C")
Building coverage ratio	62%	4%	33%
Vegetation cover and soil water content index	20%	73%	67%

4.3 Air quality of underground parking facilities

Building technical regulations in Taiwan specify that underground parking areas should have at least two vents or equivalent ventilation equipment. They should also incorporate opening windows, and the airflow in cubic metres per hour cannot be less than one-fifth of the floor area in square meters. Automatic ventilation equipment must be installed if natural air circulation is found to be insufficient.[4] As shown in Figure 5, facilities of type "A" tend to have traditional transom designs, while those of type "B" feature central patios, and type "C", both patios and floor to ceiling windows in the centre and both sides.

Figure 5. Sectional Views of Ventilation

Given the inclusion of basement-level patios and floor-to-ceiling windows of sufficient size, CMH will increase to the point that auxiliary ventilation equipment is needed just one or the other of these things, thus achieving significant energy savings, as shown in Table 3.

Table 3. Comparison of Percentage of Ventilation

	Type "A"	Type "B"	Type "C"
Floor	B1	B1-B2	B1-B2
Basement Floor Area/m²	1,113	43,000	6,393
5% Floor Area / m²	56	2,150	320
Ventilation Rate / CMH	464	17,917	2,664
Size of Window / m²	24	360	465
Patio Area / m²	0	560	470
Ventilation Rate Area / m²	34	920	935
	34 < 56	920 < 2,150	935 > 320
Percentage of Ventilation	7%	5%	35%

4.4 interior lighting

The interior lighting of parking facilities is often neglected, and underground structures in particular are stereotyped as dark and gloomy. This illumination problem can be addressed at the design stage, using solid angle project theory, and factoring in the angles of incidence of sunlight at various times of day and sizes, shapes, and precise placement of windows (Figure 6).

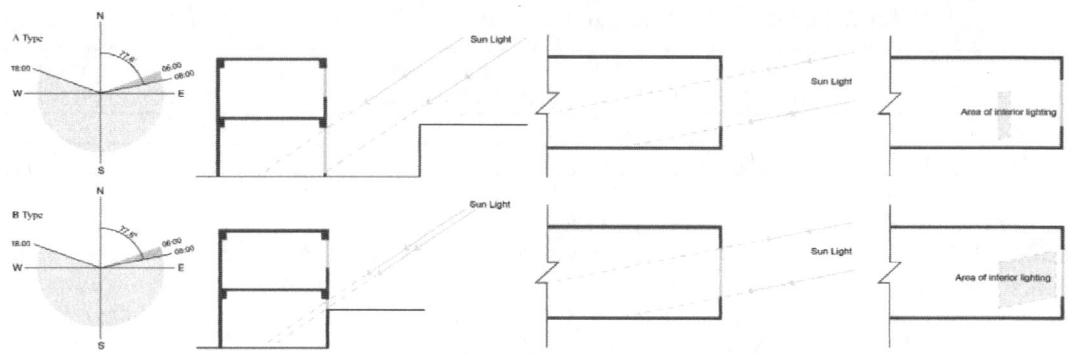

Figure 6. Analysis of Building Interior Lighting

Quantitative solid-angle analysis of type "A" and type "B" parking facilities shows that, when structures include patios and sufficiently large windows, interior lighting rates and illumination can be up to five times greater than in traditional designs with fanlight windows. And in general, increased access to natural light will decrease a given facility's day-to-day need for artificial lighting (Table 4).

Table 4. Analysis of Window Size and Area of Interior Lighting

Dimensions	Type "A"	Type "B"
Window Size	H:60XW:300CM	H:280XW:300CM
Area of Interior Lighting / m^2	2	10

5. Conclusion

Most traditional parking facilities are dark, humid, expensive to maintain, and wasteful of energy, and are increasingly deemed undesirable and scheduled for demolition. The present research has pointed out a range of measures that could render parking facilities more suited to modern environmental needs, and provided preliminary analysis showing that all of these measures – comprising planting, water management, ventilation, and lighting – would have beneficial effects. As such, this research will be of considerable value to policymakers, construction-industry practitioners, and others responsible for the planning and implementation of parking-facility designs.

6. References

[1] Riggs, William (EDT)(2017) Disruptive Transport: Driverless Cars, Transport Innovation and the Sustainable city of Tomorrow.
[2] Shiau, Tzay-An (2018). Evaluating the sustainability of constructing public off-street parking lots.
[3] Sarachanga et al. (2016). Evaluation of existing sustainable infrastructure rating systems for their application in developing countries.
[4] Building Technical Regulations, Taiwan(2017).

IOP Conf. Series: Earth and Environmental Science **291** (2019) 012043 doi:10.1088/1755-1315/291/1/012043

Implementation of a UAV-sensory-system-based hazard source estimation in a chemical plant cluster

Dan Li[1], Feiran Chen[1],*, Yiping Wang[2] and Xiaodong Wang[2]

[1]Graduate School, National University of Defense Technology, 109 Deya Road, Changsha, Hunan, 410073, China

[2]The Naval 902 Factory, Shanghai, 200000, China

*Email: 13317310232@163.com

Abstract. Since there has been growing concern about the damage that atmospheric leakage and dispersion accidents may have done to human beings, researchers are dedicated to study effective and feasible source estimation methods in chemical plant clusters. In this paper, the safety monitoring of chemical production process is conducted via an unmanned aerial vehicle (UAV) monitoring system. Based on the observed data from this system, a source estimation method incorporating Bayesian inference and Particle Swarm Optimization (PSO) is proposed. Furthermore, the method is extended to discuss the source tracing algorithm using game theory in the UAV system. Finally, a practical case study is carried out to verify the feasibility and credibility of the proposed method. Results show that the method and system are helpful for safety monitoring and risk assessment in a chemical plant cluster.

1. Introduction

In chemical industry, plants tend to geographically cluster due to environmental and collaborative considerations [1]. Within such plant clusters, atmospheric by-products generated during the process of chemical production are often discharged to the surroundings without purification. Thus, technologies including air quality monitoring and source location estimation via a mobile inspection resource have been recently used to quickly locate the sources of contaminants and greatly mitigate the influence of dispersion.

The source estimation problem is the inverse of advection and dispersion, of which the methods can be basically divided into forward and backward modeling [2]. Most of them are based on Bayesian inference or optimization. Through Bayesian inference, the emission source can be estimated by calculating the posterior function. Some advanced filters (e.g., particle filter) and optimization methods (e.g., firefly optimization or particle swarm optimization) are also widely used in source estimation problems. Ma et al. [3] used the minimum relative entropy and Particle Optimization (MRE-PSO) method to locate and quantify the diffusion source with Gaussian model. However, these methods are only suitable for scenarios with a constant release rate.

Generally, the observed data of atmospheric pollutants is essential in source estimation. The traditional approaches which support such types of accurate source estimation mainly rely on the static monitoring stations. However, it can be prohibitively expensive to build an adequate number of monitoring stations, and the collection of invalid data often occurs when the limited station resources are located in the upwind of the emission source. As a result, a UAV monitoring system is thus developed to deal with this problem due to its flexibility in architecture and mobility [3].

IOP Conf. Series: Earth and Environmental Science **291** (2019) 012043 doi:10.1088/1755-1315/291/1/012043

The goal of this paper is to monitor and estimate the potential emission sources based on the UAV sensory system and source estimation methods. The UAV system is based on DJI M100 and the gas sensors is used for data collection. With the observed data, the source estimation method combining Bayesian inference and PSO is used to estimate the source parameters (i.e., the emission rate and the emission location). This work falls through providing practical patrolling routes for UAVs herein. Fortunately, according to Tambe et al. [4], game theory has the advantage of using a sound mathematical approach in modeling under limited resources deployment in a multiple stakeholders' situation. Therefore, the method is extended to discuss the source tracing algorithm (e.g., bionic algorithms) with a UAV-system-based game-theory model. Finally, a practical case of chemical plant cluster was conducted.

2. UAV monitoring systems

Due to the fixed location, the measurements by static monitoring stations may be useless. Therefore, the UAV-based sensory system is developed to collect data.

In the past, source estimation methods usually use concentration data from static monitoring stations. Thus, the accuracy of the estimation results largely relies on the distribution and the number of monitoring stations. Generally, insufficient monitoring stations and incorrect locations of monitoring stations may cause major errors, especially when the stations are not in the downwind direction of the contaminant source. The UAV monitoring system used in this paper overcomes the aforementioned demerits of static monitoring stations for its location information and flexible mobility, which significantly increases the probability of successfully finding the emission source.

The monitoring system uses MATRICE 100 as a UAV platform, which can fly around 18 minutes at max. The maximum speed is 22 m/s, and the maximum flight distance is set at 10 km, which meets the requirement of inspecting a typical chemical cluster. The system contains gas sensors, GPS module, serial-USB converter, LTE transmitter, network connection module and a microcomputer. Moreover, the microcomputer (Raspberry Pi 2b) is used for receiving and sending measurements. In terms of data transmission, the aircraft sends measurements to a cloud database via an LTE transmitter.

In order to obtain high-quality data, the flight route is determined by the following basic rules. At the beginning, the aircraft moves in a circuit around the experimental area. When a peak appears in the measurement line (which means that the aircraft has just moved through the contaminant plume), the aircraft moves back and forth to sample the concentration data once again. After turning around several times, sufficient high-quality data have been obtained and the aircraft resumes its path along the original route.

3. Methods

3.1 Bayesian inference model

This section will explain how to use the measurements $d = \{\omega_i\}_{i=1}^{m}$ from the UAV monitoring system to estimate the source parameters (the release location l and the release rate $q(t)$) will be explained. The dataset d is collected by the UAV system at location set $\mathbf{z}$ in wind field $\mathbf{W}$. We assume that the emission source (source parameter $\theta = \{l, q(t)\}$) starts release at time t_{start} and ends at time t_{end}. Thus, the theoretical concentration value $c_z(t)$ at time t and location z can be computed through a dispersion model $c_z(t) = f(t, z, l, q, t_{start}, t_{end})$. In this paper, Gaussian multi-puff model is selected to simulate the dispersion process of gaseous pollutants. In this model, the function f is used to calculate the concentration value of plume wherein the release rate and the wind are variants. Further, the ith record of measurements contains information about location of UAV z_i, time t_i and the value of observed concentration c_i ($\omega_i = \{z_i, t_i, c_i \mid z_i \in \mathbf{z}\}$). As a result, the following formula can be obtained:

$$c_i \approx c_{z_i}(t_i) = f(t_i, z_i, \theta) + e_i$$
$$\approx \sum_{j=1}^{n} q_j \cdot f(t_i, z_i, l, 1, t_{start} + (j-1)\cdot\delta) = \sum_{j=1}^{n} q_j \cdot f_{ij} \quad (1)$$

Among which, the measurement error $\mathbf{e} = \{e_i\}$ obeys the condition $e_i \sim N(0, \sigma_e^2)$; q_j denotes the release rate of the jth puff; n is the total number of puff; and δ is the interval between each puff.
For the dispersion model $c_z(t) = f(t, z, l, q, t_{start}, t_{end})$, it is impossible to obtain analytic solution through inverse derivation because it is a nonlinear equation. Fortunately, we can utilize Bayesian inference to get the posterior probability function (PDF) of θ through the following formula:

$$p(\theta \mid d) \propto p(d \mid \theta)\cdot p(\theta), \quad (2)$$

where $p(\theta \mid d)$ is posterior probability density function, $p(\theta)$ is the prior function and $p(d \mid \theta)$ denotes the likelihood function. Moreover, the likelihood function can be expressed as follows:

$$p(d \mid \theta) \propto \exp\left\{ -\frac{1}{2\sigma_e^2} \sum_i [f(t_i, z_i, \theta) - c_i]^2 \right\}. \quad (3)$$

The key of computing posterior PDF is to determine prior PDF and likelihood function. The prior PDF is usually obtained from historical data or empirical information. Though there is little information about the diffusion source in this paper, we can still use a uniform distribution to denote the prior PDF. As for the likelihood function, since the release rate is hard to estimate accurately, it may be feasible to use the maximum likelihood estimation of θ' ($\theta' = \{l\}$) to substitute θ and estimate $p(\theta' \mid d)$ instead of $p(\theta \mid d)$. According to the maximum likelihood estimation (MLE) method, we assume the release rate $q_m(t)$ that maximizes the likelihood function in formula (3) as actual release rate. In previous research, we utilize ridge regression method to estimate the release rate. As a result, location l is the only parameter that needs to be estimated here. Then we use PSO to find the optimal location with maximum posterior probability.

3.2 *PSO algorithm*

The essence of PSO is to give random solutions to a set of particles initially and find optimal solution through iteratively updating location and velocity of particles. The concrete process is illustrated as follows: (1) first, the positions x_i of each particle are initialized uniformly; (2) then, assign the best known position x_i^p of a particle as its current position x_i^0; (3) If the posterior PDF of x_i^p is greater than that of the best known position x_g, that is, the inequality $p(x_i^p \mid d) > p(x_g \mid d)$ satisfies, the best known position x_g will be updated; (4) randomly initialize the velocity of each particle; (5) keep the particle moving until the number of iteration reaches the given value. The process is shown as below,

$$\begin{cases} v_i^{(t+1)} = w \cdot v_t^{(t)} + c_1 \cdot r_1 \cdot (x_i^p - x_i^{(t)}) + c_2 \cdot r_2 \cdot (x_g - x_i^{(t)}) \\ x_i^{(t+1)} = x_i^{(t)} + v_t^{(t)} \end{cases}, \quad (4)$$

where $v_i^{(t)}$ denotes velocity of the ith particle at t step; denotes location of the ith particle at t step; r_1 and r_2 are random number obeying uniform distribution $U(0,1)$; parameter c_1 and c_2 are accelerate ratio and they are set at 2 here; parameter w is named as momentum term and is set at $2/t$.
At each step, the best-known position of each particle x_i^p would be updated if the condition $p(x_i^{t+1} \mid d) > p(x_i^t \mid d)$ satisfies. Then, if the inequality $p(x_i^p \mid d) > p(x_g \mid d)$ satisfies at each step, the best-known position x_g will be updated. When the iteration ends, the final x_g will be the best estimated location of emission source.

4. Source tracing algorithm using game-theory

The use of UAV monitoring system in aforementioned sections can also be considered as resource monitoring with mobility. However, the ability of the UAV monitoring system and the information of the measurements are not fully utilized. Therefore, we add fluid flow sensors to the UAV system and carry out study in plume-tracing algorithm in a 3D framework.

In this framework, the search strategy of UAVs follows three stages: plume searching, plume tracing and plume source location. However, most recent studies on plume-tracing algorithm neglected the prior background information. Meanwhile, they used to set the mobile vehicles in downwind direction of the source. For these reasons, we believed that there was not solid proof of the current plume tracing algorithms having the ability to locate the emission source in complicated environments or under different initial search positions by these experiments.

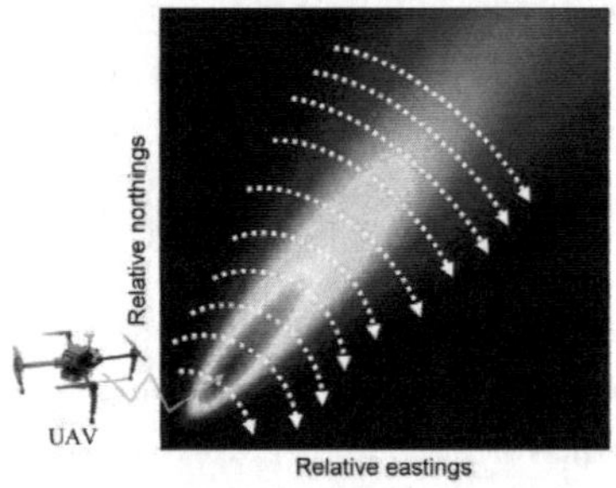

Figure 1. The illustrative tracing trajectory of the UAV.

Schedule patrolling (i.e., a traveling act of mobile resource at different locations and intervals) is one of the applications of game theory in the security domain [7], and it is also introduced in some other domains. Using game theory, the background information of the research area can be modeled into a combined algorithm (the plume-tracing algorithm and the game-theory algorithm). In this way, a robust algorithm can be proposed to fit in with different complicated environments and different initial search positions of UAVs. Figure 1 shows an illustrative tracing trajectory (i.e., the yellow line) of the UAV.

5. Practical case study

To verify and validate the feasibility of the UAV-based monitoring system and the source estimation method, a practical case study of a chemical plant cluster in Shanghai was carried out. The research area is shown in figure 2 and the yellow circles denote potential SO_2 emission spots. Information like coordinates of the potential emission sources is listed in table 1.

Table 1. Information of the potential emission sources

No.	X	Y	Height	Explanation
1	-132.575	-1317.63	50	Waste incinerator for acrylonitrile (AN)
2	-302.901	-1483.42	68	Chimney of sulfuric acid recovery (SAR) system
3	267.1415	0.359916	27	Furnace No.1
4	861.3643	147.0462	27	Furnace No.2
5	1532.017	-142.542	30	Hazardous waste incinerator

The field experiment was conducted at 11 a.m. local time in 23rd July 2017. Based on the field scene, the UAV flight path is represented as the red line in figure 2 and the concentration of SO_2 along the flight path is shown in figure 3. The flight process inspected a higher concentration of SO_2 in the south-east section of the research area, when the emission spot of SAR system was found working at the same time. This information was used as empirical estimation.

Figure 2. The GIS map of research area.

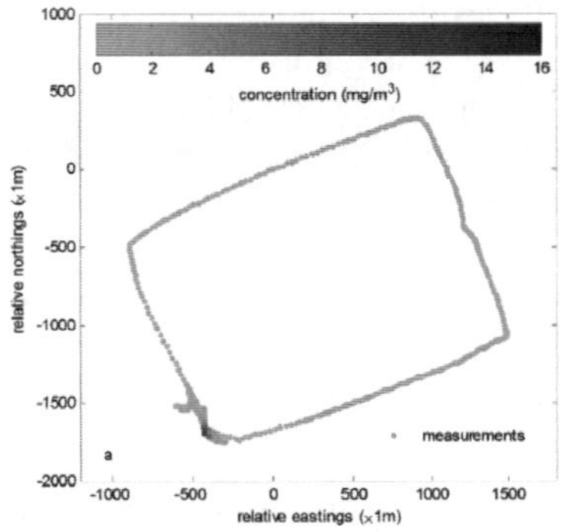

Figure 3. SO_2 concentration along the path.

Based on the SO_2 concentration data collected from the UAV sensory system, the source location was identified by PSO algorithm. Figure 4 exhibits the distribution of particles in step 10 and 15 with the observed concentration data along trajectory. The source location would be estimated by computing the weighted average sum of particles, which was (-305.6, -1445.4). Based on empirical estimation and real locations of potential emission spots, the estimated location was further confirmed as the SAR system.

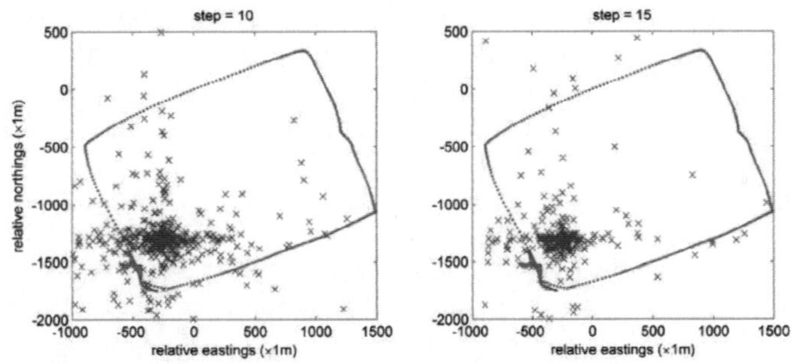

Figure 4. The distribution of particles in step 10 and 15.

6. Conclusions

This paper focuses on safety monitoring and accurate estimation of emission sources in one chemical plant cluster. To tackle these problems, this paper introduced a UAV monitoring system, an estimation method as well as an improved tracing algorithm, and implement them into a practical case. Results of the case study imply that the proposed system and method are able to estimate the source location accurately. In the future, more experiments will be conducted to verify the improved UAV-monitoring-system-based plume-tracing algorithm in more complicated environments.

7. References

[1] G Reniers et al. (2010) An external domino effects investment approach to improve cross-plant safety within chemical clusters. *J. Hazard. Mater* 177, 167-174.

[2] K S Rao. (2007) Source estimation methods for atmospheric dispersion. *Atmos. Environ* 41(33), 6964-6973.

[3] D Ma, S Wang, and Z Zhang. (2014) Hybrid algorithm of minimum relative entropy-particle swarm optimization with adjustment parameters for gas source term identification in atmosphere, *Atmos. Environ* 94, 637-646.

[4] Hutchinson M, Oh H, Chen W H. (2017) A review of source term estimation methods for atmospheric dispersion events using static or mobile sensors. *Inform. Fusion* 36, 130-148.

[5] Gatsonis N A , Demetriou M A , Egorova T. (2015) Real-time prediction of gas contaminant concentration from a ground intruder using a UAV. *IEEE International Symposium on*

IOP Conf. Series: Earth and Environmental Science **291** (2019) 012043 doi:10.1088/1755-1315/291/1/012043

[6] Tambe Milind. (2012) Security and game theory: algorithms, deployed systems, lessons learned, Cambridge University Press.
[7] L Zhang, G Reniers, B Chen, and X Qiu. (2018) CCP game: A game theoretical model for improving the scheduling of chemical cluster patrolling, *Reliab. Eng. Syst. Safe.*

Acknowledgments
This study is supported by National Key Research & Development (R&D) Plan under Grant No. 2017YFC0803300 and the National Natural Science Foundation of China under Grant Nos. 71673292 and Shanghai Special Foundation of Software and Integrated Circuit under Grant No. 150312.

IOP Conf. Series: Earth and Environmental Science **291** (2019) 012044 doi:10.1088/1755-1315/291/1/012044

Taichung City's Management Strategy for Leisure Cycleway Maintenance: A Case Study of Gaomei Cycleway

Yu-Chen Chien[1], Sue Yee Yew[2,a]

[1]Department of Landscape and Urban Design, Chaoyang University of Technology, Taiwan
[2]Graduation program in Landscape and Urban Design, Department of Landscape and Urban Design, Chaoyang University of Technology, Taiwan

[a]angel_yee0317@hotmail.com

Abstract. In an era marked by global warming, energy conservation and carbon reduction are increasingly prioritized by national and local governments, companies and individuals. The bicycle is emerging as an important tool for reducing the overall volume of traffic, protecting the environment and conserving energy, as well as improving individuals' health and relieving their stress. Therefore, environmental demand for cycleway is garnering increased attention. The cycleway built by Taichung City's government include the Gaomei Cycleway, a coastal route famous for its seascapes and sunsets, and surrounded by rich ecological resources. This study uses Gaomei as a case for exploring the city's maintenance and management strategies for its cycleway. Specifically, it looks at the varying demands placed on Gaomei and the system in general by leisure users and commuters. Importance-Performance Analysis (IPA) was used to review bikers' strategy recommendations generated from a questionnaire covering four facets: safety, continuity, service facilities, and environment. The results of IPA indicate that bike commuters give more weight of the safety and continuity facets due to the high frequency with which they use the system, whereas leisure are more attuned to the other two facets.

1. Introduction

The Intergovernmental Panel on Climate Change's (IPCC) fifth Climate Assessment Report [1] estimated that, by the end of this century, global mean temperature will rise by 4.8°C. As greenhouse-gas emissions are closely related to energy consumption and transportation, energy conservation and carbon reduction are increasingly prominent parts of national energy policies. The bicycle is emerging as an important tool for reducing the overall volume of traffic, protecting the environment and conserving energy, as well as improving individuals' health and relieving their stress. As a result, environmental demand for bicycle-related land, including cycleways and storage facilities, has also increased. In Taiwan, Taichung City's government has placed the creation of recreational cycleway at the center of its efforts to provide its citizens, and especially children, with easy access to high-quality outdoor leisure activities.

Surrounded by rich ecological resources, the Gaomei Cycleway is a famous 3.4 km coastal route, passing through Gaomei Wetland. Famed for its coastal scenery and magnificent sunsets, it has become a major tourist attraction within Taichung City, promoted by its Tourism Office and deemed a top-five bicycle route by Daily View Internet Thermometer. Because prior research has found that the quality of cycleway maintenance management is directly correlated with user satisfaction, this study (1)

IOP Conf. Series: Earth and Environmental Science **291** (2019) 012044 doi:10.1088/1755-1315/291/1/012044

explores differences in how bike commuters and general users perceive the importance and performance of the cycleway's environment and facilities; (2) uses Importance-Performance Analysis (IPA) to assess these two groups' recommendations for the maintenance and management of its environment and facilities.

2. Literature review

2.1 Bikeway design and assessment principles

2.1.1 Taiwan institute of transportation

In 2017, the Taiwan Institute of Transportation [2] published its Study of Integration and Evaluation of the Bicycle Lane Network System: a set of design reference principles for the planning of various system facilities associated with cycleways, including road-network infrastructure and ancillary facilities. The manual also contains management and maintenance mechanism for cycleway evaluation.

2.1.2 Research, development and evaluation commission, taichung city

Taichung City's 2017 Bicycle General Inspection Report [3], based on expert consultation and spot-checks of consumer opinions, aimed to understand the actual situation and facilitate further improvements to cycle safety and cycle-route quality at a strategic level. Based on its findings, the present study divides cycleway evaluation into four facets – safety, continuity, service facilities and environment – each comprising four factors, meaning that a total of 16 factors will be considered.

2.2 Performance

User performance is defined as the difference between the consumer's expectation regarding a product or service, and his or her perceptions of it after purchasing/using it.

2.3 Importance-performance analysis

IPA assesses individuals' perceptions of specific projects in terms of those projects' importance and performance, as an aid to efforts to improve user satisfaction and loyalty [4]. Its graphic depiction of users' comment content uses the overall average values of importance and performance as its separation point, with the X and Y axes cutting the space into four quadrants, as follows.

2.3.1 First Quadrant (Keep up the Good Work)

High clustering in this area indicates a highly competitive service, the quality of which should simply be maintained.

2.3.2 Second quadrant (possible overkill)

The service's quality or quantity may be excessive. Resources can be cut back somewhat without the overall experience being diluted.

2.3.3 Third quadrant (low priority)

Problems with the quality of the service are relatively unimportant, insofar as the service itself is not an especially vital one for its users.

2.3.4 Fourth quadrant (concentrate here)

This quadrant reveals the main source(s) of the service's disadvantages and drawbacks. Therefore, its content should be treated as a decisive factor for future development and improvement prioritization.

2.4 Hypotheses

Based on the foregoing literature review, this study was guided by the following two hypotheses （figure 1）:

(H1) The (a) importance and (b) performance perceptions regarding environmental quality will be correlated both with each other, and with the parallel perceptions regarding facilities.
(H2) General/leisure users and commuters will have different perceptions of the importance and performance of the cycleway's environmental quality and facilities.

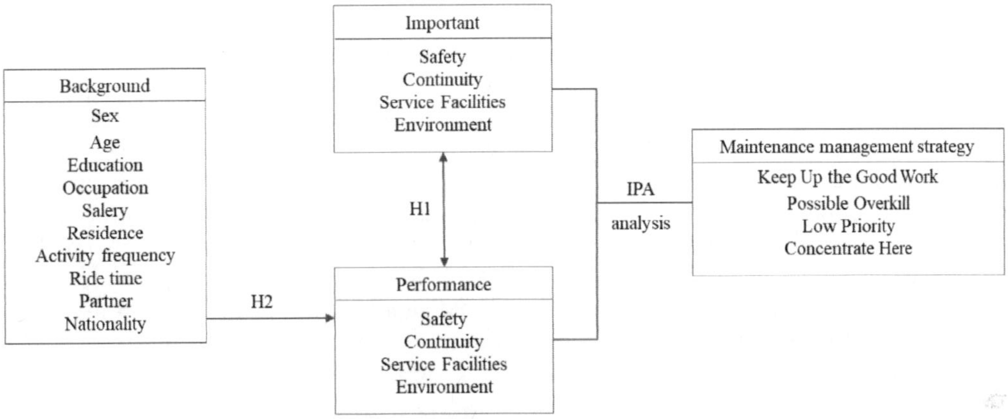

Figure 1. System Architecture

3. Research methods

Based on this study's purpose and the relevant prior literature, the researchers devised a questionnaire to explore the IPA facets of the Gaomei Cycleway. This instrument consisted of three parts, covering the respondent's socio-economic background; his/her perception for importance and performance of the cycleway environmental and facilities' quality. For purposes of analysis, socio-economic background factors were treated as independent variables, and environmental importance and performance as the dependent variables. All items were answered via the same five-point Likert scale, with 1 = "poor" and 5 = "excellent". SPSS Statistics 25 software was used to analyze the results.

Table 1. Reliability Analysis

Facets	Factors	Important Cronbach's α	Reliability	Performance Cronbach's α	Reliability
Safety	4	.784	high	.827	high
Continuity	4	.735	high	.839	high
Service Facilities	4	.761	high	.802	high
Environment	4	.642	high	.757	high
Overall	16	.847	high	.919	high

4. Results and discussion

A total of 201 questionnaires was issued via face-to-face interaction with bikeway users' between 13 and 28 October 2018. The reliability analysis overall are high(table 1).

4.1 Reliability analysis of importance and performance ratings

Reliability analysis yielded overall Cronbach's alpha coefficients of 0.847 for importance and 0.919 for performance, indicating that both parts of the questionnaire were highly reliable.

4.2 Analysis of users' background data

Male users in the sample outnumbered female ones by 12.4%. Most of the respondents were between 19 and 35 years old and highly educated, with students accounting for the single largest occupational category. It is possible that the sample was skewed in favor of individuals such as middle schoolers who felt they had ample time to answer it. Most (88.6%) of the respondents lived in the local area, and

IOP Conf. Series: Earth and Environmental Science **291** (2019) 012044 doi:10.1088/1755-1315/291/1/012044

the proportion living in the central part of Taichung was high, 46.3%. Only 77 respondents were willing to reveal their monthly income, with many of the others citing privacy concerns. In terms of their frequency of cycling, 74.1% stated they did so once per week or less, and 74.7% said they cycled on the Gaomei Cycleway was less than one hour. The majority of users and friends together accounted for 53.2% of partners.

4.3 Gaomei cycleway users' rankings of its importance and performance

The average value the respondents assigned to the Gaomei Bikeway's importance and performance were 4.199/5 and 3.754/5, respectively(figure 2). Under the general rubric of *importance*, the highest scoring factors were safety and the clarity with which the route was marked, while higher-than-average scores were assigned to the cycleway's width, the smoothness of the pavement, route indicators, the continuity of the bicycle route, route signs/marking, route network information, supply facilities, environmental cleanliness and absence of unauthorized vehicle types all scored higher than average. Under *performance*, higher-than-average scores were assigned to included most of the same factors (i.e., all except smoothness of the pavement and supply facilities), plus recreation facilities, bicycle parking space, bike rental sites, and landscape.

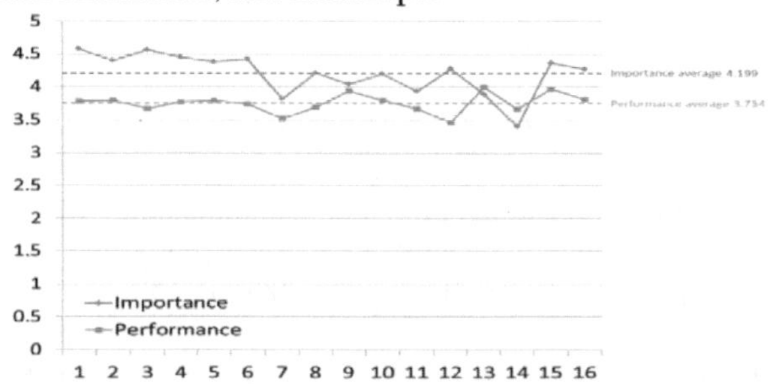

Figure 2. Gaomei Cycleway Users' Rankings of Its Importance and Performance

4.4 Importance-performance analysis: general users vs. commuters

IPA showed that the responses of general users (figure 3)and commuters (figure 4)were roughly similar in both the importance and performance categories. However, the former group tended to rate Gaomei Bikeway's security and clarity, route, route instructions, and bicycle parking spaces as "keep up the good work", whereas commuters saw all of these areas as needing improvement. Additonally, general/leisure users tended to think that bike parking spaces needs to be improved as a first priority, while commuters did not rate this concern as highly. Both groups agreed that the width of the cycleway, the continuity of its route, the cleanliness of the environment, and the exclusion of certain vehicle types were all as expected and should be kept as they were. Likewise, both groups agreed that pavement evenness, route-network information and supply facilitiesshould be improved as a matter of priority, and that route-mileage representations, bike rental sites and local specialties were in need of only minor improvement(table 2).

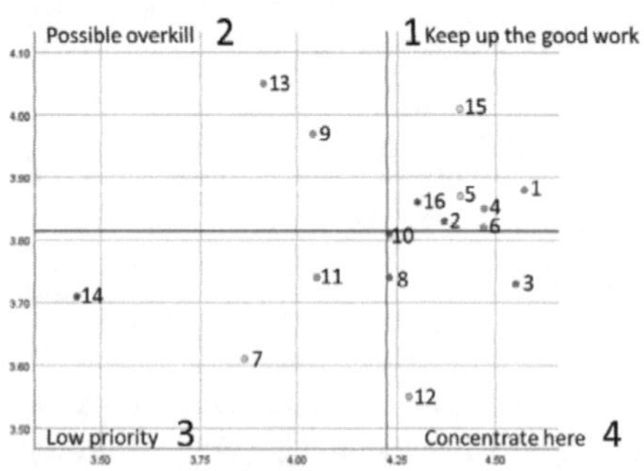

Figure 3. IPA results, general users

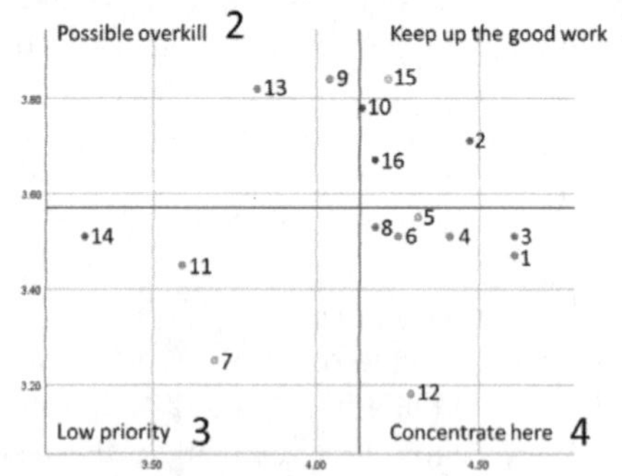

Figure 4. IPA results, commuters

Note: Factor numbers are as given in Table 2.

Table 2. Comparison of IPA strategy for general users and professional riders

Facet	Factor	General users			Bike commuters		
		Importance	Performance	Quadrant	Importance	Performance	Quadrant
Safety	1. Safety and clarity of cycleway	4.57	3.88	1	4.61	3.47	4
	2. Width of cycleway	4.37	3.83	1	4.47	3.71	1
	3. Pavement evenness	4.55	3.73	4	4.61	3.51	4
	4. Route indicators (direction indicators, warning signs, milestones, trails)	4.47	3.85	1	4.41	3.51	4
Continuity	5. Continuity of the route	4.41	3.87	1	4.31	3.55	4
	6. Bicycle route indicators/markings	4.47	3.82	1	4.25	3.51	4
	7. Mileage representations	3.86	3.61	3	3.69	3.25	3
	8. Route-network information	4.23	3.74	4	4.18	3.53	4
Service facilities	9. Recreational facilities (seats, bowers, etc.)	4.04	3.97	2	4.04	3.84	2
	10. Bicycle parking space	4.23	3.81	4	4.14	3.78	1
	11. Bicycle rental points	4.05	3.74	3	3.59	3.45	3
	12. Supply facilities (toilets, first aid equipment, etc.)	4.28	3.55	4	4.29	3.18	4
Environment	13. Landscaping	3.91	4.05	2	3.82	3.82	2
	14. Local specialties (snacks)	3.44	3.71	3	3.29	3.51	3
	15. Environmental cleanliness	4.41	4.01	1	4.22	3.84	1
	16. Exclusion of certain vehicle types	4.30	3.86	1	4.18	3.67	1

5. Conclusion

The results indicate that bike commuters had relatively greater concern about the facets "safety" and "continuity", due to the high frequency with which they used the bikeway[5,6]. Other users paid more attention to "service facilities" and "environment". Both groups agreed that the Gaomei Cycleway's width, the continuity of its route, the cleanliness of the environment, and its rules regarding what vehicles were allowed were all appropriate and should be maintained in the future. In terms of pavement evenness, route-network information, and supply facilities, on the other hand, both user groups felt that the Gaomei Cycleway had insufficient facilities and/or that its management unit should actively seek to improve user satisfaction. The fact that more attention was paid to bike rental points by general/leisure users than by commuters was probably due to the latter group mostly owning their own bikes.

6. References

[1] Intergovernmental Panel on Climate Change's (IPCC) fifth Climate Assessment Report.
[2] Institute of Transportation, Taiwan (2017). A Study of Integration and Evaluation of the Bicycle Lane Network System.
[3] Research, Development and Evaluation Commission [Taichung, Taiwan] (2017). Bicycle General Inspection Report of Taichung City.
[4] Yu-Hsuan Ho (2016). Study on Users' Satisfaction of Taiyuan Greenway in Taichung City. Unpublished master's thesis, Chao Yang University of Technology, Taiwan.
[5] Jhao-Hong Huang (2011). Research on the Service Quality of the Kaohsiung West Side Harbor Line Bike Path.
[6] Yu-Zhen Zhang (2015). Comprehensive Evaluation of Recreational Bikeway Systems.

IOP Conf. Series: Earth and Environmental Science **291** (2019) 012045 doi:10.1088/1755-1315/291/1/012045

A Design Optimization Index for Two Types of Cycleways

Yu-Chen Chien[1], Cheng-Hung Wu[2,a]

[1] Department of Landscape and Urban Design, Chaoyang University of Technology
[2] Graduation program in Landscape and Urban Design, Department of Landscape and Urban Design, Chaoyang University of Technology
[a]ko840813@gmail.com

Abstract. Amid rising environmental awareness, sustainable and low-carbon traffic tools are attracting considerable attention. Increasing worldwide use of bicycles for both leisure and basic transportation has increased the importance of public bicycle-related infrastructure, notably cycleways. This paper proposes a Design Optimization Index for two types of cycleways, i.e., those that are used primarily for commuting, and those that are mostly for leisure. Based on a literature review, the researchers initially developed 27 impact factors. Then, through the fuzzy Delphi method of collecting and refining advice from a group of experts, these were integrated into a five-part structure consisting of "security", "convenience", "environmental friendliness", "comfort" and "attractiveness", which can be further broen down into 16 factors in the case of commuter cycleways, and 15 factors in the case of leisure cycleways.

1. Introduction

Due to the greenhouse effect, global warming, rising oil prices and other related issues, numerous countries are devoting increasing attention to environmental protection and sustainable environmental management. Low-carbon cities and green transportation combine the promotion of energy conservation, carbon reduction, environmental protection and human health. Bicycles, being environmentally friendly and pollution-free, have re-emerged as a popular means of transportation. In Taiwan in recent years, the government has actively promoted a range of pro-bicycle policies, and according to 2017 Ministry of Transport statistics, 54.1% of households own bicycles. The 2006 Qianli Bike Road and Wanli Trail Implementation Plan aimed to promote the active involvement of local governments in the development of local cycleway networks and the integration of these networks into a national system. As of 2015, the total length of Taiwan cycleways reached 5,889 km .

With the aim of aiding the ongoing development of the cycleway system and bicycle-friendly cities in Taiwan, and improving the overall service level experienced by cycleway users, this study provides a detailed examination of the existing and optimal design characteristics of two distinct types of cycleways: i.e., one type tailored for people who commute to work or school, and another designed especially for leisure use by residents and tourists. To understand the environmental and facilities requirements of these different types of bikeway, and to provide for the optimal design and construction of such bikeways in the future, this study will apply the Fuzzy Delphi method to arrive at a design index that includes impact factors for both cycleway types. The research results are expected to support better-informed future bikeway policy and maintenance decisions.

2. Literature review

2.1 Taiwan bikeway types

Bikeways can be divided into three types according to their settings/functions: commuting, leisure, and competition [1]. Among these, the competition type has never been systematically introduced in Taiwan, and races carried out on the existing bikeways would undoubtedly be dangerous [2]. Therefore, this paper excludes competition bikeways from consideration.

2.1.1 Commuter bikeways

For the most part, Taiwan's commuter bikeways are closely integrated into the wider transportation system, and consist of convenient cycle routes aimed at improving urban green transportation networks while meeting the day-to-day travel needs of workers and students. Most of these routes are planned in conjunction with railways, streets, and sidewalks to connect core metropolitan areas with their residential hinterlands.

2.1.2 Leisure bikeways

Leisure bikeways are intended to provide a relaxation function and vital links in Taiwan's sightseeing/tourism network. As a system, they connect various attractions, scenic spots and monuments; encourage people to experience multiple tourist attractions that are in close proximity to one another; and construct a linear space with ecological, cultural, leisure, sightseeing and landscape functions.

2.2 Design index and impact collection

This study reviewed all the current policy literature relevant to the planning, design, and construction of bicycle lanes and bicycle road systems in Taiwan, and the Taichung Cycling Road in particular [1,3,4,5] This yielded a five-part index of such policies' key factors, comprising "safety", "convenience", "environmental friendliness", "comfort" and "attractiveness". This first level can be further subdivided into 27 impact factors relating to the two types of cycle lanes' design, updating, and post-use evaluation.

Table 1. Optimal Cycleway Design Index and Impact Factors

Design Index	Interpretation and Impact Factors
1. Safety	The impact of cycleway planning, design and protection measures on the safety of bicycle users. Its eight impact factors are: 1-1 Route clarity, 1-2 Safety-barrier facilities, 1-3 Lighting, 1-4 Warning flags, 1-5 Emergency contact methods, 1-6 Lane-net width, 1-7 Safety-net height, and 1-8 Drainage.
2. Convenience	The degree to which cycleway planning and design and the surrounding route-information facilities make users' experience easier. Its five impact factors are: 2-1 Mileage signs, 2-2 Guide boards, 2-3 Bike-rental services, 2-4 Accessibility of mass transit, and 2-5 Bicycle route interconnectedness.
3. Environmental friendliness	The extent to which a cycleway's environmental planning and design enable it to coexist with nature. Its five impact factors are: 3-1 Shade/greening, 3-2 Clean environment, 3-3 Bike parking space, 3-4 Exclusion of certain vehicle types, and 3-5 Low-impact development.
4. Comfort	The impact of cycleway planning and design decisions on users' ride comfort. Its five factors impact factors are: 4-1 Bicycle rest station, 4-2 Avoidance of delays, 4-3 Flatness, 4-4 Reduced slope interference, and 4-5 Reduced climate disturbance.
5. Attractiveness	The effect of cycleway planning and design features on cyclists' willingness to ride. Its four factors are: 5-1 Surrounding landscape, 5-2 Peripheral tourism resources, 5-3 Landscape diversity, and 5-4 Air quality.

3. Research methods

In this study, the Fuzzy Delphi method was used to screen the factors. The advantages of this approach to brainstorming are that the experts consulted can remain anonymous, which reduces the potential impact of their group membership and enables them to think independently, and that it has no space constraints. According to Delbecq, Van de Ven and Gustafson (1975) [6], the ideal number of Delphi members is five to nine, with having more or fewer indirectly affecting the accuracy of the results. Therefore, this study conducted an expert questionnaire survey with six respondents, whose professional expertise covered urban planning, urban design, engineering, architecture and landscape design.

The questionnaire items used a 11-point Likert scale; the higher the score from 0 to 10, the more important the impact factor is, according to the respondent's professional experience of each of the two types of cycleway. The statistical data were collected using Microsoft Excel 2010 software according to double-triangle fuzzy and grey-area test methods, yielding Gi values for the degree of expert consensus. The arithmetic means of these Gi values were then used as threshold values for the screening of evaluation factors, and any factors below those thresholds removed from further consideration. The average Gi value of the commuter bikeway factors was 6.55, and of the leisure bikeway factors, 6.92. The post-screening factors and their Gi values are ranked in Table 2, below.

Table 2. Experts' Ranking of Commuter Bikeway Impact Factors

Design index	Impact factor	Expert consensus value (Gi)
Safety	1-4 Warning flags	9.14
	1-6 Lane-net width	8.00
	1-3 Lighting	7.75
	1-8 Drainage	7.67
	1-7 Safety-net height	7.35
	1-2 Safety-barrier facilities	6.92
Convenience	2-4 Accessibility of mass transit	7.75
	2-3 Bike-rental services	7.08
Environmental friendliness	3-2 Clean environment	7.17
	3-3 Bike parking space	6.75
	3-4 Exclusion of certain vehicle types	6.73
	3-1 Shade/greening	6.58
Comfort	4-3 Flatness	8.08
	4-4 Reduced slope interference	7.33
Attractiveness	5-4 Fresh air	7.18
	5-1 Surrounding landscape	6.67

Table 3. Experts' Ranking of Leisure Bikeway Impact Factors

Design index	Impact factor	Expert consensus value (Gi)
Safety	1-4 Warning flags	8.00
	1-6 Lane-net width	7.75
	1-1 Route clarity	7.67
	1-7 Safety-net height	7.33
	1-8 Drainage	7.10
Convenience	2-1 Mileage signs	7.08
	2-3 Bike-rental services	6.92
Environment friendly	3-1 Shade/greening	7.67
	3-2 Clean environment	7.25
	3-5 Low-impact development	7.18
Comfort	4-3 Flatness	7.58
	4-1 Bicycle rest station	7.42
Attraction	5-2 Peripheral tourism resources	7.83
	5-3 Landscape diversity	7.65
	5-4 Fresh air	6.92

4. Results and discussion

4.1 Safety

For both types of bikeway, factors 1-4 (Warning flags) and 1-6 (Lane-net width) were deemed the first and second most important safety factors by the expert panel. Factor 1-5 (Emergency contact methods) was removed from both types of bikeway after screening. In terms of the difference between the two types of bikeway, a higher priority was assigned to commuter bikeways' factors 1-3 (Lighting), 1-8 (Drainage), 1-7 (Safety-net height) and 1-2 (Safety barrier facilities). Due to their frequent use by foreign tourists, meanwhile, leisure bikeways' highest safety priorities included 1-1 (Route clarity).

4.2 Convenience

For commuter bikeways, the experts' consensus was that factor 2-4 (Accessibility of mass transit) was the most important, indicating that the planning of such bikeways should be integrated with mass-transit planning to miximize the efficiency of the overall transportation. Somewhat surprisingly, second most important factor for commuter bikeways was 2-3 (Bike-rental services); in other words, despite 54.1% of Taiwan's households (and a higher percentage of regular commuters) owning their own bicycles, the inconvenience of carrying bikes on mass transit means that bike-rental facilities still have an important role to play for some of these users. For leisure bikeways, the experts ranked 2-1 (Mileage signs) first, followed by 2-3 (Bike-rental services).

4.3 Environmental friendliness
In the environmental friendliness category for commuter bikeways, the expert consensus ranked 3-2 (Clean environment) first, followed by 3-3 (Bike parking space), 3-4 (Exclusion of certain vehicle types) and 3-1 (Shade/greening). For leisure bikeways, the experts placed the highest value on 3-1 (Shade/greening), followed by 3-2 (Clean environment) and 3-5 (Low-impact development).

4.4 Comfort
For both types of bikeway, the experts agreed that 4-3 (Flatness) had the highest importance, i.e., that all bikeways should be solid and smooth, including in the transitional areas between different types of pavement. The comfort factors for commuter bikeways also included 4-4 (Reduced slope interference); and for leisure bikeways, 4-1 (Bicycle rest station). Factors 4-2 (Avoiding delays) and 4-5 (Reduced climate disturbance) were both excluded from both types of bikeways' factors following screening, as being either unimportant or too difficult to achieve.

4.5 Attractiveness
On the attractiveness index for commuter bikeways, the expert consensus was that 5-4 (Fresh air) should be ranked first, on the grounds that such routes should be as free as possible of busy motor vehicle traffic; followed by 5-1 (Surrounding landscape). In the leisure bikeway category, the experts ranked 5-2 (Peripheral tourism resources) most highly, followed by 5-3 (Landscape diversity) and 5-4 (Fresh air).

5. Conclusion
Having established the five main categories of factors critical to the planning and design of Taiwan's two main types of cycleway via a literature review, this study's Fuzzy Delphi approach arrived at 16 key sub-factors for the design and maintenance of commuter bikeways, and 15 for leisure bikeways. These results can usefully inform the design, renewal, and evaluation of both these types of bikeway, notably by reducing construction and maintenance costs through de-emphasizing features of relatively low importance. It is therefore hoped that this work will contribute to the overall improvement of Taiwan's urban green-traffic networks.

6. References
[1] Sports Affairs Council, Executive Yuan (2002). Planning and Construction of Bicycle Road Systems in Taiwan.
[2] Chang, C.-H., Yeh, C.-L., Yang, H. C., and Yang, C.-H. (2012). Recreational and Sport Positioning of Bicycle Paths in Taiwan. *Journal of Sport and Recreation Management*, *9*(1), pp. 155-69. ISSN: 1815-4093
[3] Construction and Planning Agency, Ministry of the Interior (2009). Reference Manual on Cycling Route Planning and Design Principles.
[4] Institute of Transportation (2017). Guide to Planning and Design of Bicycle Lane Systems.
[5] Research, Development and Evaluation Commission, Taichung City Government (2017). Taichung Cycling Road General Inspection Report.
[6] Delbecq, A. L., Van de Ven, A. H., and Gustafson, D. H. (1975). Group Techniques for Program Planning: A Guide to Nominal Group Technique and Delphi Processes. Glenville, IL: Scott Foresman.

IOP Conf. Series: Earth and Environmental Science **291** (2019) 012046 doi:10.1088/1755-1315/291/1/012046

Synthesis and physicochemical properties of epoxidized oleic acid-based palm oil

Mohd Jumain Jalil[1], Mohd Saufi Md Zaini[2], Aliff Farhan Mohd Yamin[1], Intan Suhada Azmi[1], Siu Hua Chang[1], Norhashimah Morad[3], Abdul Hadi[1*]

[1]Faculty of Chemical Engineering, Universiti Teknologi MARA Cawangan Pulau Pinang, 13500 Permatang Pauh, Pulau Pinang, Malaysia
[2]Faculty of Chemical Engineering, Universiti Teknologi MARA Cawangan Terengganu, Kampus Bukit Besi, 23200 Dungun, Terengganu, Malaysia
[3]School of Industrial Technology, Universiti Sains Malaysia

[*]Email: hadi9598@salam.uitm.edu.my

Abstract. Epoxidized oleic acid is often regarded as a high value oleochemical due to the wide range of industrial applications including cosmetics, personal care and pharmaceutical products. Epoxidized oleic acid also extensively chosen by industry due to its advantages on the good performance as intermediate chemicals and can be obtained from renewable sources. Synthesis and physicochemical properties of oleic acid palm oil was analyzed for further reaction of epoxidation. In this study, the epoxidation of oleic acid was investigated with performic acid generated in situ from hydrogen peroxide in H2O2 (30%) with formic acid. The characterization of oleic acid through iodine value, viscosity, acid value, saponification number and refractive index were 130, 146. 190.0, 206.1 and 1.47 respectively. On the other hand, the characterization of epoxidized oleic acid were 109.8, 156, 192.6, 219.9 and 1.38 of the same analysis. The chemical structures of the products were confirmed by 1H-NMR and 13C-NMR spectral analysis.

1. Introduction

Chemical modification of vegetables oil is necessary to overcome the weakness and the limitation of the compound. Epoxidation of palms oils is an important reaction because the epoxides products obtained from these renewable raw materials are widely used every day[1]. Recently, renewable material such as vegetable oil has attracted a great interest due to the environment pollution concerns and increase cost of fossil fuels [2]. The traits of vegetable oil, such as cost-effective [3] and eco-friendly to environment [4], are favourable for the production of epoxide. The epoxidized vegetable oils can be used as a raw material for a broad range of products, from pharmaceutical and plastics industries to paint and adhesives sectors [5]. They are obtained from renewable sources materials and considered as biodegradable, non-corrosive and non-toxic. Due to these reasons, they are useful as alternative products to replace petroleum-derived products that has high level of toxicity. Malaysia has high potential to produce epoxide from crude palm oil since it is abundantly available. The crude palm oil production in Malaysia in the year of 2017 had increases from 17.32 million to 19.92 million[6]. Epoxidized vegetable oil product contains carboxylic acid, which is required in industrial applications such to supply plasticizers and polymers in manufacturing of polyvinyl chloride (PVC). Similarly, the degradation of vegetable oil epoxide (oxiranes) also important process which is significant to industry. Palm oil is one of the vegetable oils that abundantly available in Malaysia and has relatively low in price compare to other vegetable oils. Furthermore, epoxidized palm oil is utilized in the production of polyurethane foam and polyester[7]. Epoxidized palm oil has become a promising reactive intermediate, since the epoxy group can be converted into other functional groups through degradation process due to high reactivity of the oxirane ring. Epoxidation may be carried out to attain many conversions, mainly if the product is to be used for synthetic chemical [8]. On the other hand, epoxidation can be employed to obtain a product in the most efficient way, such aiming at the highest conversion rate but not necessarily at the utmost conversion[9].From the most previous research, it is evidenced that the in situ peracid epoxidation of vegetable oils and consecutive ring opening reactions both having opposite reaction conditions [10]. Rangarajan et al.(1995) [11] in agreement with Campanella et al.(2008) [12] suggested that the optimum reaction condition at moderate temperature (40 – 60 °C) and sulphuric acid as catalyst. On the other hand, Zaher et al.[13] reported that the consequent ring opening took place at temperature more than 80 °C. This contrary behaviour due to the reaction take placed different for both reactions. Besides, Vaibhav et al [14] already cover of epoxidation of Jatropha oil by peroxyacids with containing 44.7% oleic acid and 31.4% linoleic acid. Other than that, studies in the area of reaction condition, optimization for improvement in yield, and selectivity of epoxidized oleic acid are lacking. In addition to that, the investigation on maximizing the selectivity of epoxidized oleic acid still remains a main concern as it will be made as a reference for the future use. The epoxidation of vegetables oil is more complicated than the epoxidation of unsaturated fatty acid vegetable oil since the epoxidized product can be more diverse (i.e., mono-, di- or tri-epoxides) depending on the seed oil origin [15]. Therefore, in the present study studies on epoxidation palm oleic acid was examine for future references.

2. Materials and Method

2.1. Materials

Oleic acid-based palm oil in this study was obtained from Quality Reagent Chemical (QReC)-Rech in its raw material. Formic acid, hydrogen peroxide and sulphuric acid were also obtained from QReC. The expoxidation procedure used was similar to those described in our previous work [16]. Initially, a mass of 100 g of palm oleic acid was mixed with specific molar ratios of formic acid to oleic acid (0.5-1.5) in a 500-mL three-neck round-bottom flask equipped with magnetic stirrer, thermometer, and reflux condenser and then it was immersed in a thermostatic oil bath. Then the mixture was stirred at stirring speed ranging between 50-300rpm at temperature of 75 C. Next, 0.5g of sulfuric acid acts as catalyst was added to the mixture, followed by dropwise addition of 30% hydrogen peroxide for a particular molar ratio of hydrogen peroxide to palm oleic acid.

3. Results and Discussion

Studies on the characteristics of raw material used and the final products obtained in this work are important in selecting suitable feedstocks, reaction system design and epoxide storage requirements. Major physical properties analyzed for both crude oleic acid and epoxidized oleic acid include acid value, viscosity, iodine value, total acid number, refractive index, saponification test and flash point. The results are tabulated in Table 1.

Table 1. Comparison of properties between the raw material and final product.

Properties	Crude Oleic Acid	Epoxidized Oleic Acid
State of Matter	Clear Liquid	White Semi-solid
Iodine value	130	109.8
Viscosity (mPa.s)	7.22	11.1
Acid number (mgKOH/g)	190.0	192.6

Saponification value (mgKOH/g)	206.1	219.9
Refractive index	1.47	1.38

In term of visual analysis, the oleic acid is a crystal-clear liquid, however after epoxidation process took place, the epoxidized oleic acid turned to be white semi solid or slurry at room condition as shown in figure 1. The change of the colour of oleic acid is due to addition of formic acid, sulphuric acid and hydrogen peroxide to form epoxidized oleic acid-based palm oil as shown in Equation 1 and Equation 2

$$FA + H_2O_2 \rightleftarrows PFA + H_2O \quad (1)$$

$$PFA + OA \dashrightarrow Epoxy + FA \quad (2)$$

The viscosity of the epoxidized oleic acid was significantly higher compared to the crude oleic acid by approximately a factor of 2. A The change in viscosity of the final products may be explained by the fact of its physical appearance which is semi solid behavior. Viscosity is an essential parameter to be considered during production, transportation and storage. Viscosity increases the molecular weight and this was proving in the data obtained for viscosity 7.22 to 11.2 after modification. For large organic molecules formed by combining a number of smaller molecules, viscosity is generally higher with higher molecular weights, where bigger molecules lead to higher viscosity. Another important property is the total acid number. It is a chemical measurement which represents the acidity of the fatty acid calculated from the mass of base material used to neutralize the acid. In this study, the total acid number is determined by the amount of potassium hydroxide (measured in milligrams) required to neutralize the acid in one gram of fatty acid. It serves as important measurement to the extent of degradation and reaction. Epoxidized oleic acid has higher acid number than crude oleic acid due to the addition of carboxylic group (peracid) in a chemical compound. Overall, the physical properties of presented above for both crude oleic acid and epoxidized oleic acid are comparable to the work of Derawi and Salimon (2010) [17], Jamaludin (2012) [18] and Yee (2017) [19] found that the iodine value palm oil is 98.64 and 96.44 gI_2/100g sample. While for epoxidized palm olein reported by Derawi and Salimon (2010) the values for iodine value and total acid number were 17 gI_2/100g sample, and 170 mgKOH/g respectively.

The iodine value analysis provides an insight view of the reactivity and degree of unsaturation of the fatty acids. The iodine value (IV) is measure of the number of double bonds, while the oxirane value is an indication of the percentage content of epoxide group [20]. The quality of the epoxidized oil is better when there have higher oxirane value and lower iodine value. Higher iodine value results in more reactive and less stable of the fatty acid. From a theoretical viewpoint, higher iodine value indicates a higher probability for the iodide ions to attach to the unsaturated carbon chains [21]. The analysis showed that iodide ion tends to attach to oleic acid is quadruple higher than epoxidized oleic acid. Thus, It is hypothesized that there were less unsaturation sites available in the former structure. It can thus be suggested that most of the available unsaturated C=C double bond in the crude oleic acid has been broken and converted into an oxirane ring structure. The saponification value of the oil was found to be 219.9 saponification value indicates the average molecular weight of the oil. A high saponification value indicates that the oil contains higher proportions of low molecular weight fatty acid [22]. The saponification value of the epoxidized palm oil of 219.9 is higher than oleic acid of 206.1. This might be due to increase in molecular weight of epoxidized palm oil, since the molecular weight decreases with increasing saponification value. This argument is consistent with other study made by Ikyenge [23].The refractive of oil was found to be 1.47 all conform values the values given by literature for others vegetable oil. This is an indication that the oil extracted is high degree of purity.

In this study refractive index of epoxidation of palm oil is 1.38 (Table 1) which is lower that oleic palm oil acid due to decrease in degree of unsaturation caused by epoxidation. The refractive index increases with degree of unsaturation [24].

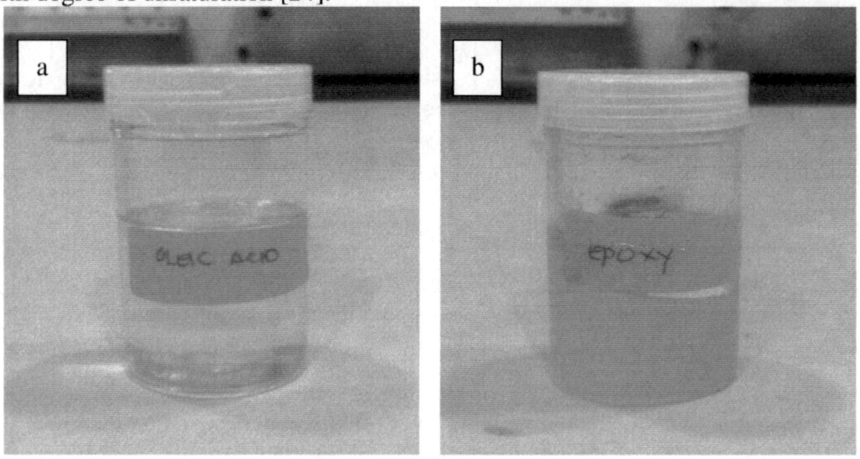

(a) Oleic acid-based palm oil (b) Epoxidized oleic acid-based palm oil

Figure 1. Physical changes of oleic acid

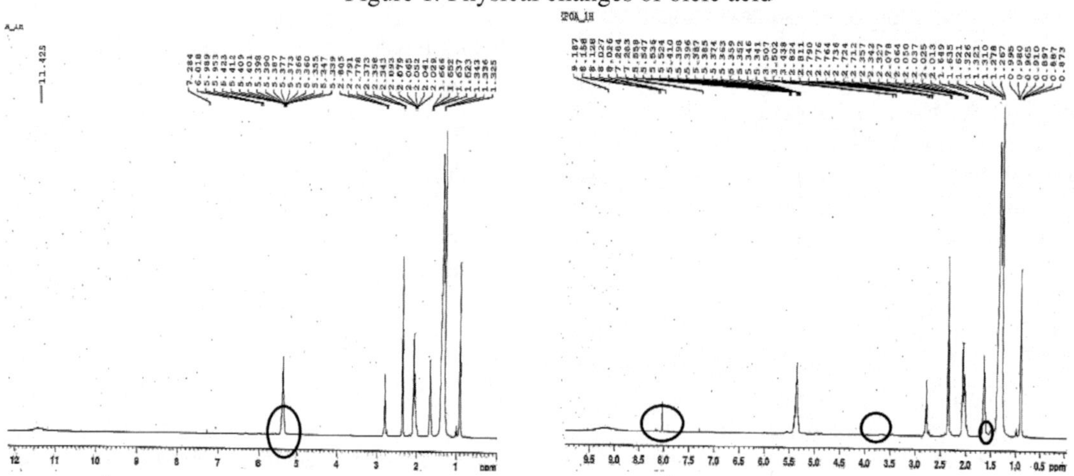

(a) ^{1}H-NMR for oleic acid (b) ^{1}H-NMR for epoxidized oleic acid

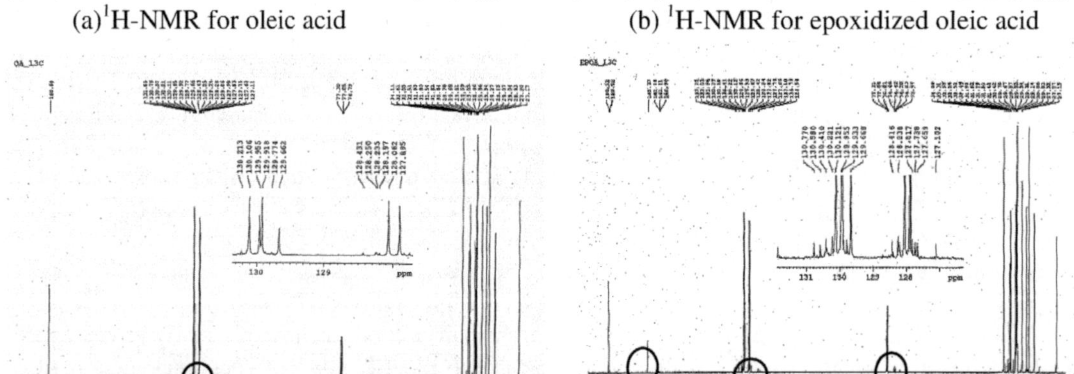

(c) ^{13}C-NMR for Oleic acid (d) ^{13}C-NMR for Epoxidized Oleic acid

Figure 2. Nuclear Magnetic Resonance (NMR) Spectral Analysis.

The formation of the epoxide adduct of oleic acid was confirm by H-NMR and C-NMR spectra, as shown in Fig. 2 H-NMR measurement for epoxidized oleic acid indicate that epoxy group is present at the 1.4-1.5 ppm (-CH_2-CHOCH-CH_2) region which are not present in spectrum for oleic acid. The signals of epoxidized oleic acid at 3.5 correspond to protons on the oxirane ring (-CHOCH-). From the Figure, the signal at 5.31-5.32 ppm (-CH=CH-) corresponding to double peak in oleic acid. It can be seen clearly a decrease of the peak area at 5.4-5.32 ppm (double bond) whereas peaks arising at 3.4-3.5 ppm correspond to CH proton attached to the oxygen atoms of both the epoxy group in Fig. 2. (c). This result indicates that double bond in oleic acid has been converted to epoxy group during epoxidation reaction. The apparent of multiple peak at 8.1-9 ppm region (HCOOC-) show the formation of new formula ester due to ring cleavage by formic acid. As shown in Fig. 2. (d), the peaks at 130.2-129.73 ppm representing the double bond of oleic acid. On the other hand, the epoxy peaks in epoxidized oleic acid were appeared at 76-77 ppm and 75 ppm. According to the C-NMR spectrum, less amount of epoxy group has been formed due to low signal intensity of epoxy peak. The peak at 161 ppm is attributed to the carbon of new ester whereas, multiple peaks at 70 and 73 ppm are the peaks of hydroxyl group. All these products show the side reaction of formic acid with the epoxy group which produces unwanted new ester.

4. Conclusion
The epoxidized yield from oleic acid-based palm oil has huge potentials in the manufacture of oleochemicals. Based on the physical properties results in this study, it is proved that the oil which is renewable raw material has a potential in replacing the non-renewable petroleum-based material as chemical precursors.

5. References
[1] Y.X. Miao, J.P. Liu, Epoxidation of Soybean Oil under Acid-Free Condition, Adv. Mater. Res. 881–883 (2014) 140–143. doi:10.4028/www.scientific.net/AMR.881-883.140.
[2] M.J. Jalil, A. Farhan, M. Yamin, S.H. Chang, I.S. Azmi, N. Morad, Selective Epoxidation of Crude Oleic Acid-Palm Oil with in situ Generated Performic Acid, 7 (2018) 152–155.
[3] A. Farhan, M. Yamin, M.J. Jalil, I. Adnan, V.G. Vsiduru, Numerical Modeling of Epoxidation Palm Kernel Oil Based Oleic Acid, 7 (2018) 123–126.
[4] M.J. Jalil, A.H. Ahmad, I.S. Azmi, Recovery of Gold in Solution from Electronic Waste by Di (2-Ethylhexyl) Phosphoric Acid, 20189 (2018). doi:10.1063/1.5066830.
[5] A. Rafizan, M. Daud, N. Mohamed, J. Jalil, S.K. Jamaludin, Formation of Dihydroxystearic Acid from Hydrolysis of Palm Kernel Oil Based Epoxidized Oleic Acid, J. Appl. Sci. Agric. J. Appl. Sci. Agric. 9 (2014) 86–92. www.aensiweb.com/JASA (accessed March 1, 2018).
[6] POINTERS: Malaysia's Palm Oil Supply and Demand for 2018, (n.d.).
[7] Z. Salem, A. Razak, A. Arsad, Mechanical Properties of Epoxidized Palm Oil / Epoxy Resin Blend, (n.d.) 1–4. doi:10.4028/www.scientific.net/AMM.695.655.
[8] M.J. Jalil, I.S. Azmi, A.R. Mohammad Daud, An Overview of Epoxidation of Vegetable Oils with Peracid- Reaction Mechanism, Recent Innov. Chem. Eng. (Formerly Recent Patents Chem. Eng. 10 (2017). doi:10.2174/2405520410666170614113317.
[9] S. Sun, G. Yang, Y. Bi, H. Liang, Enzymatic epoxidation of corn oil by perstearic acid, JAOCS, J. Am. Oil Chem. Soc. 88 (2011) 1567–1571. doi:10.1007/s11746-011-1820-1.
[10] M.J. Jalil, S.K. Jamaludin, A. Rafizan, M. Daud, DEGRADATION OXIRANE RING KINETICS OF EPOXIDIZED PALM KERNEL OIL- BASED CRUDE OLEIC ACID, 12 (2018) 296–299.
[11] B. Rangarajan, A. Havey, E.A. Grulke, P.D. Culnan, Kinetic Parameters of a Two-Phase Model for in situ Epoxidation of Soybean Oil, 72 (1995) 1161–1169.
[12] A. Campanella, C. Fontanini, M.A. Baltanás, High yield epoxidation of fatty acid methyl esters with performic acid generated in situ, 144 (2008) 466–475. doi:10.1016/j.cej.2008.07.016.
[13] F.A. Zaher, M.H.E.M.M. Ei-hefnawy, Kinetics of Oxirane Cleavage In Epoxidized Soybean

Oil, 66 (1989) 698–700.

[14] Epoxidation of Jatropha (Jatropha curcas) oil by peroxyacids, (2010) 346–354. doi:10.1002/apj.

[15] M. Jumain, N. Mohamed, S. Khatijah, A. Som, A. Rafizan, M. Daud, S. Alam, Epoxidation of Palm Kernel Oil-based Crude Oleic Acid, 906 (2014) 125–130. doi:10.4028/www.scientific.net/AMR.906.125.

[16] V. V Goud, A. V Patwardhan, N.C. Pradhan, Studies on the epoxidation of mahua oil (Madhumica indica) by hydrogen peroxide, 97 (2006) 1365–1371. doi:10.1016/j.biortech.2005.07.004.

[17] D. Derawi, J. Salimon, Optimization on Epoxidation of Palm Olein by Using Performic Acid, E-Journal Chem. 7 (2010) 1440–1448. doi:10.1155/2010/384948.

[18] S. Khatijah, J. Universiti, T. Mara, H. Abu, H. Malaysian, P. Oil, S. Universiti, T. Mara, M. Palm, O. Board, A.P. View, S.M. View, Application of Taguchi Optimization Method in Improving the Selectivity of Dihydroxystearic Acid, (2015). doi:10.4028/www.scientific.net/AMR.1113.703.

[19] A.H.N. Armylisas, M.F.S. Hazirah, S.K. Yeong, A.H. Hazimah, Modification of olefinic double bonds of unsaturated fatty acids and other vegetable oil derivatives via epoxidation : A review, 68 (2017) 1–11.

[20] A.M. Shoaib, E.A. Ismail, F. Modather, Optimum Operating Conditions for Epoxidation Reaction of Jojoba and Castor Oils, 4 (2014) 816–822.

[21] M.J. Jalil, S.K. Jamaludin, I.S. Azmi, Characterisation of Epoxidation of Palm Kernel Oil based on Crude Oleic Acid, (2017).

[22] S. Leveneur, A. Ledoux, L. Estel, B. Taouk, T. Salmi, Chemical Engineering Research and Design Epoxidation of vegetable oils under microwave irradiation, 2 (2014) 1495–1502.

[23] B.A. Ikyenge, J.T. Ageh, B.A. Anhwange, D.T. Nyiatagher, Synthesis and Characterization of Vegetable Oil-Based Polyol from (Thevetia Nerrifolia) Seed Oil, 1 (2012) 55–59.

[24] C. Young, Studies on the Synthesis and Characterization of Epoxidized Soybean Oil Studies on the Synthesis and Characterization of Title : Studies on the Characterization of Epoxidized Soybean Oil Epoxidized Soybean Oil Authors :, (2017).

Acknowledgments

The authors also gratefully acknowledge use of services and facilities Universiti Teknologi MARA, Malaysia.

Start-Up and dynamic processes simulation of SOFC-MGT hybrid power system

Kangjie Deng[1,2], Chen Yang[1,2,*]

[1]Key Laboratory of Low-grade Energy Utilization Technologies and Systems, Chongqing University, Ministry of Education, Chongqing 400030, P.R. China
[2]School of Energy and Power Engineering, Chongqing University, Chongqing 400030, P.R. China

[*]Email: yxtyc@cqu.edu.cn

Abstract. For the start-up process of a hybrid system consisting of a Micro Gas Turbine (MGT) and a Solid Oxide Fuel Cell (SOFC), the surge of the MGT is largely destructive. The MGT, which has a small thermal inertia, can start quickly. However, due to the large thermal inertia of SOFC, starting up SOFC quickly is easily malfunction. Therefore, fast and safe start-up process investigation of the coupling hybrid system is of great challenging and significance. This paper developed a dynamic model of the SOFC-MGT hybrid power systems and studied the rapid start-up process in the case of avoiding gas turbine surge and protecting SOFC by adding other afterburning, auxiliary systems, and bypass valves. The result shows that the bypass from compressor exit to atmosphere can effectively avoid system unstable and the strategy of starting the MGT and SOFC respectively is reasonable. The start-up process simulation provides a reference to the actual SOFC-MGT hybrid power systems.

1. Introduction

Solid oxide fuel cell (SOFC) is a promising technology for generating power with low environment influences. When using hydrogen as fuel, the SOFC system can generate power with zero pollutant emissions. Numerous studies[1-5] have proven that the efficiency of pressurized SOFC is higher than the one works at atmospheric pressure. For improving the efficiency of SOFC, an additional compressor, which is used to raise the pressure of SOFC cathode, is needed. Simultaneously, the SOFC works in a high temperature condition and the fuel utilization is generally below 100%. The high temperature exhaust of the SOFC can be utilized to push the turbine after burning. Therefore, the MGT, which has a compressor, a combustor, and a turbine, is considered to coupling with SOFC[6]. The MGT can provide pressurized air to the cathode of SOFC and can utilize the high temperature exhaust of SOFC. The efficient of SOFC-MGT hybrid power system is higher than independent SOFC power system[7]. However, the thermal inertia difference between MGT and SOFC is huge. The time constant of MGT is second order, but the time constant of SOFC is minute order, sometimes even hour order. The coupling characteristic of the hybrid power system decides the start-up process be complicated.

Some researchers and institutions have done a lot of work about the SOFC-MGT hybrid power system dynamic characteristic and performance analysis[8-12]. U.S. Department of Energy (DOE)/National Energy Technology Laboratory (NETL) has done numerous work about SOFC-MGT[6, 12-14]. The start-up process of the hybrid system also has been investigated[12]. However, the hybrid power system is developed using hardware in the loop simulation. The MGT is an actual system and the SOFC is simulated by a virtual model and some actual tanks and valves. The start-up process is starting up MGT to steady state, adjusting the SOFC model to a nearby boundary condition, and shifting the control signal of fuel valve from MGT to the SOFC model. Therefore, the start-up process of this hardware in the loop simulation was different from the actual SOFC-MGT hybrid power system. Petruzzi[15] developed a three-dimensional geometry code of SOFC to investigate the safe start-up process of small scale SOFC stacks. The result showed that small scale SOFC stacks can start up in 30 minutes without destructing the fuel cell. Some literature[9, 16, 17] have developed the model of gas turbine, including the Capstone MGT C30. This model can investigate the dynamic process of the MGT and can simulate the start-up process of the MGT. Liu[18] built a model of MGT/molten carbonate fuel cell (MCFC) hybrid power system and investigated the start-up process of MCFC-MGT hybrid power system by adding some auxiliary systems. However, this work assumed that the MGT had been started up. Only the start-up process of MCFC and the coupling characteristic was studied. However, when the micro gas turbine starts up, the compressor surge is extremely easy to happen. Some researchers[19, 20] have done a lot of work to avoid surge. The results prove that fast start up is an effective strategy and reducing the pressure ratio of MGT can also keep the MGT safe.

This paper developed a dynamic model of the SOFC-MGT hybrid power system. The MGT subsystem was built based on the prototype of Capstone MGT C30. The SOFC model was developed by referring to the NETL SOFC model. For the SOFC-MGT hybrid power system, in order to avoid the compressor surge during the starting process, to avoid the overheating of the gas turbine inlet and to prevent the temperature of the fuel cell from rising too fast, a supplemental combustion device and an air bypass are added. Then the steady-state validation was carried out. Finally, the safe and fast start-up process simulation of MGT-SOFC was investigated.

2. Mathematical model

2.1 Unit description

The model presented in this paper, the major components of the SOFC-MGT hybrid power system were simulated, including the compressor, heat exchanger, SOFC, combustor, turbine, bypass, etc. The diagram of the SOFC-MGT hybrid power system is shown in Fig.1. When the system is normal operation, the atmospheric air is pressurized by the compressor. Then the pressurized air goes through the heat exchanger, in which the heat of exhaust is retrieved. The heated air flows to the SOFC cathode. The exhaust of the SOFC, which include large heat and part available fuel, goes to the combustor. Next, the flue-gas flows to the turbine to generate power and drive the compressor. Then the exhaust of turbine transfers heat to the low temperature air at the heat exchanger.

In order to start up the hybrid power system, the bypass system is critical. At the initial stage of the start-up process. The MGT and the SOFC should be started up respectively. If all components are started at the same time, the temperature of the SOFC increases quickly. Therefore, the service life of SOFC stack decreases rapidly. So, the bypass valve1 and the SOFC cathode valve can separate SOFC from MGT. And the SOFC temperature can be adjusted by the bypass valve 1. The bypass valve 2 can make the hybrid system more stable by discharging pressurized air to atmosphere.

2.2 Basic equations

The SOFC-MGT hybrid power system model is developed by using the lumped parameter method. The conservation equations of the gas, which are mass, momentum, and energy conservation equations, are simulated by the homogeneous model. The model considers velocities and temperatures in pipe and heat exchanger. Therefore, it is enough that using the homogeneous model simulates the gas flow of the hybrid system.

Mass balance: $\frac{\partial \rho}{\partial t}+\frac{\partial(\rho u)}{\partial z}=0$ (1)

Momentum balance: $$\frac{\partial(\rho u)}{\partial t}+\frac{\partial(\rho u^2)}{\partial z}+\frac{\partial p}{\partial z}=\rho g_z+F_w \tag{2}$$

Energy balance: $$\frac{\partial(\rho h)}{\partial t}+\frac{\partial(\rho uh)}{\partial z}=\frac{\partial p}{\partial t}+Q_w \tag{3}$$

In these equations, Q_w represents heat transferred through walls and F_w is a friction force between the fluid and the wall surface. The space discretization and time discretization are applied to the partial differential equations. The staggered discretization scheme is utilized in the space discretization. Then, in the different space of the mesh, the different parameters are solved. For example, the state parameters are calculated in the middle of the mesh. At the junction of different mesh, the flow related parameters are solved. When calculating the enthalpy, the first order upwind scheme is utilized. Meanwhile, the implicit method is employed in the temporal discretization. The linear equations groups of pressure, void fraction and enthalpy is solved in order[21].

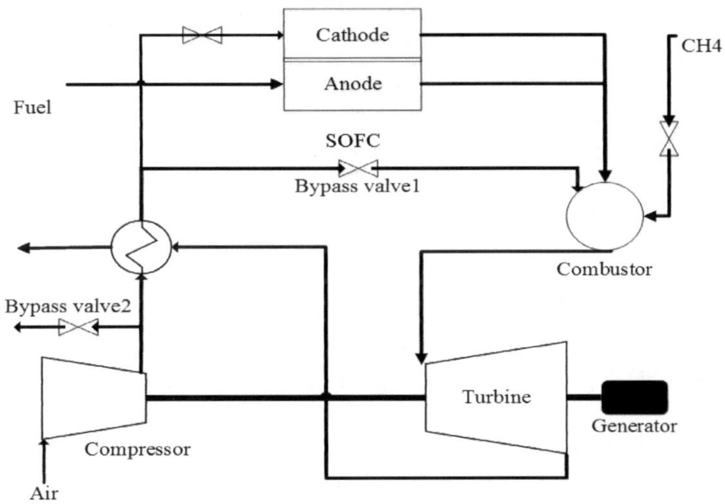

Figure 1. SOFC-MGT hybrid power system diagram.

2.3 Compressor model

The ideal process of the compressor is a definite entropy process. But due to factors such as frictional heat generation, the thermal process in the compressor is more complicated, so the entropy process needs to be corrected to some extent. The adiabatic efficiency of the compressor is introduced, and the adiabatic efficiency is the power consumption ratio of the isentropic process of the compressor adiabatic and the adiabatic entropy process:

$$\eta_{C,S}=\frac{w_{C,S}}{w_C} \tag{4}$$

The power consumption of the ideal process of the compression is:

$$w_{C,S}=\frac{1}{\gamma-1}R_gT_1\left[1-\left(\frac{p_2}{p_1}\right)^{\frac{\gamma-1}{\gamma}}\right] \tag{5}$$

Where γ is the specific heat capacity ratio and R_g is the general gas constant.

Compressor outlet temperature is calculated by:

$$T_2 = T_1\left(\frac{\pi_c^{\frac{\gamma-1}{\gamma}} - 1}{\eta_{C,S}} + 1\right) \tag{6}$$

Where π_c is the compressor pressurization ratio.

2.4 Turbine model

The turbine model mainly includes two parts: pressure calculation and enthalpy calculation.
There are different outlet pressures at different flows in the turbine. For this phenomenon, the Stodola formula is used to calculate the front and back pressure of the turbine model. The Stodola coefficient is calculated as follows:

$$K = m\cdot\sqrt{\frac{p_1 v_1}{(p_1)^2-(p_2)^2}} = m_0\cdot\sqrt{\frac{p_{10}v_{10}}{(p_{10})^2-(p_{20})^2}} \tag{7}$$

Where K is the Stodola coefficient, m is the mass flow through the turbine model, and p_1, p_2 is the model inlet and outlet pressure, respectively. v_1 is the model inlet specific volume. The subscript o is parameters under ideal conditions. From the above formula, if the inlet and outlet pressure are known, the mass flow through the model can also be obtained:

$$m_c = K\cdot\sqrt{\frac{(p_1)^2-(p_2)^2}{p_1 v_1}} \tag{8}$$

Practical form loss factor is calculated as follow:

$$k = \frac{2}{\left[K^2 r(1+\frac{p_2}{p_1})\right]} \tag{9}$$

For gas turbines, the working fluid is gas. The model enthalpy difference is:

$$H = m(h_1 - h_2) \tag{10}$$

2.5 SOFC model

For the lumped parameter SOFC model, the voltage of the fuel cell is calculated as follow:

$$V_{cell} = E_{Nernst} - \eta_{conc} - \eta_{act,an} - \eta_{act,ca} - \eta_{ohmic} \tag{11}$$

Where E_{Nernst} is the Nernst voltage. The remaining terms are the polarization losses. The Nernst voltage is given by:

$$E_{Nernst} = -\frac{\Delta G^0_{H_2O}}{2F} + \frac{R_u T}{2F}\ln\left(\frac{p_{H_2} p_{O_2}^{0.5}}{P_{H_2O}}\right) \tag{12}$$

Where the standard potential is E^0. For the model, an equation is established for E^0 as a function of temperature by fitting an equation to the standard change in Gibbs energy for formation of water reaction. The equation for the standard potential voltage is:

$$E^0 = 1.2877 - 0.0002904T \tag{13}$$

The calculation of the gas spices and the concentration polarization calculation refer to the literature[22].

3. Model validation

The model of SOFC-MGT hybrid power system was simulated with the commercial software package APROS developed by VTT Finland. To evaluate the model, the steady state of MGT was validated with the design value. The comparison between the design values and the simulation results is shown in table 1. The design and physical parameters of the fuel cell are based on articles by Aguiar[22]. And the SOFC parameters are compared with the literature[22], which is shown in table 2.

The MGT parameters comparison between the design value and the simulation results shows a good agreement. And the simulation of SOFC also is similar to the literature[22]. Therefore, the model of MGT and SOFC can be used to develop the model of SOFC-MGT hybrid power system.

Table 1. The MGT parameters comparison between the design value and the simulation result.

Parameter	Simulation result	Design value	Unit	Relative error
Compressor inlet_temperature	15	15	°C	0.00%
Compressor outlet_temperature	157.6	153	°C	3.01%
Compressor inlet_pressure	0.97	0.97	bar	0.00%
Compressor outlet_pressure	3.1	3.1	bar	0.00%
Compressor flow rate	0.31	0.31	kg/s	0.00%
Fuel inlet_flow rate	0.0024	0.0024	kg/s	0.00%
Fuel inlet_temperature	25	25	°C	0.00%
Combustor outlet_temperature	822.2	840	°C	-2.12%
Combustor outlet_pressure	2.89	2.89	bar	0.00%
Turbine inlet_temperature	822.2	840	°C	-2.12%
Turbine outlet_temperature	615	620	°C	-0.81%
Power	29.63	30	MW	-1.23%

Table 2. The SOFC parameters comparison between the literature and the simulation results.

Parameter	Simulation result	literature	Unit	Relative error
Cathode inlet flow rate	0.0013	0.0013	kg/s	0.00%
Cathode inlet temperature	800	800	°C	0.00%
Anode inlet mass flow rate	0.00012	0.00012	kg/s	0.00%
Anode inlet inlet temperature	927	927	°C	0.00%
H_2 mole fraction	0.5323	0.5323	/	0.00%
H_2O mole fraction	0.1344	0.1344	/	0.00%
CO_2 mole fraction	0.3333	0.3333	/	0.00%
Current density	0.85	0.85	A/cm^2	0.00%
Voltage	0.712	0.706	V	0.85%
Cell temperature	838	840	°C	-0.24%

4. Simulation results and discussion

The start-up of the hybrid system is a complicated process. The thermal inertia of the SOFC is different from MGT. To protect the SOFC stack, the start-up processes of SOFC and MGT are separated. Firstly, the MGT needs to be started to the state that its rotation speed reaches nominal speed. Then, the high temperature and pressurized air of heat exchanger outlet flows into SOFC cathode to heat SOFC.

The start-up process of the MGT is a process in which the unit accelerates from shutdown to rated speed, from cold state to hot state. In the start-up process, the parameters in the start-up process should be controlled to maintain a safe range. Before the gas turbine is ignited, the unit is started by the motor to gradually increase the speed of compressor and turbine. After the purge phase, the rotation speed of the gas turbine, which is driven by the starter motor, gradually increases. When the speed reaches $20\%n_0$, the gas turbine starts to ignite. Then, when the rotation speed reaches the trip speed (60% n_0),

IOP Conf. Series: Earth and Environmental Science **291** (2019) 012047 doi:10.1088/1755-1315/291/1/012047

the driving source of the MGT switches from the starter motor to the turbine. Therefore, after 60%n_0, the kinetic energy of MGT is provided by the power of the turbine. When the rotation speed of the gas turbine reaches the rated speed, the SOFC prepares to be heated. The rotation speed of the MGT start-up process is shown in Fig.2.

After the rotation speed of MGT reaches the nominal speed. The pressure rate of compressor keeps a small range variety. The SOFC cathode valve gradually open. Meanwhile, the bypass valve 1 gradually close to avoiding compressor surge and ensuring the air flow of the system keep constant. The position of the SOFC cathode valve and bypass valve 1 are shown in Fig.3.

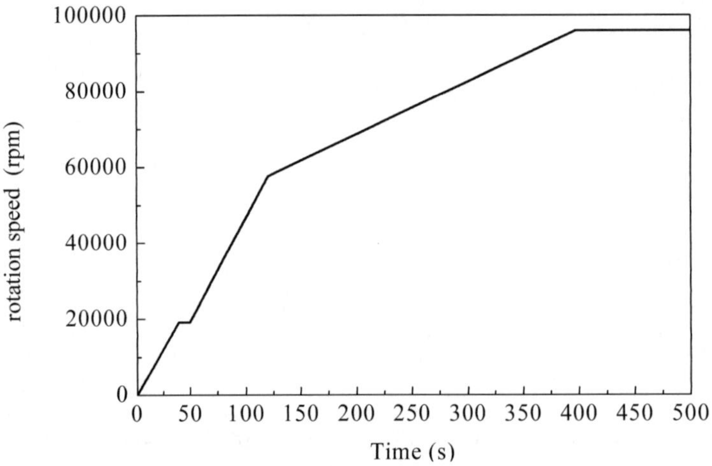

Figure 2. Rotation speed of MGT start-up process.

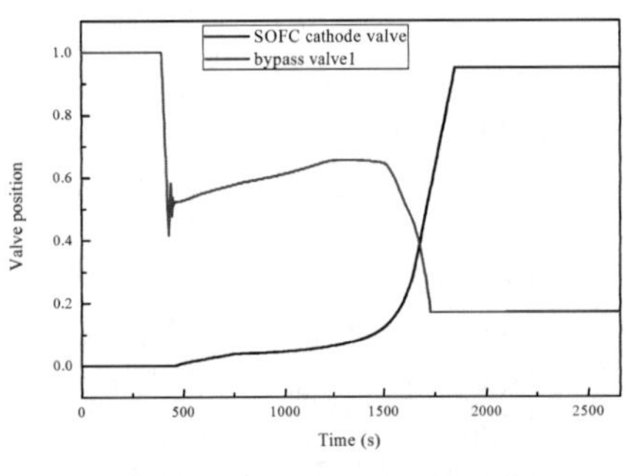

Figure 3. SOFC cathode valve position and bypass valve1 position.

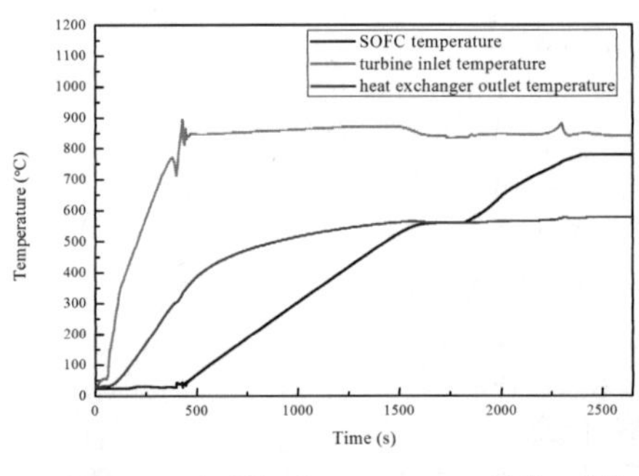

Figure 4. The temperature of turbine inlet, heat exchanger outlet and SOFC.

In order to heat the SOFC to about 600°C, improving the outlet temperature of the turbine is an effective strategy. Therefore, when heat SOFC from atmospheric temperature to start-up temperature, the turbine inlet temperature is improved to 870°C, which is still in the safe temperature range of the turbine. So, the additional heated device is not employed. About 1700s, SOFC generates power by electrochemical reaction. Then, the temperature of SOFC continues increasing with the chemical reaction heat releasing. The temperature of the turbine inlet, heat exchanger outlet and SOFC are shown in Fig.4. In 40mins, the system start-up process is ended.

As shown in Fig.5, when started the hybrid system, the CH_4 is employed to combusting to push the turbine. After MGT reaches nominal rotation speed, the temperature of heat exchanger still increases. Therefore, the CH_4 mass flow reduces slowly. About 1800s, the SOFC starts working. The fuel flow of the SOFC anode gradually increases. The temperature of the SOFC outlet also gradually increase. Therefore, the pressure drop along the pipeline increase. The pressure ratio of compressor increases and the air mass flow of the compressor reduces. The compressor running curve is close to surge line. Therefore, for the purpose of ensuring the safety of the compressor, the bypass valve 2 starts open to increasing air mass flow of compressor and reduce the pressure of the compressor outlet. At the start-up process, the compressor running curve is shown in Fig.6. At the initial stage of the MGT's start-up process, the compressor running curve is from A to B. When the speed of the MGT reaches the nominal speed, the resistance along the pipe increasing as a result of the thermal inertia of the heat-exchanger. Therefore, the compressor running state is from B to C. However, point C, which is close to the compressor surge line, represents that the safety and stability of the system are very severe. Then, the compressor running state changed from point C to D, because the bypass valve 2 is opened. The compressor surge has been avoided.

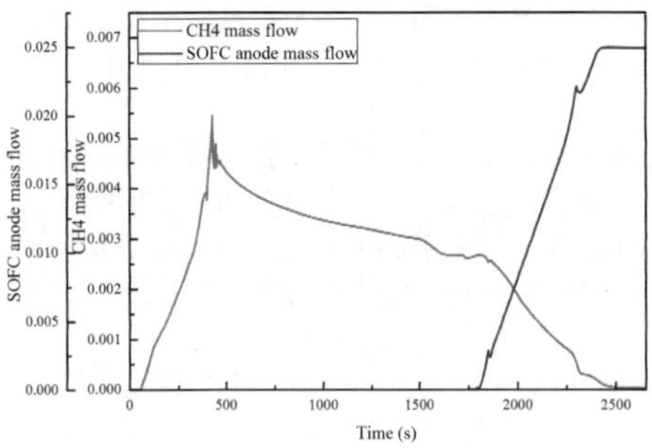

Figure 5. The mass flow of CH4 and the SOFC anode.

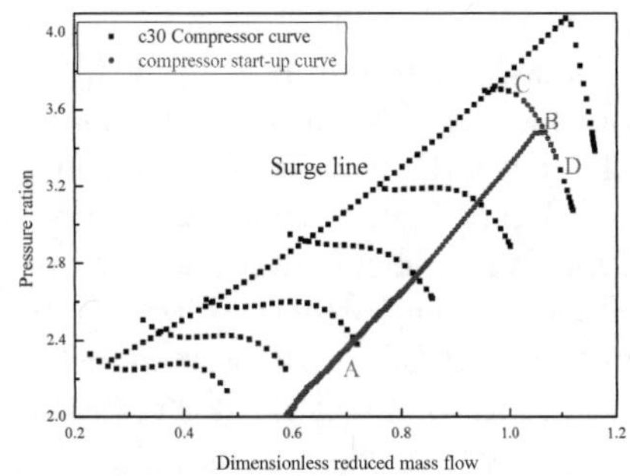

Figure 6. Compressor running

5. Conclusions

In this work, a numerical model of SOFC-MGT hybrid power system is developed. The model includes the compressor, heat exchanger, SOFC, combustor, turbine, and bypass. The MGT model and the SOFC model are validated respectively. Furthermore, the dynamic behaviour of the start-up process have been investigated by starting up the hybrid system from atmospheric temperature. Studies on the start-up process, in which protecting SOFC and avoiding compressor surge are considered, shown that this system can be started in 40 minutes. Moreover, while the temperature of SOFC outlet flue gas increases, the additional bypass valve 2 ensures the system stable. The investigation demonstrates that the hybrid power system can start up safe and fast.

6. References

[1] Y.D. Hsieh, Y.H. Chan, S.S. Shy, Effects of pressurization and temperature on power generating characteristics and impedances of anode-supported and electrolyte-supported planar solid oxide fuel cells, Journal of Power Sources, 299 (2015) 1-10.

[2] H. Xu, B. Chen, P. Tan, W. Cai, W. He, D. Farrusseng, M. Ni, Modeling of all porous solid oxide fuel cells, Applied Energy, 219 (2018) 105-113.

[3] S.S. Shy, S.C. Hsieh, H.Y. Chang, A pressurized ammonia-fueled anode-supported solid oxide fuel cell: Power performance and electrochemical impedance measurements, Journal of Power Sources, 396 (2018) 80-87.

[4] S. Seidler, M. Henke, J. Kallo, W.G. Bessler, U. Maier, K.A. Friedrich, Pressurized solid oxide fuel cells: Experimental studies and modeling, Journal of Power Sources, 196 (2011) 7195-7202.

[5] A.A. Burke, L.G. Carreiro, J.R. Izzo, Pressurized testing of a planar solid oxide fuel cell stack, International Journal of Hydrogen Energy, 38 (2013) 13774-13780.

[6] D. Oryshchyn, N.F. Harun, D. Tucker, K.M. Bryden, L. Shadle, Fuel utilization effects on system efficiency in solid oxide fuel cell gas turbine hybrid systems, Applied Energy, 228 (2018)

1953-1965.
[7] J. Pirkandi, M. Mahmoodi, M. Ommian, An optimal configuration for a solid oxide fuel cell-gas turbine (SOFC-GT) hybrid system based on thermo-economic modelling, Journal of Cleaner Production, 144 (2017) 375-386.
[8] X. Zhang, J. Li, G. Li, Z. Feng, Dynamic modeling of a hybrid system of the solid oxide fuel cell and recuperative gas turbine, Journal of Power Sources, 163 (2006) 523-531.
[9] P. Chinda, P. Brault, The hybrid solid oxide fuel cell (SOFC) and gas turbine (GT) systems steady state modeling, International Journal of Hydrogen Energy, 37 (2012) 9237-9248.
[10] C. Bao, Y. Wang, D. Feng, Z. Jiang, X. Zhang, Macroscopic modeling of solid oxide fuel cell (SOFC) and model-based control of SOFC and gas turbine hybrid system, Progress in Energy and Combustion Science, 66 (2018) 83-140.
[11] P. Saisirirat, The Solid Oxide Fuel Cell (SOFC) and Gas Turbine (GT) Hybrid System Numerical Model, Energy Procedia, 79 (2015) 845-850.
[12] N. Zhou, C. Yang, D. Tucker, P. Pezzini, A. Traverso, Transfer function development for control of cathode airflow transients in fuel cell gas turbine hybrid systems, International Journal of Hydrogen Energy, 40 (2015) 1967-1979.
[13] N.F. Harun, D. Tucker, T.A. Adams, Dynamic Response of Fuel Cell Gas Turbine Hybrid to Fuel Composition Changes using Hardware-based Simulations, in: K.V. Gernaey, J.K. Huusom, R. Gani (eds.) Computer Aided Chemical Engineering, Vol. 37, Elsevier, 2015, pp. 2423-2428.
[14] N. Zhou, V. Zaccaria, D. Tucker, Fuel composition effect on cathode airflow control in fuel cell gas turbine hybrid systems, Journal of Power Sources, 384 (2018) 223-231.
[15] L. Petruzzi, S. Cocchi, F. Fineschi, A global thermo-electrochemical model for SOFC systems design and engineering, Journal of Power Sources, 118 (2003) 96-107.
[16] A. Mehrpanahi, G. Payganeh, M. Arbabtafti, Dynamic modeling of an industrial gas turbine in loading and unloading conditions using a gray box method, Energy, 120 (2017) 1012-1024.
[17] S. Barsali, A.D. Marco, R. Giglioli, G. Ludovici, A. Possenti, Dynamic modelling of biomass power plant using micro gas turbine, Renewable Energy, 80 (2015) 806-818.
[18] A. Liu, Y. Weng, Performance analysis of a pressurized molten carbonate fuel cell/micro-gas turbine hybrid system, Journal of Power Sources, 195 (2010) 204-213.
[19] E. Munari, M. Morini, M. Pinelli, P.R. Spina, Experimental Investigation and Modeling of Surge in a Multistage Compressor, Energy Procedia, 105 (2017) 1751-1756.
[20] X. Zheng, Z. Sun, T. Kawakubo, H. Tamaki, Experimental investigation of surge and stall in a turbocharger centrifugal compressor with a vaned diffuser, Experimental Thermal and Fluid Science, 82 (2017) 493-506.
[21] T. Siikonen, Numerical method for one-dimensional two-phase flow, Numerical Heat Transfer Applications, 12:1 (1987) 1-18.
[22] P. Aguiar, C.S. Adjiman, N.P. Brandon, Anode-supported intermediate temperature direct internal reforming solid oxide fuel cell. I: model-based steady-state performance, Journal of Power Sources, 138 (2004) 120-136.

Acknowledgments
The authors gratefully acknowledge the friendly support of the National Nature Science Fund of China (No. 51576020) and Graduate Research and Innovation Foundation of Chongqing, China (No. CYB18052).

Performance prediction of zero head turbine at different water levels

I.A. Masud*, S. Yusuke and Y. Suwa

Department of Mechanical Engineering, Shibaura Institute of Technology, Tokyo, Japan

*Email: md17007@shibaura-it.ac.jp

Abstract. Hydropower technology has been focusing on conventional water turbines with performance and efficiency depending on water head without really paying much attention to the energy in flowing waters. Such energy can be harnessed by integrating zero head turbines in possibly undershoot system. The turbines are modelled to be alternative systems in renewable energy. In this research, performance of this turbines in form of a conventional water wheel is studied using both programming prediction and experimental methods. Various type of turbines was designed by altering the blade inclination angle and testing the performance at 4- water levels. The maximum and minimum efficiencies were predicted in this study at this water levels. Based on the result, the viability of this turbines are positive and can be used to harness loss potential of flowing waters in canals and rivers which can be easily installed.

1. Introduction

Hydrokinetic zero head turbines share the same mode of operation with wind turbines, and the design principles are quite similar. Both wind and hydrokinetic turbines are renewable energy sustainable systems gathering attention from researchers across the globe. [1]. Scientists have developed much interest on this type of turbines recently because it allows the use of small low-head hydro resources thereby avoiding some of the environmental, social, and economic problems associated with development of earth dams. Hydrokinetic is an alternative and a cheaper solution to remote and off-grid areas having nearby flowing streams of water [2]. Flowing water streams with little or zero head can be found in rural communities. Installation of the typical hydroelectric system is not viable in such river or water flow [3]. Developing nations can use this technique to provide electricity to remote places where transmission line cannot be connected easily or the cost becomes very high [4]. Micro hydro as a form of small hydropower is one of the most economical and environmentally friendly renewable energy technologies to be considered for powering rural communities [5]. It is often classified by the power generation potential based on head of water and discharge of water flow. A typical micro hydropower needs a substantial amount of water to be diverted from the stream/river and brought to the turbines power house without compromising the elevation which is the head. A typical micro-hydro is characterized to have high initial project capital cost and very low running costs over time [6]. In Hydrokinetic system, the turbine is placed in the water stream and flowing water enables the turbine to rotate producing electricity [7]. Hydrokinetic zero head turbine system is a new category

IOP Conf. Series: Earth and Environmental Science **291** (2019) 012048 doi:10.1088/1755-1315/291/1/012048

of hydropower energy that generates electricity by extracting kinetic energy of flowing water instead of potential energy of falling. It generates electricity without dam's storage thereby saving costs [8].
In a flowing stream, th turbines go one step further.This type of turbine has an upper hand by simply utilizing the energy in the velocity of the water in the stream [9]. Energy can be extracted from the river currents by using submerged hydrokinetic turbines, capturing energy through the processes of hydrodynamic, rather than aerodynamic, lift or drag [10].
However, the dimerit of using typical small-scale hydropower is that, in most cases it is obtained from run-of-river plants that lack storage [11]. In all this reaserch conducted by scientists, there is no any attempt to study and optimise these type of turbine performnce with respect to water levels. Our reasearch has focused on this issue which predicts the suitable water level for best performance using modelling, experimetal and simulatiion methodologies.

2. Modelling

A model was designed to study the behaviour of this turbine. In this model, a method for calculating the torque was adopted by critically studying the geometric shape. First of all, it was assumed that the water energy is completely harnessed by the upstream blades thereby shielding the downstream blades. Now suppose the flow is uniform and the blades are rotating at regular intervals. By calculating the uniform flow velocity, it seems likely that the tendency will be obtained depending on the area and angle of the blades during rotation when looking at the turbine from the upstream side. As such we have an idea of the acting force on the blades.

2.1. Modelling calculation

The force applied perpendicularly on each blade is given by equation (1) below

$$F = (1/2)\rho(u\sin\alpha)^2 \times A \qquad (1)$$

Since there are multiple blades during rotation, the force received by the front blade is not received by the blade at the back. Since only the one whose projected area of the blade is obtained as seen from the upstream side contributes to torque, we assume the behaviour in Fig. 1 which describes the cross-sectional view of the turbine model with the direction of the flow velocity coming from the left side.

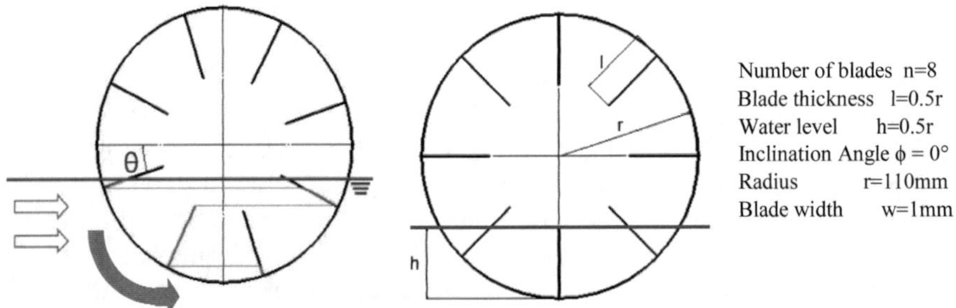

Figure 1. Turbine Cross-sectional View.

3. Experiments

In order to create a continuous flow of water around the system for the experiment, a water pool of depth 600[mm], width 600[mm] and length 1800[mm] was developed. Flow of water is influenced by a fixed waterproof 300[W] dual propeller diving pool scooter pump model BM1207 and diverted by a design curve in the pool. The turbine is located in the convergent throat of the pool at opposite side to the pump allowing it to harness energy from the relatively stable flow of water thereby rotating the turbine using free kinetic energy. The turbine 220[mm] x 220[mm] has 8 number of blades and was designed using a 1[mm]thickness plastic plate. Water level was kept at different height of the turbine blade to study the condition at different water levels.
The data targeted for this experiment is the angular velocity, time and power potential of the system. Angular velocity and time were determined using an Arduino sensor. The experiment was conducted

at 0.6m/s flow velocity determined by controlling the pump using a voltage regulator. The pump was regulated at maximum voltage of 12.2 [V]. The velocity of the system that corresponds to the voltages is 0.6[m/s] as earlier stated. The value of the output power of the turbine was determined using the principles of applied mechanics. The concept entails adding mass loads to the turbine by transferring it through a rope pulley system until the maximum power was determined at the point of no-rotation of the system turbine. The rotational speed n [rpm] and time [T] at that time were measured. The power is determined using the below expressions.

$$Pt= T \times W \text{......................................} (2)$$

$$Pin = \tfrac{1}{2}\, \rho A V^3 \text{....................................} (3)$$

Equation (2) defines the output power of the rotating turbine (Pt) and equation (2) defines the input water flow kinetic energy of the system (Pin). The efficiency of the system is determined using the two equations.

3.1 Nomenclature

ρ = Density of water (kg/m^3)
u = flow velocity (m/s)
A = cross sectional area of the blade (m^2)
α = angle of the blade relative to the flow (^{o})
r = Turbine radius (m)
T = average torque (Nm)
W = angular velocity in (rpm),
V = Average Water mode flow velocity (m/s)

4. Simulations

The research simulation was carried out to compare the laboratory experiments results. The average velocity of flow is 0.6m/s around the model. 7- type of turbines with same parameters but different inclination angles were simulated at 4 water levels. The water levels are at 1/4 the turbine radius (o.25). half the turbine radius (0.5r), 3/4 the turbine radius (0.75r) and equal the radius (h=r) . The blades incliantions are at -45°, -30°, -15°, 0°, 60°, 45°, 30°, and 15° with respect to the vertical turbine plane. Fig. 2a & 2b are the simulation of the turbine when inclined at 0° while Fig. 3a & 3b are the simulation of the turbine when inclined at -30°.

The simulation was used to determine the pressure impact on the turbine blades and velocity component around the areas of the blades at four water levels on each turbine. As shown in Figure 2 and 3 above. The above results are at water level (h) equal to 0.75 the turbine radius (h=0.75r). Based on the simulation results, the turbines flow velocity was found to be different at all water levels for wach type of tuebines, henceforth, we have a brighter idea on the reason the disperity between the modelling and experimental approach results. The turbines pressure impact is different for each turbine and the value of pressure depends on the angle of blade inclination and water levels . whereas the vector components was studied at static rotational velocity, the pattern for each turbine varies due to the shape and positioning.

IOP Conf. Series: Earth and Environmental Science **291** (2019) 012048 doi:10.1088/1755-1315/291/1/012048

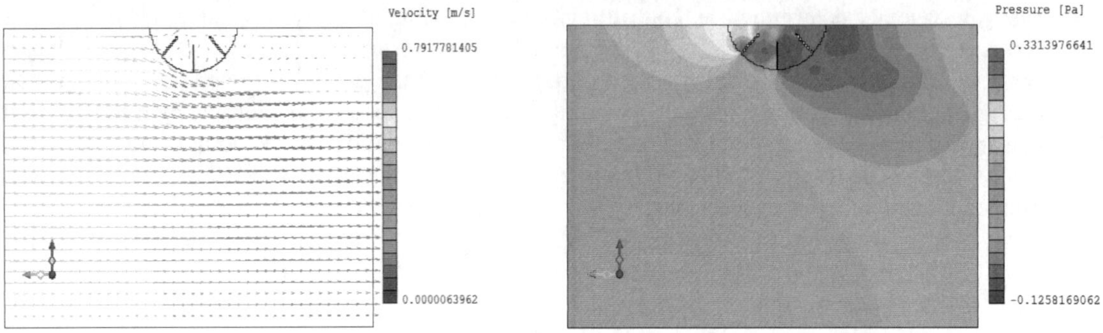

(a)Velocity vector component of 0° Turbine (b) Pressure impact of 0° inclined Turbine

Figure 2. The simulation of the turbine when inclined at 0°.

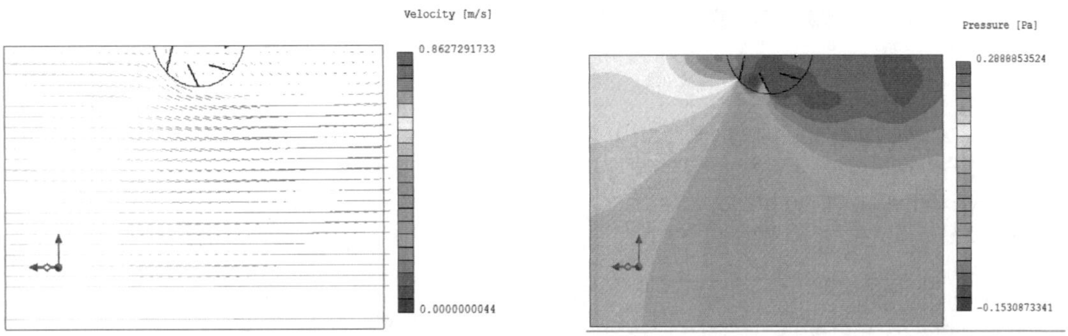

(a) Velocity vector component of -30° Turbine (b) Pressure impact of -30° inclined Turbine

Figure 3. The simulation of the turbine when inclined at -30°.

5. Results

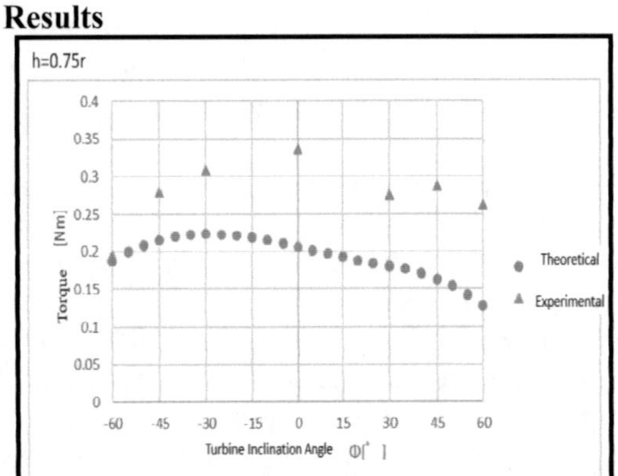

Figure 4. Theoretical and experimental comparison at h =0.75r.

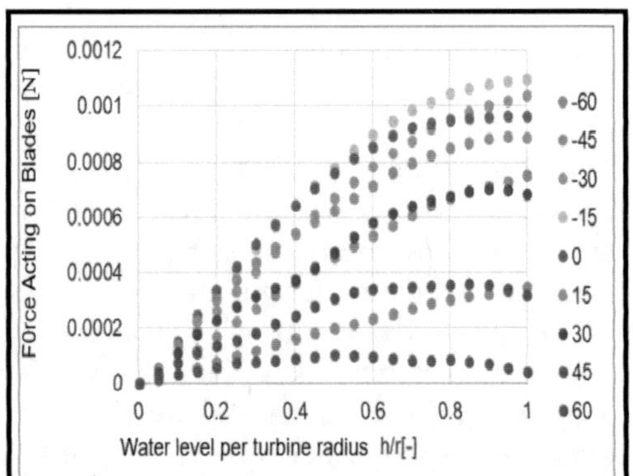

Figure 5. Impact force vs Water level.

Fig. 4 and 5 describe the result of the research. Fig.4 illustrates the comparison between the theretical and experimental values at 0.75r. we can see that the pattern of perfromance is fairly symmetrcial even though there is a little disperity. The result is equal when the turbine is inclined at -60°. In Fig. 5, variuos turbine performance are described at different water levels. From the results the turbine inclined at -15° showed a high performance at all levels and is opposite for turbine inclied at 60°.

6. Conclusion

The prototype is designed and we carried out modelling, experiments and simulations for prediction of the performance of the hydrokinetic zero head turbine. The result of the system shows that the turbine displayed a positive performance in harnessing energy in a zero head condition. Water level and blade inclination angle contributes to the efficiency of such type of turbines.

Future plan of these research will focus on improving the efficiency by carrying out a flow visualisation experiment to study the disparity between the theoretical and experimental results even though the pattern seems the same. In conclusion, this turbine has displayed a viable performance and can be used to harness loss potentials in flowing waters.

7. References

[1] M.M. Nunes, R.C.F. Mendes, T. F. Oliveira, and A.C.P. Junior, *''An Experimental Study on the Diffuser-Enhanced Propeller Hydrokinetic Turbines''*, Renewable energy, 133 PP. 840-848 (2019)

[2] G.Saini and R.P.Saini, *"A numerical Analysis to Study the Effect of Radius Ratio and Attachment Angle on Hybrid Hydrokinetic Turbine Performance"* Energy for Sustainable Development. Volume 47, December, Pages 94-106, (2018)

[3] R.P.Saini, "Performance analysis of a Savonius hydrokinetic turbine having twisted blades'', Renewable Energy, Volume 108, August, Pages 502-522, 2017

[4] M. Tanbhir, U. Nawshad., M. N. Islam, IbneaSina ,M. K. Syfullah, and R. Rahman, *''Micro Hydro Power: Promising Solution for Off-grid Renewable Energy Source''*, International Journal of Scientific & Engineering Research, Volume 2, Issue 12, December-2011

[5] Paish O. *Small hydro power: technology and current status.* Renewable and Sustainable Energy Reviews 2002;6(6):537–56

[6] Ashok S. Optimised model for community-based hybrid energy system. Renewable Energy 2007; 32(7):1155–64.

[7] Tanbhir Hoq, Nawshad UA, Islam N, Sina Ibnea, Syfullah K, Raiyan Rahman. *Micro hydro power: promising solution for off-grid renewable energy source*. International Journal of Scientific & Engineering Research 2011;2(12).

[8] Herman J. V, Kanzumba K n , Sandile P. K. *Status of micro-hydrokinetic river technology in rural applications*: A review of literature. Renewable and Sustainable Energy Reviews 29 (2014) 625–633

[9] Vince G, Clayton B. *Development and Application of a Water Current Turbine*, www.newenergycorp.ca/LinkClick.aspx?fileticket=5%2BtQK3cID%2FY%3D&tabid=84&mid=471; 2010.

[10] Mukrimin S G, *Evaluation and measures to increase performance coefficient of hydrokinetic turbines*, Renewable and Sustainable Energy Reviews 15 (2011) 3669–3675.

[11] M.J. Khan, G. Bhuyan, M.T. Iqbal, J.E. Quaicoe, *Hydrokinetic energy conversion systems and assessment of horizontal and vertical axis turbines for river and tidal applications: A technology status review*, Applied Energy 86 (2009) 1823–1835.

Acknowledgments

We gratefully wish to appreciate the effort of Japan International Cooperation Agency (JICA) for funding this research and members of fluid application and dynamics laboratory, Shibaura Institute of Technology for the advises and comments to make this research a success.

The use of faba-bean cropping as a sustainable and energy saving technology – A new protein self-sufficiency opportunity for European agriculture?

R Lybæk[*] and H Hauggaard-Nielsen

Department of Environment, Technology and Social Studies (TekSam), Institute of People and Technology (IMT), House 9.2. Universitetsvej 1. Roskilde University, Denmark.

[*]Email: rbl@ruc.dk

Abstract. European agriculture relies on large amounts of imported protein fodder, primarily soya-beans, for livestock rearing from e.g. South and North America. This has severe social and environmental impacts, and require large amount of fossil fuel to cultivate, harvest and transport. This paper investigates how faba-beans - a leguminous crop able to fixate its own nitrogen through symbiosis with soil bacteria while providing various other Eco System Services - can be adopted, as a valuable crop by farmers in EU. This could substitute import of soya-bean for animal fodder, providing global savings in pesticide, artificial fertilizer and fossil energy usage. Emphasizing transition theory and e.g. farm machinery, culture and policies, we conclude, that activities must be applied at all levels of the agricultural system. New transition pathways are e.g. to strengthen farm networks to share knowledge and farm equipment for faba-bean cultivation, and to enhance the environmental regulation to limit the amount of artificial fertilizer distributed on farmland making legumes more valuable for farmers.

1. Introduction

1.1. Impact of modern agriculture

A transition to a more sustainable and energy efficient European agricultural sector is currently challenged by a reduction in the cultivation of legume-based crops (faba-bean, pea, lupine, lenses, grass-clover, etc.), hereunder protein fodder. Despite a policy focus in countries within Europe on the cultivation of legumes, and a diversification of crops being produced, the legume area has declined during the past decades [1], [2]. Over the last 50 years the global average cereal grain yield has almost doubled, but during the same period the cultivation of legumes has dramatically declined, as for example faba-beans, with more than 56 % [3]. This is not a favorable development for the future European agriculture, either on a local or a global scale, due to the many non-harvested benefits that legumes provide for society [4]. The consequences are for example lower Eco System Services (ESS) provided by cultivation of legumes, emphasized below, as well as the need for intensified import of soya-beans from e.g. South and North America. This produce is often genetically modified spices [4], with production strategies causing deforestation and in some cases replacement of native people.

A study conducted by [5] indicate, that system internal outputs of legume nitrogen (N)-fixation of 130-153 kg N per ha can be achieved when cultivating such crops. Other studies point to the same result with variations between 100-200 kg N-fixation per ha [3]. Besides N-fixation legumes also provide crop rotation services, which limit pesticide utilization and costs by up to 25 % for e.g. cereals, and lead to yield enhancement in the following crop by up to 0.2-1.6 t. per ha [5]. ESS's provided by legumes are however highly devaluated for their contribution to establish a more sustainable, self-sufficient and energy efficient European agriculture with lower pressure on 'external' resources. In 2017 European countries for example imported 15 Mt of soya-beans for protein fodder, mainly from South America, second after China, with an import of 93 Mt. China also produced almost 14 Mt.
To put this in perspective Argentina, Brazil and the US produced 53 Mt, 87 Mt and 107 Mt soya-beans in 2017 respectively [6]. According to [7] more than 7.2 M ha forest in South America have disappeared between 1995 and 2005. The Chaco-forest, which cover areas in Brazil, Bolivia and Paraguay, have for example lost more than 10 % of its area, mainly due to cultivation of soya-beans for export and livestock farming [8]. Hence, natural and semi-natural habitats are increasingly converted into arable land, with severe impact on soil quality, climate change and biodiversity [4]. Moreover, large consumption of fossil fuels is connected to the cultivation, harvesting and transport of soya-beans to European countries for rearing of livestock like dairy cattle and pigs, with low feed efficiencies [9].

1.2. Focus of this paper

Environmental benefits and technical possibilities for deploying more legume crops, like faba-beans, have been the focus of several European and international studies and are well described in the current literature [4], [10], [11]. This paper, however, seeks to provide knowledge required for a transition of the European agriculture, with emphasis on stakeholders and organizations constituting the agricultural sector. We will identify embedded norms, cultures and traditions from e.g. Danish farmers, which currently act as obstacles for a further uptake and cultivation of protein legumes - faba-beans for animal fodder - and suggest ways of changing this. Moreover, we will provide policy suggestions to enhance the framework conditions for such changes to take place, which all together can provide more self-sufficiency in protein legume cultivation, and create a more sustainable, energy efficient, diverse and resilient agricultural system [12], largely based on 'internal' European resources.

2. Methodology

2.1. Theoretical approach and data retrieval

With point of departure in Geels' Socio-technical Transition (ST) theory [13] and emphasis on European faba-bean production in a Multi-Level Perspective (MLP) [14], [15] - constituted by the 'landscape', 'regime' and 'niche' levels - this paper develops a platform that facilitates the investigation of a 'transition pathway' for European agriculture to adapt to more legume (faba-bean) cultivation. The findings will highlight in which areas of the coherent *'regime'* (dominating mono-cultural agri-system) that the European faba-bean *'niche'* (new emerging technology or system) are challenging the existing socio-technical system, and how the *'landscape'* level (new policies, trends, consumer pressure, etc.) can impact this development. The perception of ST is to understand the conditions for changes - transitions - as being part of the three analytical levels mentioned, namely 'landscape, 'regime' and 'niche', and to identify *'champions'* [16] (stakeholders capable of influencing transition), who can impact the agricultural sector in the future to support legume cultivation and utilization. Thus, identifying future 'transition pathway's' could provide new 'windows of opportunities' within the dominating European agriculture for a future transition, in which a more sustainable and energy efficient agricultural system could be developed.
Statistical data and policy information data from e.g. public authorities and background reports and publications on European agriculture, as well as materials from organizations like FAO, WRI, etc., has been studied in this paper, and used as secondary data. Knowledge of faba-beans and legumes in

general - technical systems, cultivation practices, ESS, EU policy support schemes etc. - are also retrieved from scientific publications by means of journal articles. Primary empirical data has been collected from Danish agriculture, and utilized as examples of possible 'transition pathways' for European agriculture. Table 1 below shows the eight farmers from where primary empirical data was collected during the month of November and December 2017. Besides a tour around each farm where field notes were taken, the interviews revolved around the farmer's cultivation and utilization of faba-beans, and how different stakeholders in the value chain (farmers, seed producers, feed industry, farmer organizations and farm networks and regulatory frameworks, etc.) impact the current faba-bean situation, and how it can change to intensify faba-bean cultivation. The interviews were approached as qualitative semi-structured interviews, and recorded on tape for use as empirical research data.

Table 1. Type of farmers interviewed and their faba-bean (FB) profile.

Farm type and number	Cultivating FB (own use)	Purchase FB	Sell FB
2 Dairy cattle farmers	1	1	
2 Pig farmers	2		
1 Highland cattle/crop farm			1
1 Chicken farm			1
2 Crop farmers			2

3. Findings and Discussion

In the following section possible 'transition pathways' for adapting to more faba-bean cultivation and utilization in a European context will be outlined. We will emphasize conventional crop farmers and livestock farmers, farm networks, farmer organizations, seed producers and feed industry, as well as suggest relevant agricultural and environmental policies. This section is based on primary empirical data, as well as secondary data, as emphasized in the methodology above.

3.1. Transition pathways for European agriculture

Emphasizing ***conventional crop farmers*** more information and knowledge transfer are needed from farmers successfully working with faba-beans, as far as cultivation systems and practices, fodder quality, pest and weed control, etc. This is needed to increase the awareness and break down unfavorable myth of faba-bean cultivation. For ***conventional livestock farmers*** the question of knowledge transfer, as well as stronger cooperation, is a main issue [3]. Therefore, knowledge about the nutritional value of faba-beans and fodder mixes must increase, and farmers should to a higher extend share storage capacity and new types of farm equipment facilitating faba-bean cultivation. The latter has proved important in the Danish situation. Thus, new farms could also be established with storage capacity for fodder legumes applied already when constructed. Faba-beans can be utilized for pig fodder without any pre-treatment than grinding, but this knowledge need to be disseminated. Mobile grinders for pig fodder, and toasters for cattle feed, could eventually be rented, and the possibility for this could be disseminated more among farmers.

Farm networks play a pivotal role to achieve the above and to sustain a further development of the faba-bean 'niche', future network activities by these stakeholders must be strengthened. Enhancing farm networks in which farmers disseminate and share knowledge about legumes among each other is highly needed, and will provide good pilot cases for farm visits, sharing of machinery and storage capacity. Buying/selling faba-beans among each other could e.g. be facilitated by means of electronic platforms and increased communication between farmers in general, etc. Such pressure on the 'regime' from the 'niche' is constituted by the organic faba-bean farmers today, but must be strengthened in the future. This 'incubation center' (see [13], [15]]), composed by the 'niche', should ideally be supported by ***farmer organizations***, but this is challenging as these organizations have an embedded reluctance to change the current 'regime'. It is, however, important that farmer organizations continue and expand their existing - albeit limited - work on legumes, hereunder faba-

IOP Conf. Series: Earth and Environmental Science **291** (2019) 012049 doi:10.1088/1755-1315/291/1/012049

beans, as this shape and impact the entire 'regime' of the agricultural sector. Farmer organizations are the 'champion' most reluctant to change as they constitute the agricultural sector, being strongly impacted by traditions and culture and provide the direction for it. Changes will most likely come about slowly and require a paradigm shift from the 'landscape' level by changes in the university colloquium of researchers and staff connected to the agricultural sector, as well as for conventional farmers at farm-schools. In combination with higher environmental awareness in the society as a whole, this could impact the development.

Both ***seed producers*** and the ***feed industry*** are important 'champions' for a transition toward more legume cultivation in Europe, and do not grant legumes high interest in the agro-industrial supply chain [11]. Both stakeholders must continue their present work on developing and refining faba-bean cultivars appropriate for cultivating in Europe, as well as providing a platform for buying and selling faba-beans. The need for better cultivars supplied by these stakeholders, are also stressed by [11], which could provide higher added value and increase the interests in legumes. These 'champions' could potentially be more flexible to a changing 'regime', as they would like to sustain market shares. If protein trade among farmers intensifies, which could be facilitated by the farm networks, mentioned above, changes in the existing socio-technical system could appear relatively fast. As of now, farmers in e.g. Denmark sell and buy faba-beans among each other and achieve economic benefits, compared to selling/buying from the fodder industry. Prices of for example wet and dirty faba-beans are sold between local farmers at more favorable prices than the feed industry are willing to pay.

If protein trade among farmers intensifies, and the arable land being cultivated with faba-beans increase, seed companies could see advantages and might be more interested in participating in order to gain and sustain market shares. Stronger farm networks might also directly put pressure on the 'regime', as to develop new and more suitable faba-bean seeds appropriate for the European context, instead of farmers - like in Denmark - joining forces and buying seeds abroad. As for the farmer organizations, both seed and fodder industries could benefit from new learning colloquium at the 'landscape' level, which could facilitate changes in the 'regime', as embedded traditions and cultures - difficult to change - could be untightened. New consumer requirements at the 'landscape' level of more organic protein fodder being used for livestock's in Europe could also increase the price of soya-beans. Together with more emphasis on legumes for human consumption (food), it would make it more economically favorable to cultivate legumes for protein fodder and human food.

When it comes to ***agricultural & environmental policy*** it would benefit the balance of for example Danish agriculture, if the 'harmony rule' was re-introduced. This could potentially lead to farms becoming more balanced in the size of land and number of animals (e.g. bio-dynamic farming), as before the intensive monoculture monopolized the agricultural 'regime', where mixed farming systems were more widespread [11]. Better frameworks for becoming self-sufficient, or at least increase the faba-bean yield, could be a result of this. Stronger environmental regulatory frameworks supporting the growing of legumes for N-fixation, and increase in the cultivation of domestic protein fodder, could also be exercised. Such pressure from the 'landscape' should be adapted through agricultural and environmental policies, and would promote an intensified cultivation of legumes.

4. Conclusion

European agriculture is mainly composed of mono-cultures - primarily cereals like wheat - and a massive use of 'external' resources like artificial fertilizer N, pesticides, fossil fuels, and large import of protein fodder like soya-beans, constantly fed into the system to sustain it. This paper investigates how to increase the faba-bean cultivation and consumption within Europe to achieve ESS's and energy savings, etc. Secondary data and primary empirical data collected from farmers revile the importance of supporting activities within and create stronger farm networks at the *'niche'* level. A 'transition pathway' is thus to enhance farmer activities through farm networks, as such networks constitute a potential role as *'champion'*. Farm networks should increase in numbers and strengths and disseminate knowledge about faba-beans cultivation and utilization as fodder etc. They should also intensify the sharing of farm equipment, storage facilities and increase the trade of protein fodder among farmers.

IOP Conf. Series: Earth and Environmental Science **291** (2019) 012049 doi:10.1088/1755-1315/291/1/012049

Such *'niche'* activities could intensify the pressure on the *'regime'*, and the uptake of faba-beans in the existing agricultural system. Pressure from the *'landscape'* level - by changes in the learning colloquium of for example researchers and staff connected to farmer organizations, of seed and fodder companies, as well as of farm-schools - should also be applied. Compared to fodder and seed industries, which have a build-in flexibility as far as creating and sustaining market shares, we find that farmer organizations are the *'champion'* most reluctant to adapt to changes, and with the highest interest in sustaining the existing *'regime'*. Pressure from the *'landscape'* level could also be exercised by consumer groups requiring more sustainable and organic fodder for European livestock, as well as legumes for human consumption. Also, stronger agricultural policies where 'harmony' between farmland and the number of animals are needed. This would provide an opportunity for higher level of self-sufficiency in the protein supply in Europe. Lower amounts of N-fertilizer being used on farm soil - through stronger environmental policies - could also enhance farmers' cultivation of legumes to fixate nitrogen, and hence achieve other ESS benefits. Thus, different activities must be taken by various stakeholders connected to European agriculture, at all levels investigated here, for a transition to emerge.

5. References

[1] Davis A S, Hill, J D, Chase, C A, Johannes, A M, Liebman, M 2012 Increasing cropping system diversity balances productivity, profitability and environmental health *PloS ONE* 7(10): e47149.

[2] Lin, B B 2011 Resilience in agriculture through crop diversification: adaptive management for environmental change *BioScience* 61: 183-193.

[3] Jensen E S, Peoples M B, Hauggaard-Nielsen H 2010 Faba bean in cropping systems *Field Crop Research* 115: 203-216.

[4] Nemecek T, von Richthofen J S, Dubois G, Casta P, Charles R, Pahl H 2008 Environmental impacts of introducing grain legumes into European crop rotations *European Journal of Agronomy* 28: 380-393.

[5] Zander P, Amjath-Babu T S, Preissel S, Reckling M, Bues A, Schläfke N, Kuhlman T, Bachinger J, Uthes S, Stoddard F, Murphy-Bokern D, Watson C 2016 Grain legume decline and potential recovery in European agriculture: a review *Agron. Sustain. Dev.* 36: 26.

[6] LegValue 2017 *Fostering sustainable legume-based farming systems and agri-feed and food chains in the EU* LegValue H2020 nr. 727672. Deliverable D5.1 European Union.

[7] Forest Trends 2014 *Consumer goods and deforestation - an analysis of the extent and nature of illegality in forest conversion for agriculture and timber plantations* Washington USA.

[8] World Resource Institute WRI 2014 *5 overlooked deforestation hotspots* World Resource Institute (18.10.2014): http://www.wri.org/blog/2014/03/5-overlooked-deforestation-hotspots.

[9] Herrero M, Havlik P, Valin H, Notenbaert A, Rufino M C, Thornton P K, Blümmel M, Weiss F, Grace D, Obersteiner M 2013 Biomass use, production, feed efficiencies, and greenhouse gas emissions from global livestock systems *Proc Natl Aced Sci USA* 110 **(52)**: 20888-20893.

[10] Preissel S, Reckling M, Schläfke N, Zander P 2015 Magnitude and farm-economic value of grain legume pre-crop benefits in Europe: a review *Field Crop Research* 175: 64-79.

[11] Magrini M B, Anton M, Cholez C, Corre-Hellou G, Duc G, Jeuffroy M H, Maynard J M, Pelzer E, Voisin A S, Walrand S 2016 Why are grain-legumes rarely present in cropping systems despite their environmental and nutritional benefits? Analyzing lock-in in the French agrifood system *Ecological Economics* 126: 152-162.

[12] Raseduzzaman M D, Jensen E S 2017 Does intercropping enhances yield stability in arable crop production? A meta-analysis *European Journal of Agronomy* 91: 25-33.

[13] Geels F W 2002 *Understanding the Dynamics of Technological transition. A co-evolutionary and socio-technical analysis* PhD Thesis Twente University Press. Enschede Nederland.

[14] Geels F W 2011 The multi-level perspective on sustainable transition: responses to seven criticisms *Environ. Innov. Soc. Trans* 1: 24-40.

[15] Schot J and Geels F W 2008 Strategic niche management and sustainable innovation journeys: theory, findings, research agenda and policy *Techn. Analysis & Strategic Mangt* 20:537-554.
[16] Heves A K and Lyons D I 2008 The humanistic side of eco-industrial parks: champions and the role of trust *Reg. Stud.* 42 **(10)**: 1329-1342.

AUTHOR INDEX

Abdullah, M R 201
Abdul-Razak, Ahn 184
Abel, Bernd 230
Ahmad, N A 201, 212
Ahmad, N. A 184
Ahmad, S. S. 184
Al-Nasser, A. 65
Alrasyid, Ha 14
Al-Salem, S. M. 65
Ardiani, N A 161
Atmoko, Tu 14
Ayoub, M A 82
Azaim, Fatin Zawani Zainal 89
Azmi, Intan Suhada 275
Cai, Yi-Xin 108
Chan, Hsiao-Sung 48
Chan, K Oh 178
Chang, Siu Hua 275
Chao, Jen-Chih 20, 48
Chao, N C 101
Cheema, Taqi Ahmad 235
Chen, Feiran 259
Chen, Hsuan-Hung 108
Chen, Wei Sheng 70
Chen, Y J 101
Chen, Yen Jung 70
Chen, Yu An 70
Chen, Z 8
Chien, Yu-Chen 265, 270
Chiu, Chien-Chen 108
Danish, M S S 55
Deng, Kangjie 281
Dewi, Mbv 14
Dewi, Oc 14
Dong, L. K. 41
Dorjgotov, A 178
Du, P F 35
Duan, Yuanyuan 223
Eaktasang, N 60, 76
Ennaceri, Houda 230
Ermeey, A. K 184
Fabay, Cian Marvince 247
Faiq, A B 55
Fedayi, H. 55
Feng, S J 101
Fitriyana, Dinna 166
Ge, Zhong 223
Ghani, A A A 201
Guzy, A 129
Hadi, Abdul 275
Handayani, Tri Pratiwi 172
Hanh, N T X 195
Haron, S. N 184
Hauggaard-Nielsen, H 294
He, Liang 148
Hejmanowski, R 129
Ho, C. 82
Hsieh, Shu-Hui 254
Hsu, Che-Yu 124
Hsu, Kuo-Wei 20, 48, 119
Huang, X C 35
Indartono, Y S 240
Ismail, A 201
Jalil, Mohd Jumain 275
Jamian, J J 212
Jamil, S J T 212
Johor, Hanisah 89
Ju, G D 154
Karam, H. 65
Karimy, H 55
Kehinde, Felix 247
Kingkeaw, S 60
Koerniawan, M D 161
Laor, P. 76
Latif, A F 201
Li, Dan 259
Li, Jian 223
Li, N 35
Lin, Meng-Li 114
Liu, Zs 94
Lokman, Mohd Iqbal 89
Lybæk, R 294
Malinowska, A A 129
Manop, P. 41
Masud, I. A. 289
Mojica, Edison E 247
Morad, Norhashimah 275
Naimah, D Y N 161
Naimah, Dintani Y N 166
Ningsih, N S 240
Novitasari, D 240
Nugroho, Priyambodo Nur Ardi 217
Ou, Sheng-Jung 29, 108, 124
Pasukphun, N 76
Prabawati, Mutia 166
Priyandana, D 14
Qian, Y 1
Saat, Asmalina Mohamed 89
Sabory, N R 55
Sabri, Muhammad Irfan 89
Saleem, Abdul Samad 235
Senjyu, T 55
Shu, Z P 35
Sinutok, S. 41
Siska, D. 14
Suhendri 161, 172
Sultan, H. 65
Suma, Y 60, 76

AUTHOR INDEX

Sung, J Ok 178
Suwa, Y. 289
Tanuwijaya, Frank 166
Techato, K. 41
Tenorio, John Lambert 247
Thanh, B P P 195
Tong, Yw 189
Torii, Shuichi 217
Trang, P T T 195
Triasdian, B 240
Ullah, Rizwan 235
Ulmaniec, P. 129
Ulziidelger, B 178
Wang, Dongyan 140
Wang, Xiaodong 259
Wang, Yiping 259
Wang, Zhuoran 148
Wu, Cheng-Hung 270
Wu, Pei-Chen 48
Xie, Jhong-Ping 119
Xu, X 154
Xu, Xue-Min 29
Yamin, Aliff Farhan Mohd 275
Yang, Chen 281
Yang, Jingze 223
Yang, Rui-Xing 114
Yang, Wy 94, 189
Yang, Zhen 223
Yew, Sue Yee 265
Yi, Jianjun 148
Yi-Chien, Lee 254
Yong-Gun, S 178
Yu, S Y 1
Yu, Xl 94
Yu-Chen, Chien 254
Yunseong, J 178
Yusuke, S. 289
Zaheb, H 55
Zaini, Mohd Saufi Md 275
Zang, Liangzhen 207
Zhan, Dl 189
Zhang, Caihong 207
Zhang, Pingyu 140
Zhang, Qinchen 140
Zhang, Ty 8
Zhang, Yajun 148
Zhu, Y 8
Zong, Rh 8